Springer Series in Materials Science

Volume 296

The Springer Series in Materials Science covers the complete spectrum of materials research and technology, including fundamental principles, physical properties, materials theory and design. Recognizing the increasing importance of materials science in future device technologies, the book titles in this series reflect the state-of-the-art in understanding and controlling the structure and properties of all important classes of materials.

More information about this series at http://www.springer.com/series/856

Elena V. Levchenko · Yannick J. Dappe ·
Guido Ori
Editors

Theory and Simulation in Physics for Materials Applications

Cutting-Edge Techniques in Theoretical and Computational Materials Science

Springer

Editors
Elena V. Levchenko
CARMA, School of Mathematical and Physical Sciences
The University of Newcastle
Callaghan, NSW, Australia

Yannick J. Dappe
SPEC, CEA, CNRS
Université Paris-Saclay
Gif-sur-Yvette, France

Guido Ori
IPCMS, CNRS
University of Strasbourg
Strasbourg, France

ISSN 0933-033X ISSN 2196-2812 (electronic)
Springer Series in Materials Science
ISBN 978-3-030-37789-2 ISBN 978-3-030-37790-8 (eBook)
https://doi.org/10.1007/978-3-030-37790-8

This Springer imprint is published by the registered company Springer Nature Switzerland AG
The registered company address is: Gewerbestrasse 11, 6330 Cham, Switzerland

Preface

Nowadays, more and more industrial communities increasingly rely on atomic-scale methods to obtain reliable predictions, complement qualitative experimental analyses and circumvent the experimental difficulties. This book is a unique comprehensive overview of the latest advances, challenges and accomplishments in the field of theoretical and computational materials science. The most modern advanced simulation techniques and up-to-date theoretical approaches of a selected panel of about thirteen research teams are presented in this book. The book covers a wide range of advanced materials, their structural, elastic, optical as well as mass and electronic transport properties. This book will benefit to physicist, applied mathematicians and engineers interested in advanced simulation methods in materials science. It also can be recommended as additional literature for undergraduate and postgraduate students with majors in physics, chemistry, applied mathematics and engineering.

Newcastle, NSW, Australia — Elena V. Levchenko
Gif-sur-Yvette, France — Yannick J. Dappe
Strasbourg, France — Guido Ori

Contents

Contributors

Eduardo Alves IPFN, Instituto Superior Técnico, Universidade de Lisboa, Bobadela, Portugal

Iréné Berenger Amiehe Essomba Institut de Physique et de Chimie des Matériaux de Strasbourg UMR 7504, Université de Strasbourg - CNRS, Strasbourg Cedex 2, France

I. L. Bataronov Voronezh State Technical University, Voronezh, Russia

Mauro Boero Institut de Physique et de Chimie des Matériaux de Strasbourg, UMR 7504, Université de Strasbourg - CNRS, Strasbourg Cedex 2, France

A. S. Borsyakov Voronezh State University of Engineering Technologies, Voronezh, Russian Federation

Assil Bouzid Institut de Recherche sur les Céramiques (IRCER), Limoges, France; Institut de Recherche sur les Céramiques Centre Européen de la Céramique, Limoges, France

Alexander V. Evteev School of Engineering, Priority Research Centre for Computer-Assisted Research Mathematics and Its Applications (CARMA), The University of Newcastle, Callaghan, NSW, Australia;
School of Mathematical and Physical Sciences, CARMA, The University of Newcastle, Callaghan, NSW, Australia

Francisco C. Franco Jr. Chemistry Department, De La Salle University, Manila, Philippines

Daniel Fritsch Helmholtz-Zentrum Berlin für Materialien und Energie, Berlin, Germany

Luigi Genovese Univ. Grenoble Alpes, CEA, Grenoble, France

Miriam Izzo Solar Fuels Lab, Center of New Technologies, University of Warsaw, Warsaw, Poland

Przemyslaw Jozwik IPFN, Instituto Superior Técnico, Universidade de Lisboa, Bobadela, Portugal

Joanna Kargul Solar Fuels Lab, Center of New Technologies, University of Warsaw, Warsaw, Poland

Andreas Kromik School of Engineering, Priority Research Centre for Computer-Assisted Research Mathematics and Its Applications (CARMA), The University of Newcastle, Callaghan, NSW, Australia

Chathurangi Kumarasinghe School of Engineering, University of Warwick, Coventry, UK

Elena V. Levchenko School of Mathematical and Physical Sciences, Priority Research Centre for Computer-Assisted Research Mathematics and Its Applications (CARMA), The University of Newcastle, Callaghan, NSW, Australia

Katharina Lorenz IPFN, Instituto Superior Técnico, Universidade de Lisboa, Bobadela, Portugal;
Instituto de Engenharia de Sistemas de Computadores-Microsistemas e Nanotecnologias (INESC-MN), Lisbon, Portugal

Evelyne Martin Université de Lille CNRS Centrale Lille ISEN Université de Valenciennes, Lille, France

Carlo Massobrio Institut de Physique et de Chimie des Matériaux de Strasbourg, UMR 7504, Université de Strasbourg - CNRS, Strasbourg Cedex 2, France

Cyprian Mieszczynski National Centre for Nuclear Research, Otwock-Swierk, Poland

Tudor Luca Mitran Horia Hulubei National Institute for R&D in Physics and Nuclear Engineering, Bucharest, Magurele, Romania

George Alexandru Nemnes Horia Hulubei National Institute for R&D in Physics and Nuclear Engineering, Bucharest, Magurele, Romania;
Faculty of Physics, University of Bucharest, Magurele, Romania;
Research Institute of the University of Bucharest (ICUB), Bucharest, Romania

Neophytos Neophytou School of Engineering, University of Warwick, Coventry, UK

Lech Nowicki National Centre for Nuclear Research, Otwock-Swierk, Poland

Guido Ori Institut de Physique et de Chimie des Matériaux de Strasbourg, UMR 7504, Université de Strasbourg - CNRS, Strasbourg Cedex 2, France

Silvio Osella Chemical and Biological Systems Simulation Lab, Center of New Technologies, University of Warsaw, Warsaw, Poland

V. V. Ozyerelyev Voronezh State Technical University, Voronezh, Russian Federation

Alfredo Pasquarello Chaire de Simulation à l'Echelle Atomique (CSEA), Ecole Polytechnique Fédérale de Lausanne (EPFL), Lausanne, Switzerland

Renata Ratajczak National Centre for Nuclear Research, Otwock-Swierk, Poland

Laura E. Ratcliff Department of Materials, Imperial College London, London, UK

Sébastien Le Roux Université of Strasbourg Institut de Physique et de Chimie des Matériaux de Strasbourg, Strasbourg Cedex 2, France

Anna Stonert National Centre for Nuclear Research, Otwock-Swierk, Poland

Bartosz Trzaskowski Chemical and Biological Systems Simulation Lab, Center of New Technologies, University of Warsaw, Warsaw, Poland

Andrzej Turos National Centre for Nuclear Research, Otwock-Swierk, Poland; Institute of Electronic Materials Technology, Warsaw, Poland

V. V. Voronkov Global Wafers, Merano, Italy

V. A. Yuryev Voronezh State Technical University, Voronezh, Russian Federation

M. V. Yuryeva Voronezh State Technical University, Voronezh, Russia

N. A. Yuyukin Voronezh State Technical University, Voronezh, Russia

Part I
Development of Advanced Simulation Methods: The Predictive Power

Chapter 1
Making Computer Materials Real: The Predictive Power of First-Principles Molecular Dynamics

Carlo Massobrio, Mauro Boero, Sébastien Le Roux, Guido Ori, Assil Bouzid and Evelyne Martin

Abstract First-principles molecular dynamics (FPMD) is a well-established method to study materials at the atomic scale by taking advantage of three ingredients: the laws of statistical mechanics, the theoretical foundations of density functional theory and powerful computers. FPMD does its best when the atomic structures are unknown or poorly known and when their time trajectories are required to extract, via statistical averages, a thermodynamical evolution as a function of temperature. In this paper, key concepts of molecular dynamics are recalled and made simple, by insisting on the proper use of some definitions and by showing, via a prototypical example, the sensitivity to a crucial part (the exchange-correlation energy) of the total energy functional.

1.1 Introduction

Understanding the properties of materials at the atomic scale is a longstanding challenge of modelling approaches. The quest of a one-to-one correspondence between experimental data and theoretical counterparts rests on the establishment of realistic models and affordable techniques, keeping the pace with the increasing speed of available computers and able to complement and enrich measurements [1]. At the very beginning of modern science, computers were considered uniquely as tools making possible calculations too complex to be solved analytically [2, 3]. The notion

C. Massobrio (✉) · M. Boero · S. L. Roux · G. Ori
Université of Strasbourg Institut de Physique et de Chimie des Matériaux de Strasbourg, 23 rue du Loess, BP43, 67034 Strasbourg Cedex 2, France
e-mail: Carlo.massobrio@ipcms.unistra.fr

A. Bouzid
Institut de Recherche sur les Céramiques (IRCER), 12 rue Atlantis, 87068 Limoges, France

E. Martin
Université de Lille CNRS Centrale Lille ISEN Université de Valenciennes UMR 8520, 59000 Lille, France

E. V. Levchenko et al. (eds.), *Theory and Simulation in Physics for Materials Applications*, Springer Series in Materials Science 296,
https://doi.org/10.1007/978-3-030-37790-8_1

of time as physical parameter was completely decoupled by the computational time required to complete the calculation of a given property. For instance, the time taken to achieve the diagonalization of a Hamiltonian (a well-known procedure in many areas of theoretical physics and chemistry) did not bear any scientific significance, besides the obvious fact that it could be insightful in order to assess the performances of the computer. Looking back at the past century, it was in the fifties that theoreticians began realizing that computers could be exploited to achieve the longstanding dream of statistical mechanics. This amounts to exploring the macroscopic scale by relying on a time repetition of microscopic events produced by the equations of motion of classically perceptible degrees of freedom: the atoms. The underlying idea is as simple as plausible, even though its practical realization can be (in certain cases) of insurmountable complexity. Let us suppose that we can take reasonable guesses on a set of initial atomic positions, as it is case for a crystal or for some other ordered topology. Provided we are able to describe the interatomic forces (in a first approximation, they can be as qualitative and simple as possible, as a Lennard-Jones potential or equivalent expressions depending only on the distances between two atoms), the Newton equations of motion will drive the atoms around the configurational space [4]. In the rare and somewhat unique case that the atoms are in the absolute (or in one of the relative, if any) minimum of the potential energy surface, no appreciable movement will follow, as it should be since neither velocity or acceleration will be different from zero. However, under any other circumstance, the atoms will start to move with amplitudes directly related to the distance from the minimum energy positions. These are exactly the physical conditions allowing for the gap between computational time and real evolution time to be bridged. Once a discretization of the equation of motion is implemented in the computer, the time spent in calculating the different ingredients of the equations of motion (calculation of the forces and resulting displacement of the atomic position in an iterative manner) can be subdivided in incremental steps along a temporal trajectory. Accordingly, molecular dynamics (MD) consists in moving atoms under the action of forces on an atomic scale, thereby establishing a correspondence between the computer time (or, better, the elapsed time) spent calculating quantities depending on positions and the physical time inherent in the evolution of the coordinates [5]. These two different scales, one purely technical, the other having full physical sense, are related by the notion of step of integration for the equations of motion, chosen to describe with optimal accuracy movements occurring on frequencies typical of the system under consideration. As a natural consequence, the time needed to carry out a full molecular dynamics calculation is, to a very large extent, a multiple of the time taken by the computer to carry out all operations needed to move the atoms simultaneously under the action of the forces. This factor is essentially the same as the number of integration steps in a given trajectory, a typical number being close to a thousand for a temporal length of 1 ps. Having established that MD acts as a *transformer* of a computational time with no physical meaning into a physical time measuring actual temporal intervals, the realization of its paradigm requires the proper account of a further relationship between macroscopic and microscopic scales. Such endeavor has to do with the concept of temperature. At first sight, atoms moving around under the action of forces do not bear any clear-cut information on thermal processes, quite often associated with the presence of external constraints such as thermostats. While the idea of thermostat can also be made fully compatible with molecular dynamics, it is important to underline that any collection of moving atoms concur to its own thermal conditions [6]. This holds true via the equivalence between the temperature and the average of the kinetic energy modulated by factors readily extractable from the equipartition theorem. To move towards quantitative modelling of materials, two further issues have to be addressed: system size and reliability of the forces calculated within the selected model. Concerning system size, if we are to follow the motion of atoms individually, the number of such variables has to be limited by obvious computational constraints. However, periodic boundary conditions are a convenient tool to bypass this problem, since quite often, and especially in homogeneous disordered systems as those treated in this review, the extent of order is limited to a few interatomic distances. Therefore, the only strict requirement is to adopt simulation boxes capable of including the dimensions inherent in the problem under consideration. As an example, order in disordered network-forming systems extends well beyond the first shell of neighbors, requiring minimal periodic systems larger than those employed for describing short range order in other more conventional liquid and glasses. It remains true that, as a general hand waiving statement valid for any molecular dynamics study, the larger the number of atoms considered the better the challenge of reproducing an infinite size behavior will be met, this being more crucial for systems containing inhomogeneous regions, like extended defects. The issue of forces is conceptually the most delicate to handle in any molecular dynamics application, by marking the boundary between classical and first-principles applications of MD. What is the actual meaning of *classical* when classical molecular dynamics is invoked? To describe the movement of atoms, one is faced with the dilemma of describing the interatomic forces, on the basis of a so-called *force field* derived from the potential energy (the *effective* potential). Anytime the expression of the potential energy contains variables devoid of an intrinsic quantum nature (as the electrons) and can be obtained via an analytical expression based on atomic coordinates, the corresponding dynamics will be termed *classical*. This corresponds to the choice of neglecting the inner chemical nature of the atoms, taken as zero-dimensional objects exploring the configurational space under the laws of classical mechanics. The reliability of classical potentials employed to model materials is a longstanding subject of debate and the target of widespread efforts in the community of computational materials science. Chemical intuition has guided the choice of analytical expressions (mostly parametrized to experimental data) corresponding to well defined bonding characters by allowing for significant advances to be made ever since the first application of MD to real systems. However, with the exception of rare gases and very specific bonding situations and properties, effective potentials are bound to fail when the explored phase space differ substantially from the reference conditions employed in the fitting of the parameters. Within this context, one has to realize that the choice of modelling materials with effective potential is still appealing when the number of degrees of freedom is exceedingly high (say, more than a thousand) and the relevant temporal scales are beyond the nanosecond. A real dilemma arises when the quality of the interatomic interaction is well below the acceptable threshold separating qualitative from quantitative. An interesting textbook case is provided by liquid $GeSe_2$ [7]. This system is a prototype of a disordered network-forming material in which the regularity of a tetrahedral arrangement, based on Ge atoms fourfold coordinated to Se atoms twofold coordinated, is altered by the presence of homopolar bonds and Ge (Se) atoms departing from the main coordination motif. A few years ago, while pursuing a first-principles modelling effort of this system along the lines that will be described elsewhere in this paper, we constructed an interatomic potential based on a two-body part, predominantly Coulombic and an account of ionic polarization. Formal charges are assigned to the atoms by assuming that Ge is positively charged (4+) and Se is negatively charged (2−) as it could be tentatively extracted from their valence states. The intrinsic limitations of this model are exemplified by the value of the melting point ($T_m = 2900$ K) (experimental value of $T_m = 1025$ K) and by the fact that homopolar bonds are absent. Also, the diffusion coefficient measured at $T = 1050$ K is recovered at a temperature as high as $T = 7000$ K, demonstrating that the network atoms, rigidly connected in tetrahedra by the interatomic potential, cannot diffuse adequately unless extremely high temperatures are achieved.

We can employ the above example to define the boundaries between two kinds of description of materials, the first (*classical*) based on empirical rules (charges on atoms following the valence in the periodic table, polarization inserted *ad hoc*, Coulombic interactions between point charges combined with two body interactions) and the second (*ab-initio*) or (*first-principles*). In this approach both the ionic and the electronic structure (or, better the electronic degrees of freedom) concur to the determination of the structure and of its evolution as a function of temperature. Before taking a close look to the basics of the formalism underlying first-principles molecular dynamics (FPMD), a few points are worth to be underlined. The notions of *first-principles* and *quantum* do not have to be amalgamated and considered as equivalent. Calling FPMD as *quantum molecular dynamics* will be an overstatement driving a serious misunderstanding especially in young practitioners. FPMD calculates forces derived from a potential energy established by accounting for the quantum nature of the electronic structure [8]. This is expressed, most of the times, by resorting to density functional theory (DFT) and its associated total energy functional, playing a role of a potential energy from which the forces can be extracted. This statement is built on the Newtonian nature of the equations of motion that are by no means *quantum* since they remain fully deterministic through the evolution in time of atomic trajectories. Having established that there is nothing *quantum* in the evolution of trajectories since the quantum nature of FPMD lies in the potential energy, it is worth to dwell on the significance of *non empirical* commonly attached to first-principles approaches. Electronic structure methods such as density functional theory contain a certain degree of empiricism since they are based, without undermining their accurate character, on ideas and schemes that cannot be considered fully exact, essentially due to the complexity of the inherent many-body interactions. Several ingredients can appear to have an *empirical* nature since lying on initial assumptions. To this category belongs the expression of the exchange-correlation functional, for which all existing recipes have been proposed by following a specific rational constructed from plausible hypotheses and developed in the earnest search for improvements. Despite these undeniable (and somewhat undesirable) features, electronic structure methods as those employed in first-principles molecular dynamics has to be considered as *non-empirical* since they do not employ any experimental input in the expression of the total energy. In other words, regardless of the analytical expression for those parts that are not uniquely determined within DFT (as the exchange-correlation energy and the dispersion forces, if added) no parts of the total energy are inspired by results obtained out of the selected theoretical framework. The advantages of this drastic choice are numerous and much more rewarding of those quite often invoked (affordability, handiness, overall simplicity, control of tractable physical parameters) when using effective potentials. The main advantage is intrinsic transferability, this meaning the use of the same scheme regardless of the specific thermodynamic conditions of the system under consideration. In addition, one has access to chemical bonding anytime during the time evolution of the system, thereby allowing interpretation of any phenomena in terms of electronic structure features. Also, an appealing view on the chemical nature of the system is readily available, without a priori labelling on supposedly *ionic*, *covalent* or *metallic* character of bonding.

1.2 Basics of First-Principles Molecular Dynamics

This section is intended to recall the main guidelines of FPMD in the form introduced by Car and Parrinello (CP) in 1985 [9]. Despite the consistent period of time elapsed since the appearance of that historical paper, the method remains a remarkable example of theoretical strategy combining concepts of statistical mechanics, electronic structure theory and computational implementation. Let's start by considering the Lagrangian of a system composed of ions of coordinates R_i and electrons, described through a suitable set of orbitals $\psi_j(r)$:

$$\mathcal{L}_{CP} = \mu \sum_i \int |\dot{\psi}_i(\mathbf{r})|^2 d\mathbf{r} + \frac{1}{2} \sum_{I=1}^{N} M_I \dot{\mathbf{R}}_I^2 - E_{\text{tot}}[\{\psi_i\}; \{\mathbf{R}\}] - \sum_{ij} \lambda_{ij} \left(\int \mu \psi_i^*(\mathbf{r}) \psi_j(\mathbf{r}) d\mathbf{r} - \delta_{ij} \right) \quad (1.1)$$

The first difference to be underlined with respect to a classical molecular dynamics scheme rests on the dependence of the total energy on both the ionic and the electronic variables, i.e. $E_{\text{tot}}[\{\psi_i\};\{\mathbf{R}\}]$. The total energy of the FPMD method is the equivalent of the potential energy in a classical scheme, its analytical form being essential inaccessible since depending on a multitude of degrees of freedom. Obtaining a realistic total energy and exploiting it to gain information on the properties of a system is crucially dependent on the availability of a theoretical model such as density functional theory, this accounting for an entire field of knowledge by itself, regardless of any targeted use for FPMD calculations. The second member of the above equation is the total kinetic energy of the ionic part, easily recognizable since dependent uniquely on R_i. Consideration of these two terms only corresponds to a dynamical evolution in which all contribution of the electronic structure is built inside the total energy. If the Lagrangian consisted of these two terms only, we would be dealing with the so-called Born-Oppenheimer (BO) molecular dynamics. The underlying hypothesis views the electrons lying exactly, for each set of ionic coordinates, on the ground state Born Oppenheimer surface, the forces being calculated under these conditions. While this approach is feasible, and currently pursued in several implementations of FPMD, it requires optimization of the electronic structure through strict (and time consuming) convergence of the orbitals to their ground state value for each step of the time trajectory. Such procedure can be avoided by introducing, as in the above CP Lagrangian, the notion of fictitious electronic degrees of freedom, associated to the electronic wave functions (the orbitals) via a fictitious mass μ. Accordingly, the first member of the CP Lagrangian is a kinetic energy measuring the departure of the orbitals from the BO surface. Provided μ is small enough, one can safely avoid recalculating the electronic ground state, while keeping the electronic and ionic degrees of freedom on well separated range of frequencies for their movement. The idea of attributing a dynamical character to the electronic wave functions establishes a correspondence between the evolution of a time trajectory (in principle involving two families of degreed of freedom, ions and orbitals) and the concept of systems at equilibrium in the sense of statistical mechanics. This is key ingredient of the method that allows bridging the gap between two communities (statistical mechanics and electronic structure calculations). Due to this unifying view, the CP approach appears to benefit of an everlasting novelty, ensuring its unshakeable importance in theoretical condensed matter and beyond. In practice, the introduction of a variable μ taking very small values prevents (on reasonably long time scales for FPMD applications) the full dynamical systems (ions and "electronic" degrees of freedom) to attain equilibrium in the thermodynamic sense, this meaning equipartition of kinetic energy. The CP method has been conceived exactly to achieve the opposite: ions follow a dynamical evolution at the temperature of interest, while the fictitious electrons, bearing much higher dynamical frequencies due to their mass, adapt adiabatically to the ionic movement with no energy exchange between this two subsystems. In practice, at the beginning of a FPMD run and before starting the dynamics, one has to perform a single optimization of the electronic structure to make sure the BO conditions is fulfilled. Then, under the action of the equations of motion, one for each set of degrees of freedom as readily derived from the Lagrangian, the ionic motion will evolve in time while the electronic structure will adjust adiabatically (also by following its own Newton dynamics) to any change in the ionic configuration. Since μ has to be made very small (typically 50–100 times smaller than the ionic mass) and the corresponding frequencies are much higher than the ionic one, timesteps are shorter than in classical molecular dynamics, thereby causing an increased computational cost. This shortcoming has been pointed out as the main disadvantage of the method, together with the obvious observation that the account of the electronic structure increases the predictive power of molecular dynamics while drastically decreases its affordability. Before addressing one specific example of FPMD application, it is worth to underline that using DFT as a mean to describe the potential energy of a system can never be considered as a routine step in the implementation of the CP equations of motion. The typical stumbling block to be surmounted to confer a realistic character to the entire simulation is the choice of the exchange-correlation part of the total energy functional. This part stems directly from the approximate character of the Kohn-Sham formulation, in which the electrons are treated as independent particles, giving access to an electronic density that is simple superposition of their squared value, to indicate a probability. While this ansatz makes the full derivation of the total energy tractable, the quest of predictive power requires quantum effects (exchange energy) and many-body electronic effects (correlation energy) to be reinserted appropriately. This is the role of the exchange-correlation functionals, for which, over the years, several recipes going beyond the local density approximation (LDA) have been proposed. In 1999, we were able to demonstrate that LDA does not work properly for a prototypical network chalcogenide (liquid $GeSe_2$), by correlating this shortcoming to an insufficient account of the ionic character of bonding, partially restored with the introduction of the generalize gradient approximation. In what follows, these concepts are exemplified for the case of liquid GeSe, for which we moved to a step forward, by introducing three distinct sets of exchange-correlation functional (all beyond LDA) and performing a comparative study [10].

1.3 Liquid GeSe

It is worthwhile to consider the structural properties of disordered chalcogenides since there are a wealth of experimental data reporting on the existence of intermediate range order (IRO) as a function of composition, for instance in Ge_xSe_{1-x} ($0 \leq x \leq 1$) systems for the $0.15 \lesssim x \lesssim 0.40$ range [11, 12]. Interest in liquid GeSe rests on the lack of IRO attributed to the absence of a predominant structural motif [13, 14]. In one of our early investigations of the structural properties of disordered network-forming materials [13] we employed for the exchange-correlation functional (XC) the Perdew-Wang (PW91) scheme as the generalized gradient approximation (GGA) in the framework of density functional theory (DFT) [15, 16]. This choice was substantiated by early achievements (referred to in the previous section) showing that disordered chalcogenides cannot be modelled quantitatively by relying on the local density approximation for the exchange-correlation functional since the resulting structure has little in common with the experimental one and, in particular, intermediate range order is absent. Focusing on the first set of GGA data (those obtained via PW91) we found that some features in the Ge subnetwork were a clear indications of the need of further investigations in the search of the optimal choice for the XC functional or, at least, bringing substantial improvements. When looking at the Ge-Ge pair correlation function $g_{GeGe}(r)$, the two maxima located before and after a well discernible minimum were insufficiently reproduced by the calculations. In this respect, we took advantage of previous results on disordered $GeSe_2$ to ascribe this shortcoming to an exceedingly high impact of the metallic nature of bonding on the topology of the coordination shells. In fact, an overestimate of the metallic character of bonding was identified as responsible of shallow patterns of maxima and minima for a Ge-Ge pair correlation function [13]. The case of liquid and glassy $GeSe_2$, [17–19] is instrumental to substantiate these findings. The use of the scheme due to the Becke (B) for the exchange energy and Lee, Yang and Parr (LYP) for the correlation energy improved to a very large extent the atomic structure, as exemplified by the shape of $g_{GeGe}(r)$. Having established that BLYP can have a non-negligible effect in the case of Ge-Se disordered chalcogenides, we have adopted the same recipe to the case of liquid GeSe, thereby providing a revealing example of the effect of the XC functional on the structure of chalcogenides and, more general, of materials for which FPMD proved to be applicable with an unmistakable success. In what follows, FPMD data using BLYP will be presented for the case of liquid GeSe, at the same temperature as in [13, 14]. To enrich this analysis, we have also accounted for the Perdew-Burke-Ernzerhof (PBE) functional, very much popular in the area of disordered materials [20–23]. Also, new PW91 data are produced to complement those of [13].

1.4 Theoretical Model

The systems employed had $N = 120$ (60 Ge and 60 Se) atoms (BLYP and PBE cases) and $N = 108$ (54 Ge and 54 Se) (PW91 case). The FPMD simulations were performed in the NVT ensemble. In regard to the new PW trajectory, the results did not differ significantly from those of [13], while the diffusion coefficients, calculated on a more extended trajectory, were smaller, most likely because of the longer relaxation times allowing for actual diffusion at equilibrium. Concerning the PBE scheme, it is worthwhile to point out that this GGA scheme does not differ drastically from the PW91 one since it is based on an extrapolation of the XC energy pertaining to the homogeneous electron gas. On the contrary, it is expected to correct some the deficiencies of the PW91 scheme [20].

As customary in FPMD Car Parrinello calculations, valence electrons only were considered in combination with norm conserving pseudopotentials of the Trouiller-Martins type [24]. The energy cutoff of the wave functions expansion was equal to $E_c = 30$ Ry. Equilibrium trajectories as long as 100 ps were considered, with calculated errors bars (extracted from sub-trajectories) taking values of 5% at most for structural properties.

1.5 Neutron Total Structure Factor and Total Pair Correlation Function

With the purpose of describing the structure of liquid GeSe it is convenient to introduce the total neutron structure factor $S_{\mathrm{T}}(k)$ that reads

$$S_{\mathrm{T}}(k) - 1 \equiv \sum_{\alpha=1}^{n}\sum_{\beta=1}^{n} \frac{c_\alpha c_\beta b_\alpha b_\beta}{\langle b\rangle^2}\left[S_{\alpha\beta}^{\mathrm{FZ}}(k) - 1\right] \tag{1.2}$$

where α and β are the chemical species, $n = 2$ is their number, b_α is the coherent neutron scattering length related to the species α and c_α is the atomic fraction. In addition, $\langle b\rangle = c_{\mathrm{Ge}}b_{\mathrm{Ge}} + c_{\mathrm{Se}}b_{\mathrm{Se}}$ is the mean coherent neutron scattering length, while $S_{\alpha\beta}^{\mathrm{FZ}}(k)$ is a Faber-Ziman (FZ) partial structure factor. In real space one has the corresponding information via

$$\begin{aligned} g_{\mathrm{T}}(r) - 1 &= \frac{1}{2\pi^2 n_0 r}\int_0^\infty \mathrm{d}k\, k\, \left[S_{\mathrm{T}}(k) - 1\right]\sin(kr) \\ &= \sum_{\alpha=1}^{n}\sum_{\beta=1}^{n} \frac{c_\alpha c_\beta b_\alpha b_\beta}{\langle b\rangle^2}\left[g_{\alpha\beta}(r) - 1\right] \end{aligned} \tag{1.3}$$

where $g_{\alpha\beta}(r)$ is a partial pair distribution (also termed correlation) function and n_0 is the atomic number density.

In Fig. 1.1 the experimental result $S_{\mathrm{T}}^{\mathrm{exp}}(k)$ [12] is shown with $S_{\mathrm{T}}^{\mathrm{th}}(k)_{\mathrm{PW}}$, $S_{\mathrm{T}}^{\mathrm{th}}(k)_{\mathrm{BLYP}}$ and $S_{\mathrm{T}}^{\mathrm{th}}(k)_{\mathrm{PBE}}$. $S_{\mathrm{T}}^{\mathrm{th}}(k)_{\mathrm{PW}}$ and $S_{\mathrm{T}}^{\mathrm{th}}(k)_{\mathrm{PBE}}$ have very close shapes, while the BLYP approach improves the height of the peak at $k \sim 4$ Å^{-1}, and underestimates the height of the peak at $k \sim 2.2$ Å^{-1}. Overall, none of the three total neutron structure factors stands out as drastically better performing than the others, and, in particular, none of them agrees entirely with the experimental one for $k \geq 4$ Å^{-1}.

The total pair distribution functions are shown in Fig. 1.2. It appears that $g_{\mathrm{T}}^{\mathrm{PW91}}(r)$ and $g_{\mathrm{T}}^{\mathrm{PBE}}(r)$ are very close and in very good agreement with $g_{\mathrm{T}}^{\mathrm{exp}}(r)$ at the level of the first peak. Such peak is higher in the $g_{\mathrm{T}}^{\mathrm{BLYP}}(r)$ case, while the first minimum at $r \sim 3$ Å is more pronounced, showing that BLYP produces a more discernible sequence of shells of neighbors than the PW91 and the PBE schemes.

1.6 Reciprocal Space Properties: Faber-Ziman Partial Structure Factors

Focusing on the Faber-Ziman [25] partial structure factors, $S_{\alpha\beta}^{\mathrm{PW91}}(k)$, $S_{\alpha\beta}^{\mathrm{BLYP}}(k)$ and $S_{\alpha\beta}^{\mathrm{PBE}}(k)$ (see Fig. 1.3), it appears that the best agreement with experiments is found for $S_{\mathrm{SeSe}}(k)$. Also, the differences among $S_{\mathrm{SeSe}}^{\mathrm{PBE}}(k)$, $S_{\mathrm{SeSe}}^{\mathrm{PW91}}(k)$ and $S_{\mathrm{SeSe}}^{\mathrm{BLYP}}(k)$ are minimal.

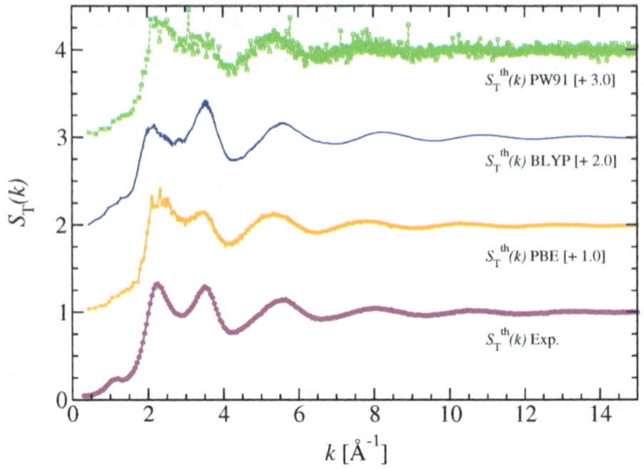

Fig. 1.1 Total neutron structure factor for liquid GeSe at T = 1000 K. The experimental result $S_T^{exp}(k)$ given in [12] (pink line with circles) is compared to $S_T^{th}(k)_{PW91}$, $S_T^{th}(k)_{BLYP}$ and $S_T^{th}(k)_{PBE}$. For clarity purposes FPMD data sets have been shifted on y axis by 3.0, 2.0 and 1.0 respectively

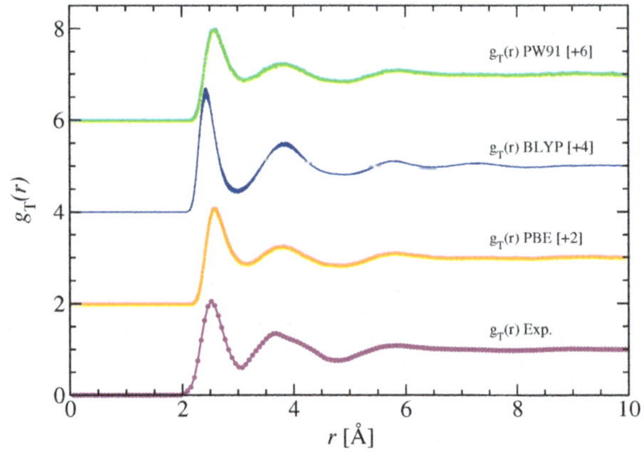

Fig. 1.2 Total pair distribution function for liquid GeSe at T = 1000 K. The experimental result $g_T^{exp}(r)$ given in [12] (pink line with circles) is compared to the calculated functions $g_T^{th}(r)$ for the PW91, BLYP and PBE models. For clarity purposes FPMD data sets have been shifted on y axis by 3.0, 2.0 and 1.0 respectively

The case of $S_{GeSe}(k)$ deserves some more comments. Clearly, $S_{GeSe}^{BLYP}(k)$ is substantially distinct from $S_{GeSe}^{exp}(k)$, $S_{GeSe}^{PW}(k)$ and $S_{GeSe}^{PBE}(k)$ in the range 2 Å^{-1} $< k <$ 4 Å^{-1}, thereby providing evidence that BLYP is less performing than PBE and PW to describe Ge-Se nearest neighbor interactions. Turning to $S_{GeGe}(k)$, this partial structure factor has intensities for the maxima and minima in better agreement with experiments for the BLYP case, especially for $k \geq 3.2$ Å^{-1}. There is a striking contrast between the patterns of $S_{GeGe}^{PW91}(k)$ and $S_{GeGe}^{PBE}(k)$ and the one made of clearly visible maxima and minima featured by $S_{GeGe}^{BLYP}(k)$. By confirming that liquid GeSe does not exhibit intermediate range order, none of the partial structure factors described above is characterized by the occurrence of the first sharp diffraction peak (FSDP) at 1 Å^{-1},

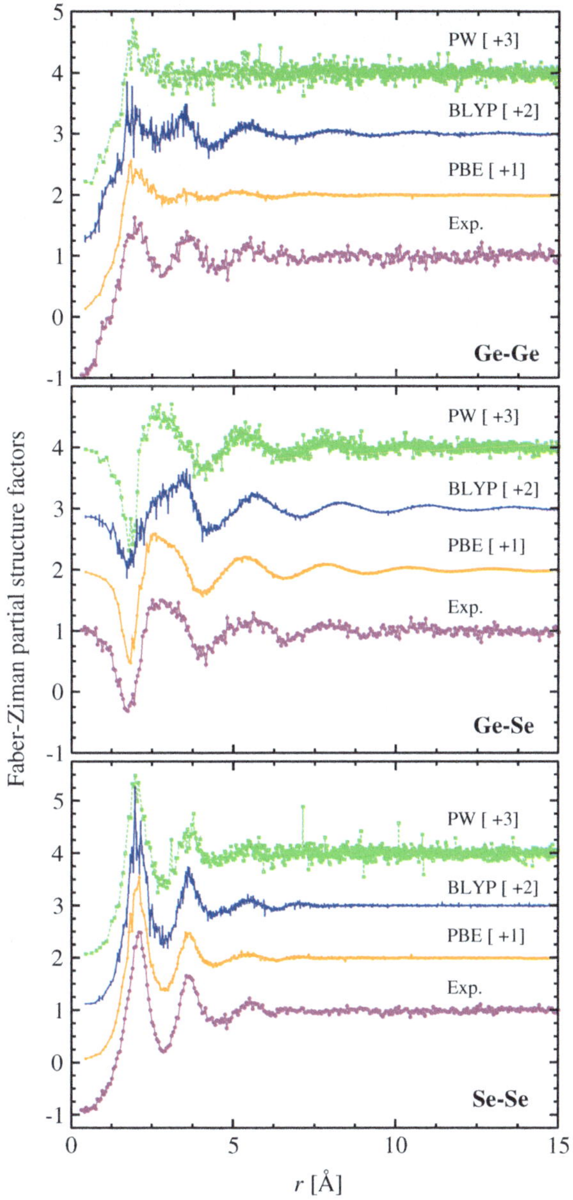

Fig. 1.3 The Faber-Ziman partial structure factors $S^{FZ}_{GeGe}(k)$ (top panel), $S^{FZ}_{GeSe}(k)$ (middle panel) and $S^{FZ}_{SeSe}(k)$ (bottom panel) for the liquids GeSe. The experimental result given in [12] (pink line with circles) is compared to the calculated functions for the PW91 (broken green lines with squares), BLYP (blue line) and PBE (orange line with crosses) models. For clarity purposes FPMD data sets have been shifted on y axis

1.7 Real Space Properties

1.7.1 Pair Distribution Functions

In the case of $g_{\alpha\beta}^{PW91}(r)$, $g_{\alpha\beta}^{BLYP}(r)$ and $g_{\alpha\beta}^{PW}(r)$ (see Fig. 1.4), the most striking difference between the BLYP, PW91 and PBE schemes (these latter two being once again very close) is in the range of interatomic distances typical of homopolar bonds, found experimentally at 2.36 Å [26]. The BLYP approach performs better than PW91 and PBE since an unambiguous minimum is found at r $\sim$ 3 Å, almost as intense as the experimental one. Also, there is maximum in $g_{GeGe}^{BLYP}(r)$ at 2.49 Å, improving upon the position 2.75–2.77 Å resulting from PBE and PW91. However, the first peak in $g_{GeGe}^{BLYP}(r)$ is by far the highest among those under consideration, giving rise to a very large coordination number associated with the Ge-Ge interactions (1.2, see Table 1.1). Focusing on the case of $g_{GeSe}^{BLYP}(r)$, the main peak is more intense and sharper than in $g_{GeSe}^{PW91}(r)$ and $g_{GeSe}^{PBE}(r)$ cases, as a consequence of the the higher electronic localization induced by this XC functional. For similar reasons, there is a more distinct separation between the shells of Ge-Se neighbors, giving rise to a visible second maximum at r $\sim$ 4 Å, not appreciable in $g_{GeSe}^{PW91}(r)$ and $g_{GeSe}^{PBE}(r)$. Negligible differences are recorded for $g_{SeSe}^{PW}(r)$, $g_{SeSe}^{PBE}(r)$ and $g_{SeSe}^{BLYP}(r)$ pair correlation functions, none of them exhibiting a first maximum at $\sim$ 2.5 Å indicative of homopolar bonds.

1.7.2 Coordination Numbers

The coordination numbers $\bar{n}_{\alpha\beta}$ can be found in Table 1.1. These quantities are equivalent to the the mean number of closest neighbors of type α around an atom of type β inside a range accounting for distances up to 3 Å, corresponding to the first minimum of the total pair correlation function. For reference purposes and in line with their frequent use when studying the topology of disordered networks, in Table 1.1 we show $\bar{n}_{\alpha\beta}$ for the chemically ordered network (CON) model and the random covalent network model (RCN). Experimental values are 3.44 for Se and 4 for Ge. Calculations underestimate these values giving coordinations equal to 2.67 (BLYP) 3.07 (PBE) 3.09 (PW91) for Se and 3.74 (PBE), 3.79 (PW91) 3.82 (BLYP). This is especially true in the case of Se due to the inability of the models to provide structures with neglect of Se-homopolar bonds. To visualize the network topology of *l*-GeSe in the BLYP case we make available in Fig. 1.5 a snapshot of the structure. There is an undeniable variety of coordination motifs for both Ge and Se, none of them being predominant as the $GeSe_4$ tetrahedron in other systems of the Ge-Se family at concentrations richer in Se ($GeSe_2$ and $GeSe_4$ for instance).

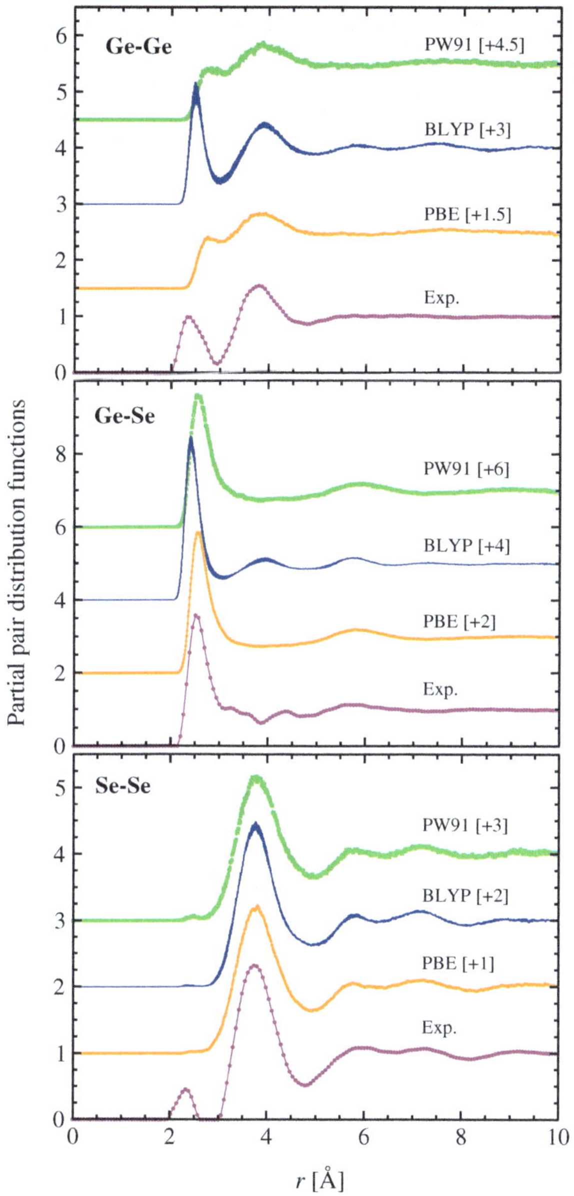

Fig. 1.4 The partial pair distribution functions $g_{\mathrm{GeGe}}(r)$ (top panel), $g_{\mathrm{GeSe}}(r)$ (middle panel) and $g_{\mathrm{SeSe}}(r)$ (bottom panel) for the liquids GeSe. The experimental result given in [12] (pink line with circles) is compared to the calculated functions for the PW91 (broken green lines with squares), BLYP (blue line) and PBE (orange line with crosses) models. For clarity purposes FPMD curves have been shifted on y axis

Table 1.1 The first (FPP) and second (SPP) peak position in $g_{\alpha\beta}(r)$ and the nearest neighbor coordination numbers $\bar{n}_{\alpha\beta}$ for the FPMD models of liquid GeSe (present work). The predictions of the CON and RCN models are also reported [12]

$g_{\alpha\beta}(r)$	Liquid GeSe	FPP (Å)	SPP (Å)	$\bar{n}_{\alpha\beta}$	$\bar{n}_{\alpha\beta}$ (CON)	$\bar{n}_{\alpha\beta}$ (RCN)
$g_{GeGe}(r)$	PW	2.77	3.87	0.8	2	2.6667
	BLYP	2.49	3.90	1.2	2	2.6667
	PBE	2.75	3.87	0.7	2	2.6667
	exp	2.36	3.81	0.8		
$g_{GeSe}(r)$	PW	2.57	–	2.99	2	1.3333
	BLYP	2.41	3.96	2.62	2	1.3333
	PBE	2.57	–	3.04	2	1.3333
	exp	2.54	3.5	3.2		
$g_{SeGe}(r)$	PW	2.57	–	2.99	2	1.3333
	BLYP	2.49	3.96	2.62	2	1.3333
	PBE	2.57	–	3.04	2	1.3333
	exp	2.54	3.5	3.2		
$g_{SeSe}(r)$	PW	2.50	3.80	0.1	0	0.6667
	BLYP	2.36	3.77	<0.1	0	0.6667
	PBE	2.46	3.82	<0.1	0	0.6667
	exp	2.34	3.80	0.222		

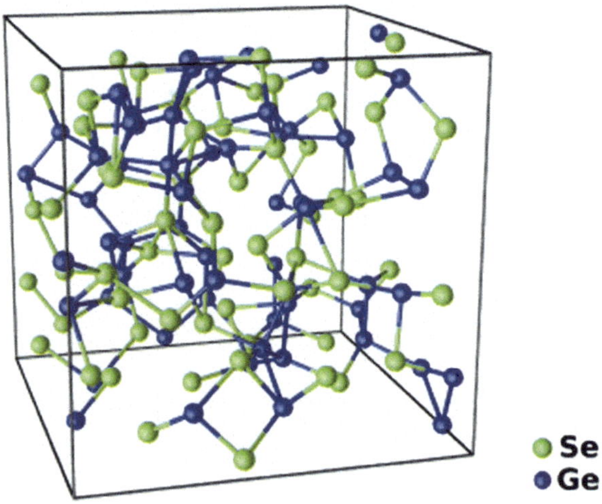

Fig. 1.5 For the BLYP model: structure of liquid GeSe at a given time along the trajectory where the Ge atoms are (dark) blue and the Se atoms are (light) green

1.7.3 Structural Units

In this section liquid GeSe is described from a different point of view, the one based on the α-l structural units in which an atom of species α (Ge or Se) is l-fold linked to its neighbors. To give an explicit example, Ge-$GeSe_2$ represents a Ge atom bound to 1 Ge atom and 2 Se atoms while Se-SeGe stands for a Se atom that is connected to 1 Se atom and 1 Ge atom (see Table 1.2, where the results are given as percentages of the total occurrences). Two main results are worth pointing out. First, by focusing on the PW91 and PBE models, it appears that none of the structural units accounts for more than half (50%) of the total coordination, this meaning that liquid GeSe does not feature a predominant topological motif that can be readily used to describe the atomic structure. Second, PW91 and PBE are quite similar when one compares the corresponding sets of data. The first of these conclusions is challenged by the BLYP results, indicating that the fourfold coordination accounts for 58.1% of the total number of linkages of Ge atoms. Interestingly, appreciable percentages of Ge-$GeSe_3$ Ge-Ge_2Se_2 units do exist also,(24.4% and 24.2% respectively). Increasing the number of fourfold coordinated Ge atoms has the effect of lowering the number of three fold and twofold units (31.3%, PW91; 33.7% PBE against 21.4% BLYP). Turning to the Se atoms, a similar effect occurs for the number of twofold Se atoms (21.5%, PW91; 18.3% PBE and 41.1% BLYP). As a general comment, a higher number of tetrahedra induces a reduction in the number of non-negligible structural units, i.e. those being present within the system in a percentage larger than 10%. The opposite is also true since, within the present PW91 and PBE calculations, fivefold units (for Ge) and fourfold units (for Se) are connected to at least 10% of the Se or Ge atoms. One can try to establish a connection between these results and the differences between the conceptual foundations of the three exchange-correlation functionals employed. This can be expressed as follows. The electron localization effects characterizing the BLYP approach promote to a smaller number of different units for the coordination of Ge ad Se, when compared to the PW91 and PBE cases.

1.8 Diffusion and Dynamical Properties

Our interest for diffusion is rooted into its key significance when seeking to understand the correlation between the topology of a network and the mobility of its atoms. Previous results have shown that whenever the predominance of the tetrahedron as main unit increases (as in liquid $GeSe_2$ within the BLYP approach), the coefficient diffusion decreases (this can be seen in comparison to the PW91 case in [19]). To establish whether the same effect is observed in liquid GeSe, we have extracted from the trajectories the statistical average of the mean square displacement of chemical species α

$$\left\langle r_\alpha^2(t) \right\rangle = \frac{1}{N_\alpha} \left\langle \sum_{i=1}^{N_\alpha} |\mathbf{r}_{i\alpha}(t) - \mathbf{r}_{i\alpha}(0)|^2 \right\rangle \tag{1.4}$$

Table 1.2 Coordination units $\bar{n}_\alpha(l)$ in liquid GeSe for the PW91, PBE and BLYP models

		Proportion $\bar{n}_\alpha(l)$ [%]		
		PW	BLYP	PBE
Ge atom				
$l = 2$	Se_2	4.8	1.8	4.0
	GeSe	0.6	0.5	0.4
$l = 3$	$GeSe_2$	9.5	9.4	8.5
	Se_3	21.8	12.0	25.2
$l = 4$	$GeSe_3$	24.7	24.4	24.5
	Ge_2Se_2	7.4	24.2	6.1
	Se_4	12.1	9.5	13.9
$l = 5$	Ge_2Se_3	5.3	4.7	4.2
	$GeSe_4$	7.0	3.2	6.7
	Se_5	2.0	0.8	2.3
Se atom				
$l = 1$	Ge	0.8	0.5	0.4
	Se	<0.1	–	<0.1
$l = 2$	Se_2	<0.1	–	<0.1
	SeGe	0.3	0.4	0.2
	Ge_2	21.2	40.7	18.1
$l = 3$	Se_2Ge	0.1	<0.1	0.1
	$SeGe_2$	2.0	1.9	1.6
	Ge_3	50.9	48.0	53.9
$l = 4$	$SeGe_3$	2.6	1.4	2.3
	Ge_4	19.7	6.5	20.9
	Ge_2Se_2	0.2	<0.1	0.1

where N_α is the total number of particles of type α and $\mathbf{r}_{i\alpha}(t)$ is the coordinate of the ith particle of chemical species α. The results are displayed in the inset to Fig. 1.6. In the diffusive regime the diffusion coefficient is:

$$D_\alpha = \frac{\langle r_\alpha^2(t) \rangle}{6t} \tag{1.5}$$

The plot of log $\langle r_\alpha^2(t) \rangle$ versus log t has a linear behavior with a gradient equal to unity extrapolated at infinite time. The values of D_{Ge} and D_{Se} obtained in this work (PW91, BLYP and PBE approaches) are reported in Table 1.3 together with those of [19] (liquid $GeSe_2$) and liquid Ge_2Se_3 [27] (BLYP calculations in the two latter cases).

Taking into account the error bars, the BLYP values of D_{Ge} and D_{Se} for liquid Ge_2Se_3 and liquid GeSe do not differ significantly. However, for both cases, we note that liquid $GeSe_2$ features smaller diffusivities. This can be rationalized as follows.

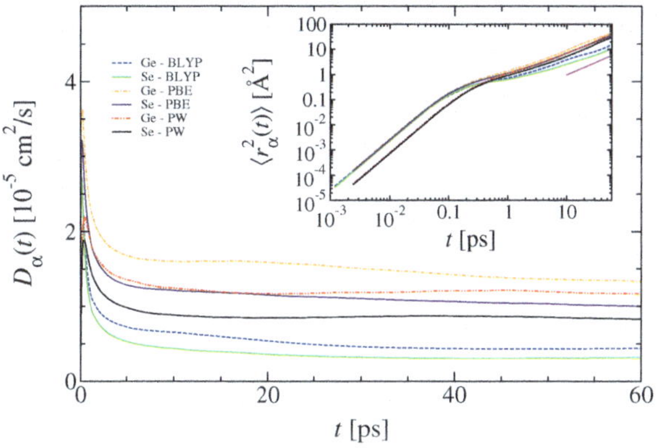

Fig. 1.6 The diffusion coefficients of Ge atoms (broken lines) and Se atoms (solid lines) at T = 1000 K extracted from the mean square displacements at time t (see 1.5). Inset: the time dependence of the mean square displacements of the Ge atoms (broken lines) and the Se atoms (solid lines). Straight line: The slope corresponding to the diffusive regime

Table 1.3 The diffusion coefficients of the Ge and Se atoms in the liquids GeSe as obtained from FPMD models

	D_α ($\times 10^{-5}$ cm^2/s)	
	Ge	Se
l-$GeSe_2$ [19]	0.2 ± 0.2	0.2 ± 0.2
l-Ge_2Se_3 [27]	0.62 ± 0.2	0.46 ± 0.2
l-GeSe PW—108 atoms (present work)	1.20 ± 0.1	0.85 ± 0.1
l-GeSe BLYP—120 atoms (present work)	0.50 ± 0.1	0.35 ± 0.1
l-GeSe PBE—120 atoms (present work)	1.48 ± 0.2	1.0 ± 0.1

Ge atoms not tetrahedrally coordinated and Ge-Ge homopolar bonds are present in both liquid Ge_2Se_3 and liquid GeSe since they are located on the Ge-rich side of the Ge_xSe_{1-x} composition range, where Ge atoms not connected within tetrahedral are more mobile that those at the center of fourfold connections. For this reason, a larger mobility of the atomic species can occur when compared to the case of the stoichiometric $GeSe_2$ network. As a second observation, the diffusion coefficient is strongly reduced for both species within BLYP due to an increase of the tetrahedral coordination and the presence of fewer structural units accounting for the shells of Ge and Se neighbors.

1.9 Conclusions

We have selected the example of liquid GeSe to provide evidence of a typical case in which the structural details of a disordered network can depend on the choice made for the exchange-correlation part of the energy functional implemented within the DFT Kohn-Sham formalism. At the origin of this postulate we would like to recall the existence of results for the case of disordered $GeSe_2$ systems that proved sensitive to the selection of the BLYP scheme as an alternative to the PW91 one. The key ingredient of this choice reside in the enhanced electronic localization of the BLYP scheme inducing a better description of short and intermediate range properties. As a preliminary result, it is important to underline that, to further test the sensitivity to the XC functional, we have also selected the PBE XC functional by finding, in the present case of liquid GeSe, only moderate changes compared to PW91. This means that, at least for this specific network, the PBE implementation is not capable of proving intrinsically better (or, merely, different) than the PW91 one. Overall, the present results confirm that BLYP improves the description of Ge-Ge short range correlations, as demonstrated by two clear maxima separated by a sharp minimum found in the $g_{\mathrm{GeGe}}^{\mathrm{BLYP}}(r)$ pair distribution function. By increasing the number of tetrahedral units, BLYP confers to the network an enhanced stability reflected by a decrease of the diffusion coefficients for both species. To complete our description in a objective fashion, these positive effects of BLYP have to be invoked together with some shortcomings also to be ascribed to the electron localization inherent in BLYP. In particular, we found Ge-Se interactions over-structured and an overestimate for the height of the first maximum in $g_{\mathrm{GeGe}}^{\mathrm{BLYP}}(r)$, leading to coordination numbers in moderate disagreement with experiments.

1.10 Perspectives

First-principles molecular dynamics is currently used in all research areas interested in acquiring knowledge on the macroscopic properties of a system from its behavior at the atomic scale. When using this technique, one is often faced to the challenge of bypassing temporal and scale limits, thereby preventing skepticism on the frequent question/comments about the reliability of results obtained for small samples and some hundredths of ps. Linear-scale methods developed by improving and extending the CP framework have been developed in recent years and yet, it appears difficult to predict for which system sizes these approaches will become less time consuming than the traditional one. For instance, in the case of the CP2K method, implemented in conjunction with the BO dynamics, the search of the electronic ground state for each ionic configuration was found to be extremely lengthy, most likely due to gap closing configurations adding additional electronic states to the relaxation. An interesting alternative consists in adopting the so-called second-generation Car-Parrinello molecular dynamics, in which the self-consistency of the electronic structure is not completely achieved but effectively controlled via the introduction of a Langevin thermostat term in the equation of motion for the wave functions. We were able to investigate a ternary material within this formalism, by achieving a convincing description of its structure [28]. Exciting perspectives have also been opened by the study of thermal transport in disordered systems via the introduction of a methodology that gives access to the calculation of the thermal conductivity on affordable simulation sample. While pointing out that the size effects might be encountered depending on the length of the phonon free paths, the methodology is fully capable of producing extrapolated results in excellent agreement with experiments [29].

References

1. T. Heine, Front. Mater. **1**, 7 (2014)
2. B. Alder, T. Wainwright, J. Chem. Phys. **27**, 1208 (1957)
3. G. Battimelli, G. Ciccotti, Eur. Phys. J. **43**, 303 (2018)
4. P. Allen, J.D. Tildesley, *Computer Simulation of Liquids: Second Edition* (Oxford University Press, 2017)
5. D. Frenkel, B. Smit, *Understanding Molecular Simulation: From Algorithms to Applications* (Elsevier, 2001, 1996)
6. P. Blochl, M. Parrinello, Phys. Rev. B **45**, 9413 (1992)
7. M. Wilson, B.K. Sharma, C. Massobrio, J. Chem. Phys. **128**, 244505 (2008)
8. R. Car, F. de Angelis, P. Giannozzi, N. Marzari, *Handbook of Materials Modeling: First-Principles Molecular Dynamics* (Springer, Dordrecht, 2005)
9. R. Car, M. Parrinello, Phys. Rev. Lett. **55**, 2471 (1985)
10. C. Massobrio, A. Pasquarello, R. Car **121**, 2943 (1999)
11. P.S. Salmon, R.A. Martin, P.E. Mason, G.J. Cuello, (London) **435**, 75 (2005)
12. P.S. Salmon, J. Non-Cryst, Solids. **353**, 2959 (2007)
13. F.H.M. van Roon, C. Massobrio, E. de Wolff, S.W. de Leeuw, J. Chem. Phys. **113**, 5425 (2000)
14. J.Y. Raty, V.V. Godlevsky, J.P. Gaspard, C. Bichara, M. Bionducci, R. Bellissent, R. Ceolin, J.R. Chelikowsky, P. Ghosez, Phys. Rev. B **64**, 235209 (2001)
15. J.P. Perdew, Y. Wang, Phys. Rev. B **45**, 13244 (1992)
16. J.P. Perdew, J.A. Chevary, S.H. Vosko, K.A. Jackson, M.R. Pederson, D.J. Singh, C. Fiolhais, Phys. Rev. **B 46**, 6671 (1992)
17. A.D. Becke, Phys. Rev. A **38**, 3098 (1988)
18. C. Lee, W. Yang, R.G. Parr, Phys. Rev. B **37**, 785 (1988)
19. M. Micoulaut, R. Vuilleumier, C. Massobrio, Phys. Rev. B **79**, 214205 (2009)
20. J.P. Perdew, K. Burke, M. Ernzerhof, Phys. Rev. Lett. B **77**, 3865 (1996)
21. J. Akola, R.O. Jones, Phys. Rev. Lett. **100**, 205502 (2008)
22. G.C. Sosso, S. Caravati, R. Mazzarello, M. Bernasconi, Phys. Rev. B **83**, 134201 (2011)
23. G.C. Sosso, G. Miceli, S. Caravati, J. Behler, M. Bernasconi, Phys. Rev. B **85**, 174103 (2012)
24. N. Troullier, J.L. Martins, Phys. Rev. B **43**, 1993 (1991)
25. Y. Waseda, *The Structure, of Non-Crystalline Materials* (McGraw-Hill, New-York, 1980), describes the relationship between the three sets of partial structure factors commonly used (Faber-Ziman, Ashcroft-Langreth and Bhatia-Thornton)
26. I. Petri, P.S. Salmon, H.E. Fischer, J. Phys. **11**, 7051 (1999)
27. S. Le Roux, A. Zeidler, P.S. Salmon, M. Boero, M. Micoulaut, C. Massobrio, Phys. Rev. **84**, 134203 (2011)
28. A. Bouzid, S. Gabardi, C. Massobrio, M. Boero, M. Bernasconi, Phys. Rev. B **91**, 184201 (2015)
29. A. Bouzid, H. Zaoui, P.L. Palla, G. Ori, M. Boero, C. Massobrio, F. Cleri, E. Lampin, Phys. Chem. Chem. Phys. **19**, 9729 (2017)

Chapter 2
Assessing the Versatility of Molecular Modelling as a Strategy for Predicting Gas Adsorption Properties of Chalcogels

Iréné Berenger Amiehe Essomba, Carlo Massobrio, Mauro Boero and Guido Ori

Abstract Modelling gas adsorption of porous materials is nowadays an undeniable necessary in order to complement experiment findings with the purpose to enrich our fundamental understanding of adsorption mechanisms as well as develop better performing materials for gas mixture separation. In this contribution, we explore the possibility to use first-principles molecular dynamics (FPMD) and grand canonical Monte Carlo (GCMC) simulations to target the gas adsorption of disordered nanoporous chalcogenides (i.e. chalcogels). This computational scheme allows us to take advantage of the ability of FPMD to accurately describe the structure and bonding of the disordered nature of chalcogels as well as the potential of GCMC to model the adsorption mechanisms of porous networks. We assess the versatility of such scheme by evaluating the role of pore size, chemical stoichiometry and composition for multiple chalcogenide-based systems on nitrogen adsorption isotherms.

2.1 Introduction

The development of porous materials for gas adsorption, separation and purification is central to many energy and environmental applications [1]. A great deal of progress has been made over the last decades in this field and many different porous materials have been studied for gas adsorption including metal-oxides [2–5], zeolites [6–9], porous carbon [10–13], and metal-organic frameworks (MOFs) [14, 15]. Keeping into consideration that the ideal sorbent material needs to show enhanced uptake capacity, high selectivity, easy recyclability, as well as very good structural and chemical stability, the development of porous materials able to satisfy all these requisites still represents a challenge. Porous gels [16–19] and zeolitic chalcogenides [20, 21] have recently attracted increasing attention in the field of gas adsorption because of their high surface area and the soft Lewis base nature

I. B. Amiehe Essomba · C. Massobrio · M. Boero · G. Ori (✉)
Institut de Physique et de Chimie des Matériaux de Strasbourg UMR 7504,
Université de Strasbourg - CNRS, 23 rue du Loess, BP43, 67034 Strasbourg Cedex 2, France
e-mail: guido.ori@ipcms.unistra.fr

E. V. Levchenko et al. (eds.), *Theory and Simulation in Physics for Materials Applications*, Springer Series in Materials Science 296,
https://doi.org/10.1007/978-3-030-37790-8_2

of chalcogen elements (S, Se, and Te) [22]. Unlike nanocrystalline chalcogenides, glassy chalcogels feature amorphous networks similar to those of amorphous silica. While their properties depend on the synthesis conditions, most chalcogels exhibit a high surface area owing to their mesoporous structure (pores from 2 to 50 nm) along with a smaller amount of micropores (<2 nm) and macropores (>50 nm) [16–18, 23–25]. As a result, chalcogels are considered promising alternatives as adsorbent materials for gas transport and separation. However, despite the growing interest for this class of porous materials, the study of their adsorption properties is still limited [26–29]. While adsorption and transport in porous materials such as zeolites, porous carbon and MOF are relatively well understood, the specific case of chalcogels remains unclear with many questions left unexplained such as the role of the surface chalcogenide chemistry and the specific interactions playing at the gas/solid interface in the confined environment. This arises from a general limitation of the experimental studies of gas adsorption of chalcogels as most methods are unable to detect the features of their buried gas/solid interface. In this respect, molecular modelling has proven to be an efficient technique to investigate the gas adsorption of such porous materials. In few recent works, we resorted on a computational scheme based on first-principles molecular dynamics (FPMD) and grand canonical Monte Carlo simulations (GCMC) to identify the microscopic mechanisms of N_2, CO_2, H_2 and CH_4 adsorption in a prototypical chalcogel made of glassy GeS_2 (g-GeS_2 hereafter) [30, 31]. With the present contribution we aim to widen this first work and to assess the versatility of such computational scheme to gain insights on the role of pore size, chemical stoichiometry and composition for multiple chalcogenide-based systems on nitrogen adsorption isotherms. Gas adsorption of simple fluids such as N_2 at 77 K is a routine technique which allows characterizing the specific surface, porous volume, and pore size distribution of microporous and mesoporous media [32, 33]. This contribution is organized as follows. In a first section, the main advantages of employing FPMD and GCMC simulations within the field of modelling chalcogels are presented. In a second section, as case studies, the role of pore size, chemical stoichiometry and compositions are analyzed and discussed. We critically assess our results by invoking available experimental data before drawing a final set of conclusions and perspectives.

2.2 Computational Methodology

The computational scheme proposed here is based on the employment of FPMD simulations where the electronic structure is described within the framework of the density functional theory (DFT) and it allows to accurately describe the chemistry and bonding of amorphous chalcogenides and GCMC simulations to model the adsorption mechanisms of porous chalcogels. In Fig. 2.1 a schematic cartoon of the four main building blocks of this methodology is presented:

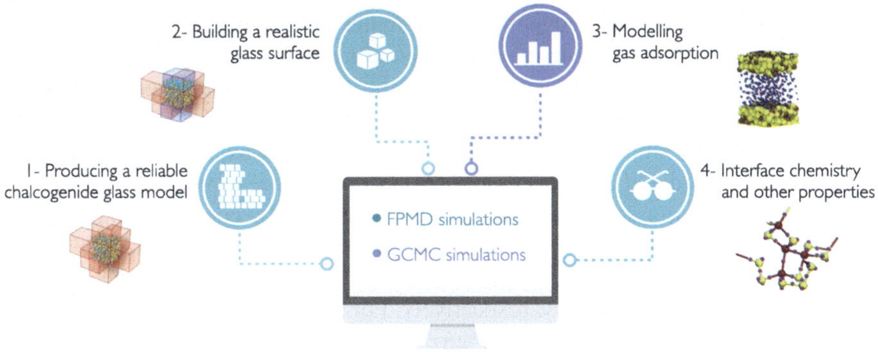

Fig. 2.1 Schematic representation of the computational methodology for modelling gas adsorption of chalcogel materials

1. Producing of a chalcogenide glass structural model in quantitative agreement with experimental findings;
2. Building and refinement of chalcogenide surface models at finite conditions;
3. Modelling gas adsorption mechanisms of chalcogel with defined porosity;
4. Detailed analysis of the gas/solid interface chemistry and computation of other properties.

Three out of the four building blocks resort to FPMD simulations in order to produce realistic and reliable structural models of glassy chalcogenides as well as being able to accurately describe the chemical interactions involved at the gas/solid interface and compute other electronic properties that might be of interest for specifics cases. GCMC simulations are used to model the gas adsorption and isotherms within the porous chalcogenide network.

2.2.1 *First-Principles Molecular Dynamics: Quantitative Prediction of Structure and Bonding of Bulk Disordered Chalcogenides and Their Surfaces*

In this section we tentatively summarize the reasons of FPMD [34] combined with DFT as the computational method of choice for the proper description of the chemistry and structure of disordered chalcogenides. Empirical interatomic potentials, routinely employed in classical MD simulations, are found to generally fail in reproducing the correct chemical behavior for systems other than those characterized by chemically ordered topologies (typical of oxide-based materials) as well as the presence of structural defects such as miss-coordinations and homopolar bonding. Thanks to a proper description of the local electronic structure, FPMD allows instead to face the hurdle of describing the complex interplay between different bonding interactions particular of disordered chalcogenide-based materials, whether liquids or glasses (such as ionocovalency [35] and metavalent bonding [36] among others). Typically, glassy phases are created by means of FPMD mimicking the melt-to-quench process used in the laboratory. Although employing a different time scale with respect to the experiments, FPMD simulations allow nowadays to achieve a quantitative structural description of disordered chalcogenides. This performance, accurately reproducing experimental data, arises from a proper choice of the FPMD computational scheme details such as the type of exchange-correlation (XC) functional, the formalism of dispersion forces (such as van der Waals (vdW)) and the size and time scales of the system and simulation trajectory, respectively. For the class of Ge-X based chalcogenide glasses with X being the chalcogen counter-ion (S, Se or Te), the generalized gradient approximation (GGA) scheme together with the XC functional developed by Becke [37] Yang and Parr [38] (BLYP) has been found to offer quantitative description (>96%) for multiple structural properties (such as neutron and X-ray structure factors, total and partial pair correlation functions). This capability arises from the ability of BLYP to enhance the valance electron cloud localization of the local atomic arrangements with respect to other XC (Perdew, Becke and Ernzehof (PBE) [39, 40] for instance) [41]. Table 2.1 shows the details of the systems considered in the present work and for which a quantitative agreement with experimental data was found over last few years. The specifics of the computational methodology employed for each system can be found in the references reported in Table 2.1. The role of vdW interactions has also be taken into attentive account, especially if the interest is towards the study of the interactions of chalcogenide surfaces with other chemicals (such as gas or organic molecules). Multiple mathematical formalisms are available to describe the vdW interactions based on empirical corrections or on first-principles calculations making use of the electronic structure provided by DFT. However, regarding the specific case of disordered chalcogenides, a thoughtful choice has to be undertaken with respect to the vdW formalism used depending on the type of system and chalcogen under study [35, 45, 46]. Typically a glass model is obtained by means of FPMD simulations within about $\sim$100–300 ps of thermal treatment plus few tens of ps of equilibration at the final finite temperature of interest. This time frame is often statistically adequate to obtain a realistic structural model of the chalcogenide glass of interest. Longer equilibration time can be eventually needed for the computation of other properties (electronic, vibrational and thermal). Once the bulk chalcogenide glass model is obtained, the corresponding surface model can be built by opening to a vacuum space the cell box along one direction and properly relax its structure at finite temperature. This thermal treatment is needed in order allow surface local atomic rearrangements and structurally relax the atoms at the surface. In so doing, the presence of dangling bonds is reduced to a minimum. Table 2.2 shows the local and average coordinations around the atomic species for the surface models of the chalcogenides targeted in this work. The reported values are evaluated by considering the whole systems and have been calculated including neighbors separated by a cutoff corresponding to the first minimum in the partial pair correlation functions. All the systems considered are mainly made of fourfold coordinated Ge and twofold coordinated X atoms. However, these surface models show an increase in Ge and

X atoms under- and over-coordinations contents at the expenses of the fourfold and twofold coordinations with respect to the bulk counterparts. These results show that the surface models show a slightly lower chemical order than the bulk counterparts. It is important to highlight that the decrease of the chemical order degree characterizes the external layers of the surface models (about ∼7–8 Å from the surface), whereas in the inner-region the atom species keeps and/or recover coordination distributions typical of bulk models. As an example, the inner-region region of g-$GeSe_2$ surface model shows a ∼73% of Ge atoms fourfold coordinated and ∼86% of Se atoms twofold coordinated, which is inline with the contents found in the parent bulk model (∼76% and ∼93%, respectively). Regarding the analysis of chemical bonding, both the electron localization function (ELF) and the maximally localized Wannier functions (MLWF) schemes are nowadays two widely used methodologies to study the valence electron (de)localization and structure and infer on the type of chemical interactions involved in amorphous chalcogenides. As an archetypal example, herein we report the typical information that can be extracted resorting to the MLWF formalism, based on the notion of Wannier functions, the MLWF centers (WFCs) and their spread (ω). This scheme allows to seek further insight into the interplay between atomic structure and electronic properties with respect to the standard information based on the electronic density of states. Recently, we have employed these quantities to rationalize the extent of covalent versus ionic nature of bonding for g-GeS_4, g-$GeSe_4$ and g-$GeTe_4$. This has ben achieved by relying on the correspondence between the distances identified in the atom-WFCs pair correlation functions ($g_{\alpha\beta}(r)$, with α = X and β = WFCs). The data ($g_{XW}(r)$ and WFCs spread) considered in the following discussion can be found in [41, 42]. In particular, three types of WFCs can be distinguished in this series of binary glassy chalcogenides. The first type, labeled W_b, represents a clear fingerprint of Ge-X heteropolar bonding and it is identified as the $g_{XW}(r)$ second peak found at about ∼0.9 Å, ∼1 Å and ∼1.25 Å for g-GeS_4, g-$GeSe_4$ and g-$GeTe_4$, respectively. The second type, labeled W_h, refers to homopolar X-X bonding and it corresponds to the typical $g_{XW}(r)$ third peak: W_h is found at about ∼1.02 Å, ∼1.18 Å and ∼1.4 Å for g-GeS_4, g-$GeSe_4$ and g-$GeTe_4$, respectively. The third type, labeled W_{lp}, refers to lone pair valence electrons not participating directly to chemical bonds but remaining localized in the proximity of the X atoms. Given the above definitions, the locations of the W_b centers with respect to the X atoms can be used to compare the covalency versus ionicity degree of bonding that characterizes these glasses. The relative position of the $g_{XW}(r)$ second peak and the closer location to its origin allow to quantify the greater ionic character of the Ge-X covalent bonds along the serie: g-GeS_4 > g-$GeSe_4$ > g-$GeTe_4$. This stems from the fact that the centers of valence electronic localization (related to heterogeneous bonding) are systematically closer to the S sites with respect to Ge sites than to the case of Se and Te sites, respectively. Furthermore, the inspection of the electronic (de)localization degree (spread, ω) of the WFCs with respect to the X-WFCs distance allows to further quantify the degree of polarity of the G-X bonding, which results following the serie Ge-S (higher polarity) > Ge-Se > Ge-Te (lower polarity). These conclusions are found in sound agreement with the expected trends on the basis of Pauling electronegativity scale and polarizability values (see Table 2.2)

Table 2.1 Details of the bulk chalcogenide glasses produced by FPMD and considered in this work

Chemical composition	Stoichiometry	n. atoms	Box size L (Å)	Density ρ (atoms Å^{-3})	References
g-Ge_xS_y	GeS_4	480	24.845	0.0313	[42]
g-Ge_xSe_y	$GeSe_2$	480	24.495	0.0327	[43]
	$GeSe_4$	480	24.871	0.0312	[42]
	$GeSe_9$	260	19.900	0.0330	[44]
g-Ge_xTe_y	$GeTe_4$	215	19.234	0.0302	[41]

Table 2.2 Local atomic coordinations and charges of the glassy chalcogenide surfaces considered in this work. For completeness, polarizability and electronegativity values are also reported

Species	Coordination	Local coordination distribution (%)[a]/charge q [b]				
		g-GeS_4	g-$GeSe_2$	g-$GeSe_4$	g-$GeSe_9$	g-$GeTe_4$
Ge	II	6.3/0.99	6.9/0.57	10.4/0.76	11.5/0.68	–
	III	11.5/1.02	19.4/0.62	21.9/0.78	15.4/0.79	18.6/0.58
	IV	66.7/1.09	58.8/0.72	56.3/0.78	73.1/0.94	62.8/0.59
	V	15.6/1.01	10.0/0.73	9.4/0.84	–	14.0/0.59
X	I	8.3/−0.29	12.5/−0.30	12.5/−0.19	7.3/−0.10	14.5/−0.13
	II	91.1/−0.26	67.8/−0.34	62.0/−0.20	88.9/−0.10	52.9/−0.14
	III	0.5/−0.51	18.4/−0.38	22.9/−0.19	3.0/−0.13	29.1/−0.17
$\bar{n}/\bar{q}$[b]	Ge	3.92/1.06	3.57/0.66	3.58/0.77	3.62/0.89	3.77/0.56
	X	1.92/−0.27	2.03/−0.34	2.05/−0.19	1.94/−0.10	2.08/−0.14
	Ge	S	Se			Te
α (a.u.)[c]	42	19	27			40
E.N.[d]	2.01	2.5	2.4			2.01

[a] Atomic coordinations are calculated including neighbors separated by a cutoff corresponding to the first minimum of the corresponding partial pair correlation functions
[b] The atomic charges q are calculated by using the Qeq method developed by Rappé and Goddard et al. [47]
[c] Polarizability values are obtained from [48]
[d] Electronegativity values are obtained from [49]

[48, 49]. In particular, the ω values and the extension of its distribution with respect to X-WFCs distances allow to infer about the more softer (more polarizable, i.e. higher polarizability) nature of Te atoms with respect to Se and S atoms.

Once the surface model of the amorphous chalcogenide system of interest is produced by means of FPMD simulations, its optimized structure can be used to build a slit-like pore by introducing a proper vacuum space with respect to the pore width of interest. However, before switching to GCMC simulations for the study of gas adsorption and isotherms, a suitable set of atomic charges have to be assigned to the atoms of the host porous network. This step, together with the refinement of the chalcogenide glass' structure, represent the main challenge of the proposed computational procedure. Indeed, classical potentials are often unable to describe deviations from chemically ordered topologies occurring, in specific network structures made of interconnected tetrahedra (as glassy chalcogenides). In the present work, we tentatively propose a set of atomic partial charges dependent on the local coordination. This approach is based on the exploitation of the charge equilibration method developed by Rappé and Goddard [47]. This method is derived from atomic ionization energies and electron affinity values in order to compute partial atomic charges of the atoms with respect to covalent radii. Thanks to the fact that the charge on each atom is distributed over a Slater orbital having the size of the atom, the advantage of the Qeq method is to predict charges as a function of the local coordination environment [47]. This charge assignment results particularly suitable for the present case since it is shown to provide a physical picture of the glassy chalcogenide surfaces by describing the charge distribution as a function of the chemical order. The chemical order for glassy chalcogenides is not only sensitive to the composition but it can substantially differ from the perfect chemical order found in the crystalline parent phases. For instance, in the case of glassy g-$GeSe_2$, we recall the perfect chemical order corresponds to the absence of any under- or over-coordinated Ge or Se atoms. To be noted that the trend of partial charges obtained with the Qeq method are found to be in agreement with those obtained with the EQeq method [50] as well as with the Bader method [51, 52] on the basis of charge densities computed by FPMD-DFT data. Whereas the Lowdin [53] and Mulliken [54] and the electrostatic potential-based (ESP) [55] methods lead to a nonzero net charge or to charges nearly insensitive to the coordination number [30]. The Qeq (EQeq) and Bader methods capture the effect of local coordination on the partial charges in chalcogenide materials. These methods provide reasonable partial charges for both the Ge and X atoms as their absolute charge increases with the coordination. However, the absolute charge values obtained by the Bader method are too large to be employed in classical simulations, conferring to the Ge and X atoms a nearly pure ionic character [30]. The charge-coordination correlation found in the Qeq(EQeq) method seems more appropriate to be employed for classical simulations such as GCMC or classical MD, presuming the potential parameters are consistent. Table 2.1 shows the atomic partial charges obtained by means of Qeq method for the g-GeS_4, g-$GeSe_4$ and g-$GeTe_4$ systems as well as g-$GeSe_2$ and g-$GeSe_9$. With this methods over-coordinated Ge (X) atoms possess a large positive (negative) charge with respect to that corresponding to stoichiometric coordination. This is directly related to their higher valence state which, in a formalism purely based on formal ionic charges [cations (Ge) and anions (X)], result in an increased charge localization. Likewise, the absolute charge value decreases descending along the element of the VI group of the periodic table for the counter-ion X ($q_S > q_{Se} > q_{Te}$). Along the Ge-Se serie, the absolute charge value for Ge sites increases with the Se:Ge ratio and it is counterbalanced by a decrease of the charge value of Se sties. Such method seems to be the best suited technique to describe changes in the valence (charge) state for different coordinations (e.g. structural order) as well as different chemical compositions and stoichiometry. However, it has to be underlined that it has not been demonstrated whether Qeq is as accurate as DFT-based methods reproducing QM energies nor that the predicted changes in polarization during dynamics agree with QM. Moreover, problem using this method could occur when high temperatures come into play or when extreme compositions are studied. With all these cautions in mind, the charges determined by Qeq method, in the presence of rationally physical situations, are reasonable and can be used to calculate the Coulomb interaction between atoms in (semi)empirical approaches.

2.2.2 Grand Canonical Monte Carlo Simulations: Quantitative Prediction of Gas Adsorption Isotherms

Once a proper model of chalcogel pore is optimized by means of FPMD simulations, its structural model (i.e. atoms coordinates) and the corresponding atomic charge distribution (coordination dependent) can be employed to perform GCMC simulations in order to model the gas adsorption isotherm. The GCMC technique is a stochastic method applicable to a system having a constant volume V (the pore with the adsorbed phase) in equilibrium with an infinite reservoir of molecules imposing a chemical potential μ for each species (N_2 for this work) and temperature T. The absolute adsorption isotherm is given by the ensemble average of each number of adsorbate molecule as a function of the fugacity f_{N_2} of the reservoir (the latter is determined from the chemical potential μ_{N_2}). For the adsorbate, given the fugacities and temperatures considered in this work, the gas pressure P is assumed to be equal to the fugacity f_{N_2} (i.e. ideal gas assumption). We performed GCMC simulations of N_2 adsorption at $T = 77$ K for chalcogel pores made of: g-GeS_4, g-$GeSe_4$ and g-$GeTe_4$ as well as g-$GeSe_2$ and g-$GeSe_9$. We model the N_2 adsorption isotherm between zero pressure and the N_2 saturation pressure P_0. Nitrogen adsorption at low temperature is a routine laboratory characterization technique of porous materials. For instance, the specific surface area of porous materials is usually assessed from adsorption experiments. For the purpose of this work, nitrogen was described using the model of Potoff and Siepmann (Trappe forcefield) [56]. In this model, each N atom of the rigid N_2 molecule is a center of repulsion and dispersion interactions via Lennar-Jones potential. In addition, each N atom bears a partial charge with $q_N = -0.482e$, charges interacting through Coulombic forces. At the center of the N–N bond a partial charge $q = +0.964e$ compensates the negative charge on the N atoms. Such charge distribution mimics the measured quadrupole moment of the N_2 molecule. All the interactions between the atoms of the N_2 molecules and the Ge and X atoms of the chalcogels were calculated by considering the intermolecular energy $U_{ij}(r)$ between two sites (i and j) as the sum of a Coulombic contribution and a pairwise-additive Lennard-Jones (LJ) 12-6 potential. In our simulations, the LJ cross interaction parameters (σ_{ij}, ϵ_{ij}) between unlike sites are calculated using the Lorentz-Berthelot mixing rules. The LJ parameters (σ_{ij}, ϵ_{ij}) for the Ge and X of the chalcogels were taken from [57]. The dispersive interactions were neglected beyond a cutoff of 10 Å. The electrostatic interactions were computed using the Ewald summation technique (the parameters were chosen so that the relative accuracy in the Coulomb energy calculation is 10^{-5}). More details about the GCMC procedure and potential parameters employed in this study can be found in [30].

2.2.3 Detailed Analysis of the Gas/Solid Interface Chemistry and Computation of Other Properties

In order to refine the output configurations of the GCMC calculations of the systems made of N_2 adsorbed in the chalcogels' pores, FPMD simulations can be employed with the purpose to further optimize the systems structure and chemistry as well as compute other properties of interest. As an example, once the new chalcogels systems are obtained containing different contents of gas molecules, MLWFs can be further computed to investigate into the details the interactions involved between the gas molecules and the solid glassy surfaces. Furthermore, MLWFs can be also employed to compute molecular dipoles to measure the induced molecular polarization due to the interaction with the glassy solid surface. This exploitation of MLWFs allows to obtain a deeper insight into the interactions between the adsorbed gas molecules and the chalcogenide surfaces. Even for the case of apolar gas molecules for which the dipole moment averaged over time is expected to be zero, the instantaneous molecular distortions can result in non-zero instantaneous dipole moments (μ) and, in this respect, this analysis gives access to the local molecular polarization induced by the solid surface. Recently, we investigated the case of CO_2 molecules adsorbed in a g-GeS_2 chalcogel slit-like pore [58]. In this case, for the system made of a layer of CO_2 molecules physisorbed on a g-GeS_2 surface, the magnitude of the induced μ_{CO_2} is found to be strongly dependent on the distance from the g-GeS_2 surface. Close to the solid surface, the CO_2 molecules show a large μ_{CO_2} of about ~0.51 D, while departing from the surface μ_{CO_2} reaches values close to those expected for the CO_2 in the gas phase (~0.1 D). The large variation of the dipole moments for the CO_2 molecules in contact with the solid surface can be explained by the high polarizability of both Ge and S atoms of the g-GeS_2 surface. CO_2 molecules departing from the surface towards the empty space in the center of the pore restore almost completely the μ_{CO_2} value typical of CO_2 molecules in the gas phase. Similarly, Karseemeijer et al. [60] and Sun et al. [61] found values of μ_{CO_2} of about ~0.5 D for CO_2 molecules adsorbed on top of solid water. Interestingly, our calculated dipole moments values are similar to those of induced μ_{CO_2} found for CO_2 molecules adsorbed in a hydrated Ca-exchanged Montorillonite (in the range ~0.4–1.0 D) [62]. While the oxygen dipole moment coming from the electronic polarization in glassy silica was found weakly dependent from the local melt composition [63].

2.2.4 Models and Methods Details Relevant to This Work

The bulk structural models of the chalcogenide glasses studied in this work were obtained by means of FPMD simulations following the typical melt-quenching technique. The details about the chalcogenide systems considered here are reported in Table 2.1. The electronic structure was described within DFT by using as generalized gradient approximation (GGA) the Becke, Lee, Yang, and Parr (BLYP) exchange-correlation functional [37, 38]. The BLYP functional was combined with the empirical Grimme dispersion correction for van der Waals (vdW) interactions [59]. This theoretical scheme has been successfully validated on a wide range of glassy chalcogenides. More specifics about the methodology used to obtain the different chalcogenide glasses can be found in the references reported in Table 2.1. The surfaces models were initially built from the parent bulk models by removing the periodic boundary conditions along the z direction. By inserting the slab having a thickness Δ_z in a simulation box of a size h_z, one defines a slit pore of width $H = h_z - \Delta_z$ via the use of periodic boundary conditions. More precisely, we took the pore width (H) as the distance between the mean positions of the chalcogen and Ge atoms on the two opposite surfaces. We prepared systems with H in between 2 and 4 in order to obtain Ge_xCh_y (with Ch: S, Se, Te) slit-like nanopores with different widths. A selection of the output configurations obtained from the GCMC simulations were further refined by means of FPMD simulations with applying a friction force on the atom dynamics (ion velocities scaled by a factor 0.95 at each step) in order to optimize the systems at $T = 0$ K. The surface models were then equilibrated by FPMD at finite temperature and volume. The above calculations were performed with the CPMD code [34].

2.3 Case Studies

2.3.1 *Pore Size Effect*

Figure 2.2 shows the N_2 adsorption isotherm for g-$GeSe_4$ nanopores with $H = 2$, 3 and 4 nm computed at 77 K. The data obtained for the three pores confirm the typical behavior observed in the experiments of adsorption/condensation in solid nanopores. At low pressures, the adsorbed amount increases in a continuous fashion upon increasing the pressure with the N_2 forming an adsorbed film at the g-$GeSe_4$ pore surface. Beyond adsorption of the first layer, the slope of the adsorption isotherm decreases once the pore gets filled along the multilayer adsorption regime. Then, at a pressure lower than the bulk saturating vapor pressure P_0, capillary condensation occurs. As expected on the basis of capillary condensation theories such as Derjaguin and Derjaguin-Broekhoff-De Boer model [3, 5] the capillary condensation pressure increases with increasing the pore width H: the condensation pressures for the nanopores with $H = 2$, 3, and 4 nm are $\sim$0.10 P/P_0, $\sim$0.59 P/P_0, and $\sim$0.89 P/P_0, respectively. Figure 2.2 also shows typical molecular configurations of N_2 molecules adsorbed at different relative pressures upon adsorption in the nanopore with $H =$ 4 nm. Regardless of the pore width, the surface of g-$GeSe_4$ nanopores is covered with a homogeneous film at the onset of capillary condensation. A discontinuous transition (i.e., capillary condensation) between the partially filled and completely filled configurations occurs when the adsorbed film becomes unstable, in line the results found for g-GeS_2 and typical experimental data [31]. Note that for a proper quantitative comparison with experimental data this has to be normalized to the cor-

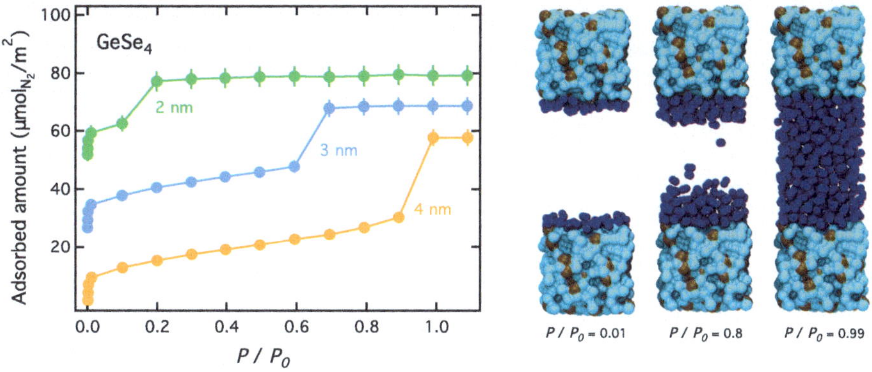

Fig. 2.2 Left: N_2 adsorption isotherms at 77 K for porous g-$GeSe_4$ with slit-like pores of different width H (indicated next to each adsorption isotherm). Adsorbed amounts are in μmol/m^2. For the sake of clarity, adsorption isotherms have been shifted up. Pressures are in relative units with respect to the bulk saturating pressure P_0 for N_2 at 77 K. Right: Typical molecular configurations for N_2 adsorbed in g-$GeSe_4$ pore with H = 3.6 4: (from left to right) $P/P_0 = 0.01$, $P/P_0 = 0.80$, and $P/P_0 = 0.99$. The ochre and cyan spheres are the Ge and Se atoms of the chalcogenide surfaces, respectively. The blue spheres are the atoms of the N_2 molecules

rected Brunauer-Emmett-Teller (BET) [32] surface area. Generally, it is appropriate to correct the BET surface to account for the overestimate of the true geometrical surface areas by about 25–30% [4, 33].

2.3.2 *Chemical Stoichiometry-Se:Ge Ratio Effect*

Figure 2.3 shows the N_2 adsorption isotherm for nanopores with H = 2 nm computed at 77 K for Ge_xSe_y chalcogels with different Se:Ge ratio $\phi_{Se/Ge}$: 2, 4 and 9 for g-$GeSe_2$, g-$GeSe_4$, and g-$GeSe_9$ respectively. The N_2 adsorbed amount is normalized with respect to the pore surface area in order to take into account the different size of the simulated system (see Table 2.1 for details). We recall that the pore surface areas considered in this work allows to rule out any possible size effect, in accordance with the previous results obtained for the g-GeS_2 system [31, 58]. The data obtained for the three systems show a similar trend that follows the typical behavior observed for adsorption/condensation experiments in solid nanopores as discussed in the previous section. The onset of capillary condensation within the pore occurs at similar relative pressure for systems with $\phi_{Se/Ge}$= 2 and 4 (P/P_0 ~0.1) whereas for the $\phi_{Se/Ge}$= 9 system it occurs at larger relative pressure (P/P_0 ~0.2). Figure 2.3 (inset) shows the comparison between the N_2 adsorption isotherms in the range $0.0 < P/P_0 < 0.25$ for the three systems. The non-negligible difference found corresponds to a lower adsorbed amount of N_2 at low pressure (< 0.15 P/P_0) for the g-$GeSe_9$ system with respect to g-$GeSe_2$ (~ −11% and ~ −9% at P/P_0 ~0.01 and P/P_0

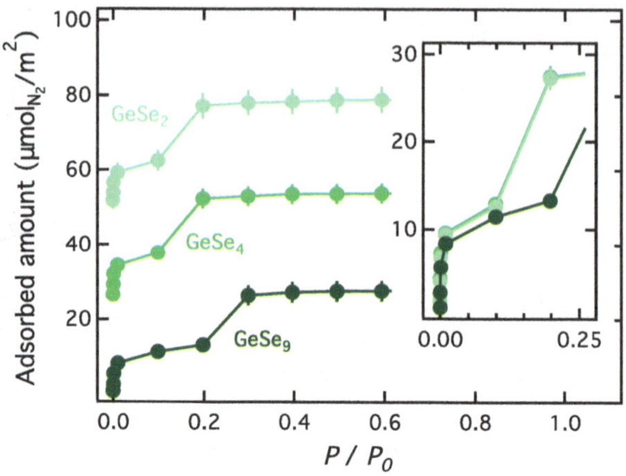

Fig. 2.3 N_2 adsorption isotherms at 77 K for slit-like pores with $H = 2$ nm made of Ge_xSe_y chalcogels with different Se:Ge ratio $\phi_{Se/Ge} = 2$, 4 and 9 for g-$GeSe_2$, g-$GeSe_4$, and g-$GeSe_9$ respectively. For the sake of clarity, adsorption isotherms have been shifted up. Insets: zoom-in within the same $\mu mol_{N_2}/m^2$ scale at the low relative pressure P/P_0 range ($0.0 < P/P_0 < 0.25$)

$\sim$0.1 respectively). This result suggests a weaker interaction between N_2 molecules and the Se atoms of the chalcogel surface with respect to the interaction between N_2 molecules and the Ge atoms. The stronger interaction between N_2 and chalcogels with lower $\phi_{Se/Ge}$ arises from the interplay between the electronegativity and electropositive charge of Ge atoms and its larger polarizability with respect to Se ($\alpha_{Ge} \sim 42$ vs. $\alpha_{Se} \sim 27$) [48]. The specific gas interactions with these surfaces is a key factor determining the behavior of gas mixtures through porous structures. Our data is found in fair agreement with the experimental behavior for adsorption/condensation obtained by Armatas et al. [18] for Ge_xSe_y chalcogels, where a system with a greater Se:Ge ratio is found to promote lower N_2 adsorption at the same relative pressure ($\phi_{Se/Ge} \sim 0.44$ vs. $\phi_{Se/Ge} \sim 0.21$).

2.3.3 *Chemical Composition-Chalcogen Effect*

Figure 2.4 shows the N_2 adsorption isotherm for nanopores with $H = 3$ nm computed at 77 K for chalcogels made of different chalcogen counter-ions: g-GeS_4, g-$GeSe_4$, and g-$GeTe_4$. The data obtained for the three systems show a similar trend that follows the typical behavior observed for adsorption/condensation experiments in solid nanopores as discussed in the previous section. The onset of capillary condensation within the pore occurs at similar relative pressure for the sulfide and selenide systems ($P/P_0 \sim 0.6$) whereas for the telluride system it occurs at larger relative pressure ($P/P_0 \sim 0.8$). Figure 2.3 (right panel) shows the comparison between the N_2 adsorption isotherms in the range $0.0 < P/P_0 < 0.25$ for the three systems. At a given relative pressure, the amount of N_2 adsorbed on the chalcogel surface is found to be strongly related to the nature of the chalcogen counter-ion and following the series g-GeS_4 > g-$GeSe_4$ > g-$GeTe_4$. For example, at relative pressure of $P/P_0 \sim 0.3$, g-$GeSe_4$ and g-$GeTe_4$ show a lower adsorbed of N_2 with respect to g-GeS_4 (−5.9%

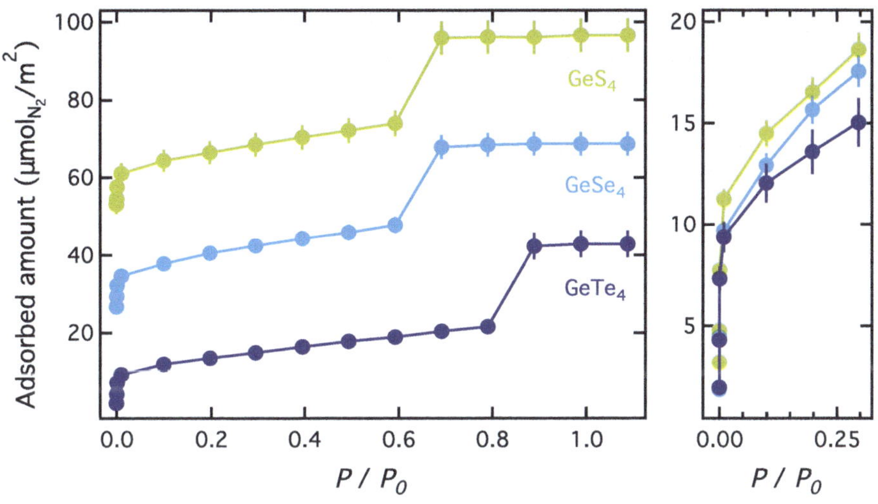

Fig. 2.4 Left: N_2 adsorption isotherms at 77 K for slit-like pores with $H = 3$ nm made of chalcogels based on different chalcogen element: g-GeS_4, g-$GeSe_4$, and g-$GeTe_4$ respectively. For the sake of clarity, adsorption isotherms have been shifted up. Right: zoom-in within the same $\mu mol_{N_2}/m^2$ scale at the low relative pressure P/P_0 range ($0.0 < P/P_0 < 0.25$)

and -19.4%, respectively). This behavior can be ascribed to the interplay between the electronegativity, charge and polarizability of the chalcogels constituents elements. It remains to be investigated if the interactions of different gases (such as CO_2, CH_4, and H_2) with these porous glassy chalcogenides would follow the same trend.

2.4 Conclusions and Perspectives

Realistic models of porous glassy chalcogenides were used to probe the viability of such a class of materials for adsorption applications. Using atomic-scale simulation based on first-principles molecular dynamics and grand canonical Monte Carlo simulations, we gained insights into the adsorption mechanisms for N_2 molecules. Both the effects of pore size and chalcogenide chemistry (composition and stoichiometry) were investigated in order to gain fundamental understanding on the chalcogels gas adsorption properties and assess the versatility of computational modelling approach proposed. Our results are found to be in good agreement with available experimental data, confirming the ability of our methodology to produce realistic chalcogel models and predict their N_2 gas adsorption performance. The present work shows that glassy porous chalcogenides are a valuable class of materials for gas adsorption. While further work is needed to clarify gas adsorption for different gas probes, these results shed light on the adsorption mechanisms for porous chalcogenides. Such an approach can be used for *a priori* evaluation of their gas adosorption performances and pave the way for the design of chalcogenide-based adsorbents towards specific applications.

Acknowledgements We acknowledge the Pôle HPC and Equipex quip@Meso at the University of Strasbourg and the Grand Equipement National de Calcul Intensif (GENCI) under allocation DARI-A0060807670. G.O. acknowledges the Seed Money program of Eucor—The European Campus (project MEDIA) for financial support.

References

1. *The Strategic Energy Technology Plan 2017*, https://bit.ly/2kqMR6W
2. C.H. Lau, S. Liu, D.R. Paul, J. Xia, Y.-C. Jean, H. Chen, K. Shao, T.-S. Chung, Adv. Energy Mater. **1**, 634–642 (2011)
3. B. Coasne, A. Galarneau, R.J.M. Pellenq, F. Di Renzo, Chem. Soc. Rev. **42**, 4141–4171 (2013)
4. B. Coasne, P. Ugliengo, Langmuir **28**, 11131–11141 (2012)
5. B. Coasne, New J. Chem. **40**, 4078 (2016)
6. X. Xu, C. Song, J.M. Andresen, B.G. Miller, A.W. Scaroni, Energy Fuels **16**, 1463–1469 (2002)
7. N.R. Stuckert, R.T. Yang, Environ. Sci. Technol. **45**, 10257–10264 (2011)
8. G. Li, P. Xiao, P. Webley, J. Zhang, R. Singh, M. Marshall, Adsorption **14**, 415–422 (2008)
9. J. Merel, M. Clausse F. Meunier, Ind. Eng. Chem. Res. **47**, 209–215 (2008)
10. N. Du, H.B. Park, M.M. Dal-Cin, M.D. Guiver, Energy Environ. Sci. **5**, 7306–7322 (2012)
11. J. Lee, J. Kim, T. Hyeon, Adv. Mater. **18**, 2073–2094 (2006)
12. P. Billemont, B. Coasne, G. De Weireld, Langmuir **29**, 3328–3338 (2013)
13. G.P. Hao, Z.Y. Jin, Q. Sun, X.Q. Zhang, J.-T. Zhang, A.H. Lu, Energy Environ. Sci. **6**, 3740–3747 (2013)
14. J.-R. Li, R.J. Kuppler, H.-C. Zhou, Chem. Soc. Rev. **38**, 1477–1504 (2009)
15. C.E. Wilmer, O.K. Farha, Y.-S. Bae, J.T. Hupp, R.Q. Snurr, Energy Environ. Sci. **5**, 9849–9856 (2012)
16. V. Stanić, A.C. Pierre, T.H. Etsell, J. Am. Ceram. Soc. **83**, 1790–1796 (2000)
17. K.K. Kalebaila, D.G. Georgiev, S.L. Brock, J. Non-Cryst, Solids **352**, 232–240 (2006)
18. G.A. Armatas, M.G. Kanatzidis, Nat. Mater. **8**, 222–271 (2009)
19. B.J. Riley, J. Chun, W. Um, W.C. Lepry, J. Matyas, M.J. Olszta, X. Li, K. Polychronopoulou, M.G. Kanatzidis, Environ. Sci. Technol. **47**, 7540–7547 (2013)
20. Q. Lin, X. Bu, C. Mao, X. Zhao, K. Sasan, P. Feng, J. Am. Chem. Soc. **137**, 6184–6187 (2015)
21. H. Yang, M. Luo, X. Chen, X. Zhao, J. Lin, D. Hu, D. Li, X. Bu, P. Feng, T. Wu, Inorg. Chem. **56**, 14999–15005 (2017)
22. R.G. Parr, R.G. Pearson, J. Am. Chem. Soc. **105**, 7512–7516 (1983)
23. M.G. Kanatzidis, Adv. Mater. **19**, 1165–1181 (2007)
24. M. Shafai-Fallah, A. Rothenberger, A.P. Katsoulidis, J. He, C.D. Malliakas, M.G. Kanatzidis, Adv. Mater. **23**, 4857–4860 (2011)
25. E. Ahmed, A. Rothemberger, J. Mater. Chem. A **3**, 7786–7792 (2015)
26. K.S. Subrahmanyam, C.D. Malliakas, D. Sarma, G.S. Armatas, J. Wu, M.G. Kanatzidis, J. Am. Chem. Soc. **137**, 13943–13948 (2015)
27. B.J. Riley, D.A. Pierce, W.C. Lepry, J.O. Kroll, J. Chun, K.S. Subrahmanyam, M.G. Kanatzidis, F.K. Alblouwy, A. Bulbule, E.M. Sabolsky, Ind. Eng. Chem. Res. **54**, 11259–11267 (2015)
28. S. Murugesan, P. Kearns, K.J. Stevenson, Langmuir **28**, 5513–5517 (2012)
29. G. Leyral, M. Ribes, L. Courthéoux, D. Uzio, A. Pradel, Eur. J. Inorg. Chem. **31**, 4967–4971 (2012)
30. G. Ori, C. Massobrio, A. Bouzid, M. Boero, B. Coasne, Phys. Rev. B **90**, 045423 (2014)

31. G. Ori, C. Massobrio, A. Pradel, M. Ribes, B. Coasne, Phys. Chem. Chem. Phys. **18**, 13449–13458 (2016)
32. S. Brunauer, P.H. Emmett, E. Teller, J. Am. Chem. Soc. **60**, 309–319 (1938)
33. A. Galarneau, H. Cambon, F. Di Renzo, F. Fajula, Langmuir **17**, 8328–8335 (2001)
34. R. Car, M. Parrinello, For this contribution as FPMD simulation method we adopted the Car-Parrinello approach. Phys. Rev. Lett. **55**, 2471 (1985). Using the CPMD code [see http://www.cpmd.org/, copyright IBM Corp. 1990–2013, copyright MPI für Festkörperforschung Stuttgart 1997–2001.]
35. G. Ori, A. Bouzid, E. Martin, C. Massobrio, S. Le Roux, M. Boero, Solid State Sci. **95**, 105925 (2019)
36. J.-Y. Raty, M. Schumacher, P. Golub, V.L. Deringer, C. Gatti, M. Wuttig, Adv. Mater. **31**, 1806280 (2019)
37. A.D. Becke, Phys. Rev. A **38**, 3098 (1988)
38. C. Lee, W. Yang, R.G. Parr, Phys. Rev. B **37**, 785 (1988)
39. J.P. Perdew, K. Burke, M. Ernzerhof, Phys. Rev. Lett. **77**, 3865 (1996)
40. J.P. Perdew, K. Burke, M. Ernzerhof, Phys. Rev. Lett. **78**, 1396 (1997)
41. A. Bouzid, C. Massobrio, M. Boero, G. Ori, K. Sykina, E. Furet, Phys. Rev. B **92**, 134208 (2015)
42. A. Bouzid, S. Le Roux, G. Ori, M. Boero, C. Massobrio, J. Chem. Phys. **143**, 034504 (2015)
43. A. Bouzid, S. Le Roux, G. Ori, C. Tugene, M. Boero, C. Massobrio, in *Molecular Dynamics Simulations of Disordered Materials*, vol. 12 (Springer Series in Materials Science, Cham, 2015), pp. 313–344
44. S. Le Roux, A. Bouzid, K.Y. Kim, S. Han, A. Zeidler, P.S. Salmon, C. Massobrio, Chem. Phys. **145**, 084502 (2016)
45. E. Lampin, A. Bouzid, G. Ori, M. Boero, C. Massobrio, J. Chem. Phys. **147**, 044504 (2017)
46. C. Massobrio, E. Martin, Z. Chaker, M. Boero, A. Bouzid, G. Ori, Front. Mater. **5**, 1–5 (2018)
47. A.K. Rappé, W.A. Goddard III, J. Phys. Chem. **95**, 3358–3363 (1991)
48. P. Schwerdtfeger, J.K. Nagle, Mol. Phys. **117**, 9–12 (2019)
49. L. Pauling, *The Nature of the Chemical Bond*, th edn. (Cornell University Press, Ithaca, 1960)
50. C.E. Wilmer, K.C. Kim, R.Q. Snurr, J. Phys. Chem. Lett. **3**, 2506 (2012)
51. R.F.W. Bader, *Atoms in Molecules: A Quantum Theory* (Oxford University Press, New York, 1990)
52. C. Gatti, P. Macchi, *Modern Charge-Density Analysis* (Springer, Dordrecht, 2012)
53. P.-O. L'owdin, J. Chem. Phys. **18**, 365 (1950)
54. R.S. Mulliken, J. Chem. Phys. **1955**, 23 (1833)
55. S.R. Cox, D.E. Williams, J. Comput. Chem. **2**, 304 (1981)
56. J.J. Potoff, J.I. Siepmann, AIChE J. **47**, 1676–1682 (2001)
57. S.L. Mayo, B.D. Olafson, W.A. Goddard, J. Phys. Chem. **94**, 8897 (1990)
58. Z. Chaker, A. Bouzid, B. Coasne, C. Massobrio, M. Boero, G. Ori, J. Non-Cryst, Solids **498**, 288–293 (2018)
59. S. Grimme, J. Compt. Chem. **27**, 1787–1799 (2006)
60. L.J. Karssemejier, G.A. Wijs, H.M. Cuppen, Phys. Chem. Chem. Phys. **16**, 15630 (2014)
61. Z. Sun, D. Pan, L. Xu, E. Wang, Proc. Natl. Acad. Sci. **109**, 13177–13181 (2012)
62. M.-S. Lee, B.P. McGill, R. Rousseau, V.-A. Glezakou, J. Phys. Chem. C **122**, 1125 (2017)
63. R. Vuilleumier, N. Sator, B. Guillot, J. Non-Cryst, Solids **357**, 2555 (2011)

Chapter 3
Exploring Defects in Semiconductor Materials Through Constant Fermi Level Ab-Initio Molecular Dynamics

Assil Bouzid and Alfredo Pasquarello

Abstract We focus on the determination of point defects in semiconductor materials through constant-Fermi-level ab initio molecular dynamics and demonstrate that this technique can be used as a computer-based tool to reveal and control relevant defects in semiconductor materials. In this scheme, the Fermi level can be set at any position within the band gap during the defect generation process, in analogy to experimental growth conditions in the presence of extra electrons or holes. First, the scheme is illustrated in the case of GaAs, for which we generate melt-quenched amorphous structures through molecular dynamics at various Fermi levels. By a combined analysis that involves both the atomic structure and a Wannier-function decomposition of the electronic structure, we achieve a detailed description of the generated defects as a function of Fermi level. This leads to the identification of As–As homopolar bonds and Ga dangling bonds for Fermi levels set in the vicinity of the valence band. These defects convert into As dangling bonds and Ga–Ga homopolar bonds, as the Fermi level moves toward the conduction band. Second, we investigate defects at the InGaAs/oxide interface upon inversion. We adopt a substoichiometric amorphous model for modelling the structure at the interface and investigate the formation of defect structures upon setting the Fermi-level above the conduction band minimum. Our scheme reveals the occurrence of In and Ga lone-pair defects and As–As dimer/dangling bond defects, in agreement with previous studies based on physical intuition. In addition, the present simulation reveals hitherto unidentified defect structures consisting of metallic In–In, In–Ga, and Ga–Ga bonds. The defect charge transition levels of such metallic bonds in Al_2O_3 are then determined through a hybrid functional scheme and found to be consistent with the defect density measured at InGaAs/Al_2O_3 interfaces. Hence, we conclude that both

A. Bouzid (✉)
Institut de Recherche sur les Céramiques Centre Européen de la Céramique,
12 Rue Atlantis, 87068 Limoges, France
e-mail: assil.bouzid@unilim.fr

A. Pasquarello
Chaire de Simulation à l'Echelle Atomique (CSEA),
Ecole Polytechnique Fédérale de Lausanne (EPFL), 1015 Lausanne, Switzerland
e-mail: alfredo.pasquarello@epfl.ch

E. V. Levchenko et al. (eds.), *Theory and Simulation in Physics for Materials Applications*, Springer Series in Materials Science 296,
https://doi.org/10.1007/978-3-030-37790-8_3

In and Ga lone pairs and metallic In–In bonds are valid candidate defects for charge trapping at InGaAs/oxide interfaces upon charge carrier inversion. These two studies demonstrate the effectiveness of constant-Fermi-level ab initio molecular dynamics in revealing and identifying semiconductor defects in an unbiased way.

3.1 Introduction

Silicon is the most used material in the microelectronic industry. Its wide use is due to the facility of doping silicon with both donors and acceptors, allowing thereby the realization of both p-type and n-type field effect transistors. In particular, silicon is highly advantageous thanks to its native oxide SiO_2, which gives rise to well-passivated interfaces and good dielectric behavior. Despite these excellent properties, the scaling of this system is hindered due to fundamental limits associated with quantum tunneling [1, 2].

As an alternative, attention has recently been devoted to III–V semiconductor materials. While these compounds offer higher carrier mobilities compared to silicon [3, 4], their integration in metal-oxide-semiconductor field-effect transistors [5–7] remains difficult due to the formation of performance degrading defects at their interfaces with other materials [8, 9]. Hence, finding appropriate routes to hamper the formation of undesired defects is a general issue that often needs to be overcome upon the introduction of novel materials in electronic devices.

In practice, it remains difficult to identify the nature of active defects experimentally. In this field, computational modelling is the best tool to achieve insight into possible defect structures. Currently, the identification of defects through computational methods relies on educated guesses based on physical intuition or results from trial and error procedures. First, candidate defect structures are imagined, then generated in atomistic models and compared to the available experimental characterization. In the absence of experimental data, the selection process generally rests on the sole calculation of formation energies, which does not correspond to an inclusive search. In addition, the calculated energetics might be flawed by the consideration of inappropriate structural models. Given these issues, it is of paramount importance to develop computer-aided simulation approaches for identifying and controlling active semiconductor defects without introducing any human bias in the search procedure.

As an alternative scheme, constant Fermi level molecular dynamics represents a very promising technique for the identification and study of defects in semiconductor materials [10–13]. Within this scheme, the system is allowed to exchange electrons with an external potentiostat set at a pre-fixed target Fermi energy. The charge transition levels of a defect separate energy ranges in the band gap, in which the defect stabilizes in different charge states. Hence, setting the Fermi-level within a given energy range of the band gap should, at equilibrium, lead to the formation of the defect in the corresponding charge state. Furthermore, the application of this technique in a dynamical fashion, i.e. in combination with ab-initio molecular dynamics, facilitates the realization of such equilibrium conditions through thermal annealing.

In this chapter, we review the application of ab initio molecular dynamics at constant Fermi level as a computer-aided tool to reveal and identify semiconductor defects. We focus on two case studies, namely defects in GaAs and at the $In_{0.53}Ga_{0.47}As$/oxide interface. GaAs and $In_{0.53}Ga_{0.47}As$ (InGaAs hereafter) alloys are among the III–V semiconductors that are considered as alternatives to silicon in metal-oxide-semiconductor compounds [14–16] owing to their high carrier mobility [17–19]. However, the occurrence of a high defect density at the interface with oxide materials hampers their widespread use and severely reduces their performance compared to theoretical expectations [20–22]. Hence, understanding the nature of these defects and their impact on the electrical properties represents an important step toward their control and passivation [23–26].

3.2 Computational Methods

The electronic structure is described in the framework of density functional theory within the generalized gradient approximation due to Perdew, Burke and Ernzerhof (PBE) [27]. Core-valence interactions are described by normconserving pseudopotentials [28]. The wave functions of the valence electrons are expanded in a plane-wave basis set defined by kinetic energy cutoffs of 40 Ry and 70 Ry for GaAs and InGaAs/oxide systems, respectively. The Brillouin zone is sampled at the Γ point. We perform Born-Oppenheimer molecular dynamics using a time step of $\Delta t = 0.48$ fs to integrate the equations of motion. The control of the temperature is ensured through a velocity rescaling method.

In order to control the Fermi level during the simulation, the system is connected to an external potentiostat at a fixed potential $\bar{\epsilon}_F$ acting like an external electron reservoir. Hence, the electronic charge N_e is considered as a dynamical variable having a fictitious mass M_e. In this way, the extended system, which comprises the physical system and the external electron reservoir, is driven by the grand canonical potential: $\Omega = E^{\text{tot}}(r_i, N_e) - N_e\bar{\epsilon}_F$, where $E^{\text{tot}}(r_i, N_e)$ represents the total energy of the considered system and $N_e\bar{\epsilon}_F$ the energy corresponding to N_e electrons in the external reservoir [10, 11]. Consequently, this formulation defines the forces acting on the atoms and the electronic charges, which are given by:

$$F_i = -\frac{\partial E^{\text{tot}}(r_i, N_e)}{\partial r_i} \quad \text{and} \quad F_e = -(\epsilon_F - \bar{\epsilon}_F), \tag{3.1}$$

where ϵ_F is the instantaneous Fermi energy. F_e drives electrons into the system when ϵ_F is lower than $\bar{\epsilon}_F$ and drives electrons into the reservoir in the opposite case. The average of the instantaneous Fermi energy ϵ_F then corresponds to the pre-set $\bar{\epsilon}_F$. The dynamical equations for the charge evolution read:

$$\dot{N}_e = \frac{P_e}{M_e} \quad \text{and} \quad \dot{P}_e = F_e = -(\epsilon_F - \bar{\epsilon}_F), \tag{3.2}$$

where P_e is a fictitious momentum associated to the dynamical variable N_e. Upon the exchange of electrons with the reservoir, the charge neutrality in the supercell is ensured by the use of a uniform neutralizing background, as customary in the study of charged defects in periodically repeated supercells [29]. Further details about the implementation of this technique can be found in [11–13, 30].

In practice, we set the fictitious electronic mass to $M_e = 1000$ a.u. for GaAs and to 500 a.u. for InGaAs. The temperature of the electronic charge T_c is controlled by a velocity rescaling method and is set to 30 K. We use Gaussian-smeared occupations with widths of at most 0.13 eV to prevent numerical instabilities associated to the use of fractional occupations.

3.3 The Case of GaAs

Among the III–V semiconductor materials, GaAs is considered as a prototypical system and has received considerable attention. A large body of theoretical studies based on density-functional calculations has provided a detailed description of possible defect structures in GaAs and at its interfaces with oxide materials [22, 31–38]. In particular, the Fermi-level pinning in GaAs has been attributed to an amphoteric defect [33, 37, 39, 40]. When the Fermi energy is below the pinning level, the defect occurs in the form of a homopolar As–As bond. It transforms into two doubly occupied dangling bonds (DBs) on As atoms when the defect captures two electrons. This particular defect occurs in several GaAs systems, including interfaces [33, 39], surfaces [33], and amorphous phases [37].

In this work, we mainly focus on GaAs in its amorphous phase. This phase represents a suitable model for bonding arrangements at interfaces of the parent crystalline semiconductor. We apply the constant-Fermi-level ab initio molecular dynamics technique to investigate defects in GaAs. This scheme is expected to reveal the occurrence of the amphoteric defect and of its transformation, even in the absence of any prior knowledge. Consequently, the structure of amorphous models generated by melt-quench procedures would be highly influenced by the position of the fixed Fermi level.

3.3.1 Model Generation at Constant Fermi Level

We consider a system containing 32 Ga and 32 As atoms in a periodically repeated cubic cell of size 11.54 Å, corresponding to the experimental density [41–43]. The amorphous phase is generated through the melt-quenching technique starting from an initial crystalline structure of GaAs. Through a molecular dynamics evolution at $T = 2500$ K for a duration of 6 ps, the structure departs from the initial configuration through significant diffusion of Ga and As. Next, the temperature is gradually lowered to $T = 1600$ K within 10 ps through a stepwise decrease of the temperature. The system is annealed during 25 ps at $T = 1600$ K to ensure good diffusion in the liquid state. This state is the starting point for all the amorphous models generated in this work. The final amorphous system is achieved by quenching from the melt according to the following protocol: $T = 1100$ K for 10 ps, $T = 800$ K for 13 ps, $T = 500$ K for 10 ps, and $T = 300$ K for 17 ps. Statistical averages are collected over the last 10 ps of the trajectory at 300 K.

We first generate a neutral reference amorphous model without the use of the constant-Fermi-level technique. For this model we find a HOMO-LUMO gap of 0.53 eV with a Fermi level lying in the middle of the band gap at 0.26 eV from the valence band maximum (VBM). Despite the small value of this band gap, we are able to perform molecular dynamics simulations with Fermi levels fixed at various positions in the gap. A detailed description of the alignment procedure used to set the Fermi level during the dynamics is described in [13].

We then fix the Fermi level at various positions with respect to the band edges during the molecular dynamics. As a consequence, the system is driven out of equilibrium, exchanges electrons with the external reservoir and arranges its structure accordingly. In this manner, we can explore the dependence of the defect populations in particular and of atomistic structure in general on the position of the pre-fixed Fermi level. Four amorphous models are generated by setting the target Fermi level at different energies with respect to the band edges. For each of these four systems, we start the constant Fermi-level molecular dynamics from the liquid state equilibrated at $T = 1600$ K and adopt a quenching protocol similar to the one used for the reference system. When referred to the VBM of the neutral GaAs, the Fermi levels of these four systems are at -0.10, +0.07, +0.45, and +0.61 eV. The time evolution of the electronic charge and of the Fermi level during the molecular dynamics of the amorphous systems at 300 K are shown in Fig. 3.1.

We generate two amorphous models with Fermi levels at higher (0.45 eV) and lower (0.07 eV) energy in the band gap with respect to the Fermi level of the reference system (Fig. 3.1c). Additionally, we generate two models with Fermi levels degenerate with either the valence (-0.10 eV) or the conduction (+0.61 eV) band in order to simulate conditions of carrier degeneracy. For all the generated systems, the instantaneous Fermi levels show stable time evolutions around the pre-set values during the molecular dynamics. When the target Fermi level is set close to the VBM (Fig. 3.1a, b), the system releases electrons toward the reservoir. Hence, the total charge of the system becomes positive and stabilizes around $+16e$ and $+10e$ when $\bar{\epsilon}_F$ is fixed at -0.10 and 0.07 eV, respectively. When setting the target Fermi level close to the CBM, extra electrons are driven into the system. We find total electronic charges of about $-6e$ and $-12e$ for the simulations with $\bar{\epsilon}_F$ at 0.45 eV and 0.61 eV (Fig. 3.1d, e), respectively. In this manner, the generated amorphous models are representative of both n-type and p-type GaAs.

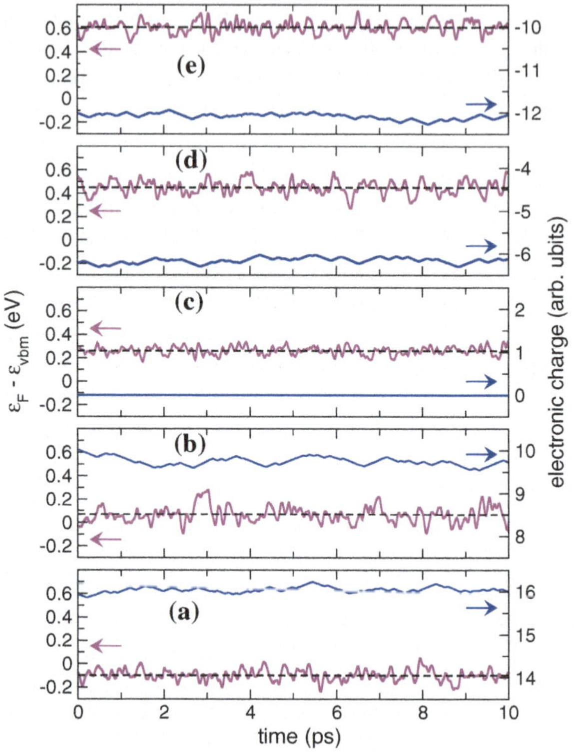

Fig. 3.1 Time evolution of the instantaneous Fermi level ϵ_F (green) and of the total electronic charge (in units of e) in the system (red) during the constant-Fermi-level molecular dynamics of amorphous GaAs at 300 K. Panel (**c**) corresponds to the reference system produced without fixing the Fermi level ($\epsilon_F = 0.26$ eV). Panels (**a**), (**b**), (**d**), and (**e**) correspond to systems generated by setting the Fermi level $\bar{\epsilon}_F$ at −0.10, +0.07, +0.45, and +0.61,eV, respectively

3.3.2 Analysis of the Atomic Structure

The partial pair correlation functions $g_{\alpha\beta}(r)$ with α, β = Ga or As, shown in Fig. 3.2, allow one to investigate the structural properties of the amorphous models of GaAs generated at various Fermi levels. The first peaks in $g_{\mathrm{GaGa}}(r)$ and $g_{\mathrm{AsAs}}(r)$ are located at about 2.50 Å and are signatures of homopolar Ga–Ga and As–As bonds, respectively. When the Fermi level moves from the conduction to the valence band, the intensity of the first peak in $g_{\mathrm{AsAs}}(r)$ shows a steady increase, which reflects the occurrence of a higher concentration of As–As dimers for Fermi levels close to the VBM. In the case of Ga–Ga correlations, the intensity of the first peak shows the opposite behavior. For Fermi levels moving closer to the VBM, this peak undergoes broadening until the first minimum is no longer discernable. Such a change indicates that the topology of the system is highly affected by holes injected at these Fermi levels. For Fermi levels moving closer to the CBM, the formation of homopolar Ga–Ga is favored as the first peak in $g_{\mathrm{GaGa}}(r)$ becomes sharper. The modification of the local environment around the Ga atoms as reflected by the Ga–Ga correlations is also mirrored in the Ga–As correlations. In particular, as the Fermi level approaches the CBM and the first peak in $g_{\mathrm{GaGa}}(r)$ gets better defined, the first peak in $g_{\mathrm{GaAs}}(r)$

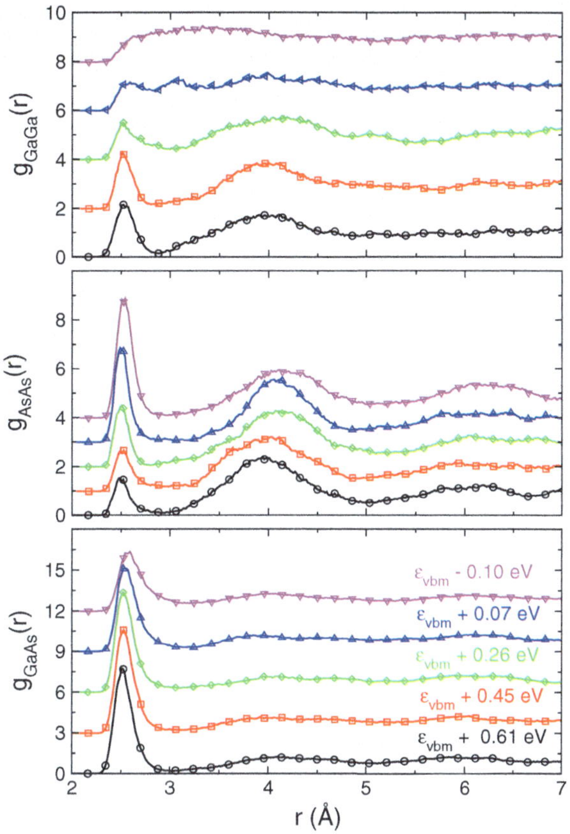

Fig. 3.2 Ga–Ga, As–As, and Ga–As partial pair correlation functions of the amorphous models of GaAs generated at various Fermi levels. The results correspond to averages over 10 ps of molecular dynamics at a temperature of 300 K

becomes sharper and its intensity higher. These variations indicate that the chemical order in the system is related to the presence of well-defined homopolar Ga–Ga bonds. In particular, the present results suggest that chemical order is favored in GaAs generated in n-type conditions. At a more general level, the real space analysis based on pair correlation functions shows that the position of the Fermi level in the band gap highly influences the atomic structures of the generated amorphous model.

3.3.3 *Defect Population Analysis*

A compact real-space representation of the electron structure can be achieved by resorting to maximally localized Wannier functions and in particular to their centers (W), which can be exploited to detect chemical bonds and lone pairs. Within this formalism, each Wannier function represents the localization of two electrons at a given average position corresponding to its center W. We distinguish three different

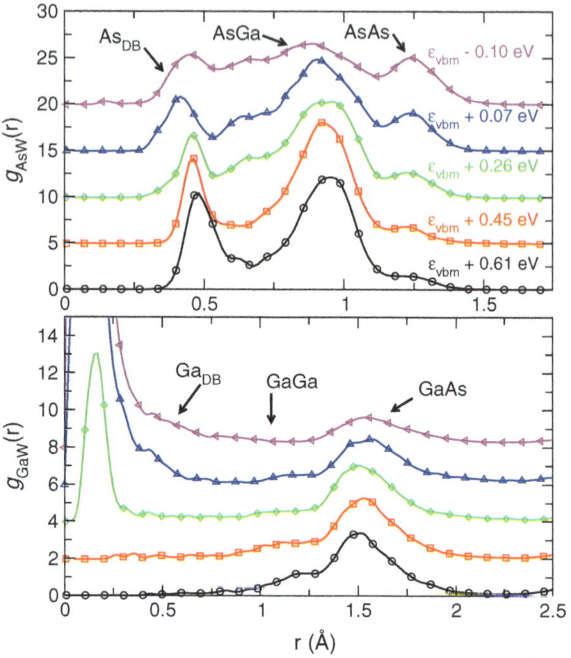

Fig. 3.3 Pair correlation functions $g_{\mathrm{AsW}}(r)$ and $g_{\mathrm{GaW}}(r)$ between As and W centers (upper panel) and between Ga and W centers (lower panel), respectively, for the amorphous models evolving at 300 K. The various curves are labeled with their corresponding Fermi energy

Wannier centers in GaAs systems. First, Wannier center (W_B) occurs along heteropolar Ga–As bonds, as it is the case in crystalline GaAs. Second, a Wannier centers (W_H) are found close to the middle of Ga–Ga or As–As bonds and correspond to homopolar linkages. Third, lone pairs attached to As or Ga, which are not participating in the chemical bonding, are also represented by Wannier functions (W_{LP}).

The pair correlation functions $g_{\mathrm{AsW}}(r)$ and $g_{\mathrm{GaW}}(r)$ displayed in Fig. 3.3 show As–W and Ga–W correlations as computed for all our structures. Three main features can be distinguished for As–W correlations [33, 37], which arise from W_{LP}, W_B, and W_H with increasing distance. The intensity of As–W_{LP} correlations increases, while that of As–W_H correlations is reduced as the Fermi level moves towards higher energies. This trend is in agreement with the observed behavior of As–As dimer/DB defects at GaAs surfaces and interfaces [33, 39]. Taking Ga–W correlations under consideration, we find W centers corresponding to W_{LP}, W_H, and W_B with increasing distance.

The intensity of Ga–W_{LP} and Ga–W_H correlations show an opposite behavior upon variation of the Fermi level. More precisely, as the Fermi level moves towards the CBM, the amount of Ga lone pairs decreases and the amount of homopolar Ga–Ga bonds increases. In addition, when the Fermi level reaches the neighborhood of the CBM, the peak corresponding to Ga–As bonds of both As–W and Ga–W correlations gains higher intensity. As for the As dangling bonds and Ga homopolar bonds, significant fractions of them are formed at the Fermi levels close to the CBM.

These defects are replaced by As homopolar bonds and Ga dangling bonds for Fermi level moving toward the VBM.

A proper description of the local environments requires definitions for the chemical and dangling bonds. First, lone pairs correspond to W centers occurring at shorter distances than the first minima in $g_{\mathrm{GaW}}(r)$ or $g_{\mathrm{AsW}}(r)$. Second, heteropolar or homopolar bonds are defined when a W center is shared between two atoms of different or of the same chemical species. In these definitions, we allow for a tolerance on the bond lengths, according to the inequality $r_{\mathrm{W}\alpha} + r_{\mathrm{W}\beta} - r_{\alpha\beta} < 0.25$ Å, where α and β indicates either Ga or As atoms. In this manner, all the W centers can be assigned to either atomic bonds or lone pairs.

Using the aforementioned bond definitions, we calculate the coordination numbers of As and Ga atoms in our amorphous models. For Fermi levels in the vicinity of the VBM, Ga atoms feature a high fraction of fourfold coordinated configurations. When the Fermi level is set in the vicinity of the VBM, the Ga atoms are found to be mainly fourfold coordinated, with a substantial fraction of undercoordinated atoms ($\ell = 0$, 1, 2, and 3). In agreement with their low electronegativity, we do not find any threefold or fourfold coordinated Ga atoms carrying lone pairs. We note that fivefold coordinated Ga atoms occur in very small fractions. When the Fermi level shifts toward the CBM, the amount of undercoordinated Ga atoms reduces in favor of fourfold coordinated ones. Focusing on As atoms, we find that they occur mainly in fourfold coordinated configurations but that they also show a sizeable fraction of threefold coordinated motifs. As the Fermi level moves from the VBM to CBM, the fraction of fourfold coordination reduces while that of threefold coordination increases. We note that almost all the threefold coordinated As atoms carry a lone pair and that fivefold coordinated As atoms only occur in small proportions.

A complete picture of the defective structures occurring in GaAs as a function of Fermi level requires that we consider unoccupied dangling bonds in the analysis. We adopt the following rational. In the case of As atoms, the Wannier-function decomposition shows the occurrence of dangling bonds in the form of lone pairs. When one excludes the negligible amount of fivefold coordinated As atoms, we find that the majority of the remaining As atoms are either fourfold coordinated without carrying any lone pair or threefold coordinated with a single lone pair. As such, one can establish a dominant preference for fourfold hybridization in amorphous GaAs. It is then reasonable to assume that such a fourfold hybridization is also preserved in the case of Ga atoms. This allows us to assign unoccupied dangling bonds to Ga atoms on the basis of their coordination number and their number of lone pairs. In this manner, the amounts of atoms with heteropolar and homopolar bonds together with those carrying occupied and unoccupied dangling bonds can be established. This counting is normalized to account for a fourfold hybridization per atom. Fivefold coordinated atoms are considered separately.

Figure 3.4 shows the fractions of As and Ga atoms forming homopolar and heteropolar bonds and carrying unoccupied and occupied dangling bonds as a function of the pre-set Fermi level. From the fraction of Ga–As bonds, we find that the chemical order is mostly favored when the Fermi level is set in the upper half of the band gap. The analysis of the local environment of the As atoms reveals that the

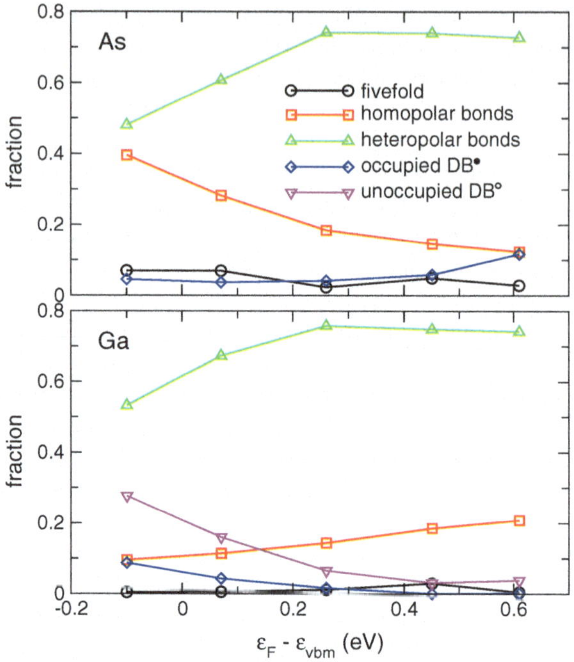

Fig. 3.4 Fractions of Ga and As atoms involved in heteropolar and homopolar bonds and carrying occupied and unoccupied dangling bonds as a function of the pre-set Fermi level. The results are obtained under the assumption of fourfold hybridization. The fractions of fivefold coordinated atoms are accounted for separately

fraction of the occupied As dangling bonds increases when the Fermi level moves from the VBM toward the CBM. On the contrary, we find that the fraction of As–As homopolar bonds shows an opposite trend.

It has been show previously [33, 37, 39, 40] that the conversion of As–As dimers into occupied As dangling bonds (As–As $+ 2e^- \rightarrow 2\,\mathrm{DB}^{\bullet}_{\mathrm{As}}$) is the key mechanism to explain Fermi-level pinning in GaAs systems. We remark that in the constant Fermi energy scheme, the key features of this reaction are captured in a straightforward manner through computer simulations without the necessity of any prior knowledge of the relevant defect structure.

In the case of Ga atoms, we find that the amount of homopolar bonds increases with Fermi energy. This behaviour is accompanied with a reduction of the fraction of unoccupied dangling bonds and to a lesser extent of occupied ones. This suggest that the following reaction mechanism occurs as a function of the Fermi level:

$$\text{Ga–Ga} \rightarrow 2\,\mathrm{DB}^{\circ}_{\mathrm{Ga}} + 2e^{-}. \tag{3.3}$$

This mechanism might lead to the formation of a defect level in the band gap. However, at interfaces relevant for microelectronic applications, GaAs is generally in contact with excess oxygen, which leads to their passivation as Ga–Ga bonds get oxidized.

3.4 The Case of the InGaAs/Oxide Interface

$In_{0.53}Ga_{0.47}As$ represents a very promising candidate for practical implementation in microelectronic devices [17]. This system is usually interfaced with high dielectric-constant materials, such as Al_2O_3, which is characterized by a wide band gap and excellent thermal stability [17, 44]. It has been revealed experimentally that, when interfaced with Al_2O_3, $In_{0.53}Ga_{0.47}As$ shows a density of defect states with a large feature in the vicinity of the valence band and a second smaller feature at mid-gap [45–48]. These interface defects have been found to be independent of the nature of the deposited oxide, which implies that they originate most likely from the semiconductor substrate [46–49].

Capacitance-voltage experiments demonstrate the presence of oxide traps centered at +1.5 and −0.5 eV with respect to the conduction band minimum (CBM) of InGaAs [20]. Furthermore, bias temperature instability experiments reveal the occurrence of a defect density distribution centered at $\approx$1 eV above the CBM [21]. Under carrier population inversion, the Fermi level moves up into the conduction band, allowing thereby the carrier population to reach high concentrations. In these conditions, the electrical properties of the device are undermined by the occurrence of oxide defects. The nature of these defects remains inaccessible to experimental techniques. We here resort to constant Fermi energy molecular dynamics to investigate these defects.

3.4.1 Model Generation at Constant Fermi Level

We model the InGaAs/oxide interface through a disordered assembly composed of the stoichiometric interfacial components at a ratio of 50:50. In agreement with the experimental characterization [50], our choice of composition results in a substoichiometric oxide representative of the composition at the interface. As in the case of GaAs, we here resort to a melt-quenching technique to generate an amorphous phase. This protocol allows the system to overcome energy barriers and to reach a reasonable structure. We note that our primary aim is to unveil typical defect structures rather than to achieve a realistic description of the atomistic network at the interface.

The suboxide model $(GaAs)_{0.5}(Ga_2O_3)_{0.5}$ generated in [37] is taken as the starting structure to build the interface model. This model consists of 48 Ga, 16 As, and 48 O atoms at a mass density of 5.42 g/cm^3. Replacing half of the 48 Ga atoms by In atoms, we then achieve a system composed of 24 Ga, 24 In, 16 As and 48 O, which is described by the formula $(In_{0.5}Ga_{0.5}As)_{0.5}(InGaO_3)_{0.5}$. We calibrate the system density to eliminate residual pressure [51] and yield a mass density of 5.98 g/cm^3. In the following, this structure is used as the initial configuration for the melt-quenching protocol.

Within our melt-quenching protocol, we set the temperature to 2500 K and allow the system to evolve towards a disordered structure for a duration of 6.4 ps, ensuring thereby complete decorrelation with respect to the initial configuration as all the atoms diffuse over several inter-atomic bond lengths. Next, we achieve the amorphous structure by gradually decreasing the temperature according to the schedule: $T = 1500$ K for 7.6 ps, $T = 1000$ K for 7.2 ps, $T = 500$ K for 13.4 ps, and $T = 300$ K for 10 ps. A similar stepwise quenching protocol has been demonstrated to achieve accurate amorphous models featuring excellent agreement with experiments [52–55]. The final results are averaged over the last 7 ps of the trajectory at 300 K. This model is generated under neutral conditions without the use of the constant Fermi-level technique and is referred to as model I in the following. Model I features a band gap of 0.30 eV with an average Fermi level lying at midgap.

Furthermore, we generate a second model, model II, that simulates the experimental condition achieved upon carrier population inversion. In practice, we set the Fermi level above the CBM of the reference model during the molecular dynamics. Starting from the liquid-like state of model I achieved at $T = 1500$ K, we activate the constant Fermi-level simulation and lower the temperature according to the following quenching protocol: $T = 1000$ K for 6.9 ps, $T = 500$ K for 11 ps, and $T = 300$ K for 10 ps.

We adopt an alignment scheme based on the average O 2*s* energy level of twofold coordinated O atoms in order to reference the energy levels of model II to those of model I. As a result of this alignment, the Fermi level of model II is found to be at +0.16 eV with respect to the CBM of model I.

3.4.2 *Defect Population Analysis*

As in the case of GaAs, we here resort to maximally localized Wannier functions to analyse the nature of the atomic environment in terms of chemical and dangling bonds. The Wannier decomposition analyses applied to model I and model II are shown in Fig. 3.5. In model I, we find about 7.5% of the In atoms carrying an occupied DB and a similar fraction forming homopolar bonds. An atom is considered to be involved in homopolar bonds when it forms at least one of such bonds. Both DBs and homopolar bonds are absent in the case of Ga, but for As they occur in fractions of 70% and 62%, respectively. In the case of O atoms, we observe a large fraction of DBs around 68%, but these atoms are not found to be involved in homopolar bonds. We note that the fractions of DBs and homopolar bonds for In and Ga atoms increase when the Fermi-level is set above the CBM. Hence, two main changes are observed upon increasing the Fermi level above the CBM. On the one hand, the formation of metallic homopolar bonds is promoted, and on the other hand, the system tends to form a larger fraction of atoms carrying DBs. The formation of occupied DBs on In and Ga atoms upon capture of two electrons has already been identified in [56] as one possible charge trapping mechanism at energies above the CBM. In [56] these

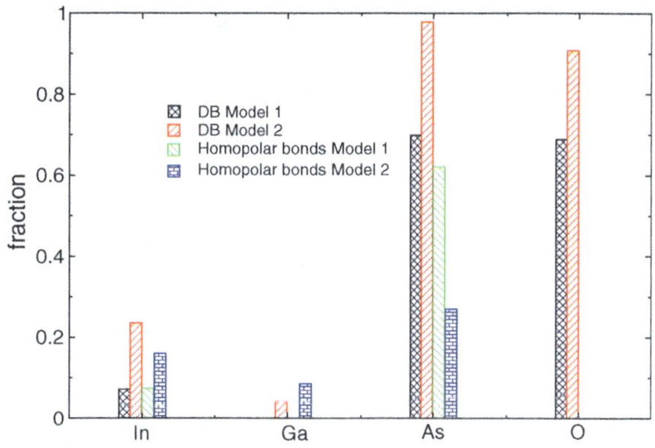

Fig. 3.5 Fractions of In, Ga, As, and O atoms involved in homopolar bonds and carrying occupied dangling bonds, as obtained for model I and model II

defects were identified by physical intuition. Interestingly, they emerge directly as the result of the applied simulation method in the present work.

Focusing on the local environment around As atoms, we find a reduction of the fraction of As atoms belonging to homopolar bonds accompanied by an increase of the fraction of As atoms carrying DBs. These two observations indicate the occurrence of As–As dimer/DB defects according to the following reaction mechanism [33, 39]:

$$\text{As–As} + 2e^- \rightarrow 2\,\text{DB}^{\bullet}_{\text{As}}. \tag{3.4}$$

Overall, our search for defects at InGaAs/oxide interfaces through constant Fermi-level molecular dynamics reveals two main points. First, we are able to reproduce defects previously identified through conventional defect studies. Specifically, we find that upon electron charging the fractions of In and Ga atoms carrying occupied DBs increase and we retrieve the mechanism associated with the As–As dimer/DB defect. Second, our simulation scheme reveals the occurrence of previously unidentified defect structures consisting of metallic In–In, In–Ga, and Ga–Ga bonds, which might potentially play a role in the charge trapping at the InGaAs/oxide interface. This possibility is examined in the next section through a study of the electronic structure at the hybrid functional level of theory.

3.4.3 Defects at the Interface

In order to evaluate the electrical activity of metallic-bond impurities in $In_{0.53}Ga_{0.47}As$ based devices, we determine their energy levels in bulk Al_2O_3. We here resort to an amorphous Al_2O_3 model generated previously through first principles molecular dynamics [57] and take advantage of the accurate band alignment at the $In_{0.53}Ga_{0.47}As/Al_2O_3$ interface achieved in [56]. Metallic bond defects are created in this system by replacing two Al atoms bonded to the same O atom in their first

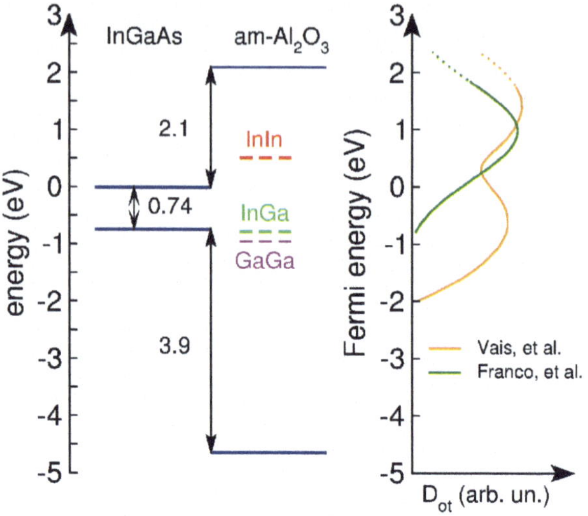

Fig. 3.6 Defect levels of In–In, In–Ga, and Ga–Ga defects as obtained from electronic-structure calculations with the hybrid functional proposed by Heyd, Scuseria, and Ernzerhof. The experimental density of oxide traps as a function of Fermi energy (from [20, 21]) is presented for comparison. The energies are referred to the CBM of InGaAs

neighbor shell. In this manner, we choose two Al atoms separated by a small distance, which favors metallic bond formation upon substitution. The defect charge transition levels are calculated on the thermodynamically equilibrated defect structure. We perform cyclical structural relaxations in the neutral and $Q = -2$ charge states until convergence is achieved [58] and determine the associated $\epsilon_{0/-2}$ defect level. As such, one ensures that the structural relaxations occurring upon defect charging are solely related to the defect rather than to defect-independent structural relaxations of the amorphous model [58]. For In–In, we find that the relaxed configuration in the charge state $Q = -2$ features a bond length of 2.7 Å, and an overall local environment which resembles the one achieved during the molecular dynamics at 300 K. In–Ga and Ga–Ga defects are created by substituting In with Ga and repeating the procedure based on cyclical structural relaxations.

Details of the calculation of the defect levels corresponding to thermodynamic charge transition levels $\epsilon_{0/-2}$ are provided in [12]. The resulting defect levels are displayed in Fig. 3.6. When referred to the CBM of InGaAs [56], the energy levels of the In–In, In–Ga, and Ga–Ga defects lie at +0.51, −0.78, and −0.95 eV. We remark that the In–In defect level falls close to peaks in the experimental densities of interface states centered around 1 eV above the CBM of InGaAs (see Fig. 3.6) [20, 21]. Hence, we infer that In–In pairs are valid candidates for the trap states observed experimentally upon carrier population inversion. This result complements the one put forward previously, in which energy levels associated to In and Ga defects carrying lone pairs were found to lie between 0.7 and 1.3 eV above the CBM [56]. At variance, the In–Ga and Ga–Ga defect levels fall below the VBM of InGaAs in a good match with the experimental peak found at 0.5 eV below the CBM [20]. However, they do not appear to play a significant role upon charge carrier inversion.

These findings consolidate the observation from constant Fermi-level molecular dynamics that metallic bonds in the oxide phase could stabilize at the InGaAs/oxide interface. The present results together with those of [56] provide a comprehensive picture of the defects occurring at InGaAs/Al_2O_3 interfaces and disclose the nature of the defects that could contribute to the measured density of interface states upon inversion.

3.5 Conclusions

Inspired by experimental studies on semiconductors, in which the Fermi level is controlled during growth, we have reviewed in this chapter the use of constant Fermi level ab initio molecular dynamics as a computer-aided simulation technique to investigate the occurrence of defects in semiconductor materials. Our approach is free from human bias and does not require prior knowledge of the defect structures. Using this simulation technique, we have demonstrated that it is possible to generate amorphous models, for which the Fermi level is fixed at a pre-set value. The system spontaneously favors the formation of defects that show higher stability at the selected Fermi level. Hence, the basic principle of our scheme is that the defect population undergoes modification when the Fermi level crosses defect charge transition levels. In particular, this offers the possibility of achieving computer-generated semiconductor structures representative of n-type or p-type semiconductors.

We have validated this scheme in the case of amorphous GaAs for which we have observed a clear anticorrelation between the occurrence of As–As homopolar bonds and As dangling bonds as the Fermi-level varies across the band gap. This clearly hints at the interconversion of these two defective structures upon charge exchange, which has been proposed as the key mechanism leading to Fermi level pinning in this material. Next, we have investigated defect structures occurring at the InGaAs/oxide interface and have found that In–In defects, in addition to In and Ga lone-pair defects identified previously, constitute valid candidates for trapping charge carriers upon inversion.

In conclusion, we have demonstrated the effectiveness of the constant-Fermi-level ab initio molecular dynamics technique in revealing charge trapping mechanisms associated to defect levels in the band gap. The method is expected to serve as an invaluable tool towards the identification of undesired defects in semiconductor compounds in view of their passivation. Especially when common practices based on physical intuition have failed, the present scheme provides an alternative search procedure, which is computer-driven and thus not subject to human bias.

References

1. P.A. Gargini, IEEE Circuits Devices Mag. **18**(2), 13 (2002)
2. J.A. Carballo, W.T.J. Chan, P.A. Gargini, A.B. Kahng, S. Nath, in *2014 32nd IEEE International Conference on Computer Design (ICCD)* (IEEE, 2014), pp. 139–146

3. L. Xia, J.B. Boos, B.R. Bennett, M.G. Ancona, J.A. del Alamo, Appl. Phys. Lett. **98**(5), 053505 (2011)
4. Y. Xuan, Y.Q. Wu, P.D. Ye, IEEE Electron Device Lett. **29**(4), 294 (2008). https://doi.org/10.1109/LED.2008.917817
5. K. Rajagopalan, J. Abrokwah, R. Droopad, M. Passlack, IEEE Electron Device Lett. **27**(12), 959 (2006). https://doi.org/10.1109/LED.2006.886319
6. K. Rajagopalan, R. Droopad, J. Abrokwah, P. Zurcher, P. Fejes, M. Passlack, IEEE Electron Device Lett. **28**(2), 100 (2007). https://doi.org/10.1109/LED.2006.889502
7. Y. Sun, E.W. Kiewra, S.J. Koester, N. Ruiz, A. Callegari, K.E. Fogel, D.K. Sadana, J. Fompeyrine, D.J. Webb, J.P. Locquet, M. Sousa, R. Germann, K.T. Shiu, S.R. Forrest, IEEE Electron Device Lett. **28**(6), 473 (2007). https://doi.org/10.1109/LED.2007.896813
8. M. Hong, J. Kwo, A. Kortan, J. Mannaerts, A. Sergent, Science **283**(5409), 1897 (1999)
9. G. Brammertz, H.C. Lin, K. Martens, D. Mercier, S. Sioncke, A. Delabie, W.E. Wang, M. Caymax, M. Meuris, M. Heyns, Appl. Phys. Lett. **93**(18), 183504 (2008)
10. N. Bonnet, T. Morishita, O. Sugino, M. Otani, Phys. Rev. Lett. **109**(26), 266101 (2012)
11. A. Bouzid, A. Pasquarello, J. Chem. Theory Comput. **13**(4), 1769 (2017)
12. A. Bouzid, A. Pasquarello, J. Phys. Cond. Matt. **29**(50), 505702 (2017)
13. A. Bouzid, A. Pasquarello, Phys. Rev. Appl. **8**, 014010 (2017)
14. I. Thayne, R. Hill, M. Holland, X. Li, H. Zhou, D. Macintyre, S. Thoms, K. Kalna, C. Stanley, A. Asenov et al., ECS Trans. **19**(5), 275 (2009)
15. E. O'reilly, J. Robertson, Phys. Rev. B **34**(12), 8684 (1986)
16. L. Lin, J. Robertson, Appl. Phys. Lett. **98**(8), 082903 (2011)
17. J.A. Del Alamo, Nature **479**(7373), 317 (2011)
18. M. Passlack, P. Zurcher, K. Rajagopalan, R. Droopad, J. Abrokwah, M. Tutt, Y.B. Park, E. Johnson, O. Hartin, A. Zlotnicka et al., in *IEEE International Electron Devices Meeting, 2007. IEDM 2007* (IEEE, 2007), pp. 621–624
19. J. Wiley, Semiconductors Semimetals **10**, 91 (1975)
20. A. Vais, J. Franco, H.C. Lin, N. Collaert, A. Mocuta, K.D. Meyer, A. Thean, Appl. Phys. Lett. **107**(22), 223504 (2015). https://doi.org/10.1063/1.4936991
21. J. Franco, A. Alian, B. Kaczer, D. Lin, T. Ivanov, A. Pourghaderi, K. Martens, Y. Mols, D. Zhou, N. Waldron et al., in *2014 IEEE International Reliability Physics Symposium* (IEEE, 2014), pp. 6A–2
22. J. Robertson, Appl. Phys. Lett. **94**(15), 152104 (2009)
23. M. Caymax, G. Brammertz, A. Delabie, S. Sioncke, D. Lin, M. Scarrozza, G. Pourtois, W.E. Wang, M. Meuris, M. Heyns, Microelectron. Eng. **86**(7), 1529 (2009)
24. M. Houssa, E. Chagarov, A. Kummel, MRS Bulletin **34**(7), 504 (2009)
25. M. Huang, Y. Chang, C. Chang, Y. Lee, P. Chang, J. Kwo, T. Wu, M. Hong, Appl. Phys. Lett. **87**(25), 252104 (2005)
26. L. Lin, J. Robertson, J. Vac. Sci. Technol. B, Nanotechnol. Microelectron. Mater. Process. Meas. Phenom. **30**(4), 04E101 (2012)
27. J.P. Perdew, K. Burke, M. Ernzerhof, Phys. Rev. Lett **77**(18), 3865 (1996). Erratum Phys. Rev. Lett. **78**, 1396 (1997)
28. N. Troullier, J.L. Martins, Phys. Rev. B **43**(3), 1993 (1991). https://doi.org/10.1103/PhysRevB.43.1993
29. H.P. Komsa, T.T. Rantala, A. Pasquarello, Phys. Rev. B **86**(4), 045112 (2012)
30. A. Bouzid, A. Pasquarello, J. Phys. Chem. Lett. **9**(8), 1880 (2018)
31. D.L. Winn, M.J. Hale, T.J. Grassman, J.Z. Sexton, A.C. Kummel, M. Passlack, R. Droopad, J. Chem. Phys. **127**(13), 134705 (2007)
32. K. Laasonen, R.M. Nieminen, M.J. Puska, Phys. Rev. B **45**(8), 4122 (1992)
33. D. Colleoni, G. Miceli, A. Pasquarello, J. Phys. Cond. Matt. **26**(49), 492202 (2014)
34. J. Robertson, Y. Guo, L. Lin, J. Appl. Phys. **117**(11), 112806 (2015)
35. S. Pöykkö, M.J. Puska, M. Alatalo, R.M. Nieminen, Phys. Rev. B **54**(11), 7909 (1996)
36. M.J. Caldas, J. Dabrowski, A. Fazzio, M. Scheffler, Phys. Rev. Lett. **65**(16), 2046 (1990)
37. D. Colleoni, A. Pasquarello, Appl. Phys. Lett. **107**(3), 031605 (2015)

38. T. Mattila, R.M. Nieminen, Phys. Rev. B **54**(23), 16676 (1996)
39. D. Colleoni, G. Miceli, A. Pasquarello, Phys. Rev. B **92**(12), 125304 (2015)
40. D. Colleoni, A. Pasquarello, Microelectron. Eng. **109**, 50 (2013)
41. M.L. Theye, A. Gheorghiu, H. Launois, J. Phys. C: Solid State Phys. **13**(36), 6569 (1980)
42. C. Ascheron, A. Schindler, R. Flagmeyer, G. Otto, Nucl. Instr. Meth. B **36**(2), 163 (1989)
43. C.M.H. Driscoll, A.F.W. Willoughby, J.B. Mullin, B.W. Straughan, in *IOP Conference Proceedings Gallium Arsenide and Related Compounds* vol. 24 ed. by J. Bok (IOP, London, 1975). p. 275
44. S. Oktyabrsky, D.Y. Peide, *Fundamentals of III–V Semiconductor MOSFETs* (Springer, 2010)
45. G. Brammertz, H. Lin, K. Martens, A.R. Alian, C. Merckling, J. Penaud, D. Kohen, W.E. Wang, S. Sioncke, A. Delabie et al., ECS Trans. **19**(5), 375 (2009)
46. G. Brammertz, H.C. Lin, M. Caymax, M. Meuris, M. Heyns, M. Passlack, Appl. Phys. Lett. **95**(20), 202109 (2009)
47. V. Djara, T. O'Regan, K. Cherkaoui, M. Schmidt, S. Monaghan, É. O'Connor, I. Povey, D. O'Connell, M. Pemble, P. Hurley, Microelectron. Eng. **109**, 182 (2013)
48. Y.C. Fu, U. Peralagu, D.A. Millar, J. Lin, I. Povey, X. Li, S. Monaghan, R. Droopad, P.K. Hurley, I.G. Thayne, Appl. Phys. Lett. **110**(14), 142905 (2017)
49. K. Tang, A.C. Meng, R. Droopad, P.C. McIntyre, A.C.S. Appl. Mater. Interfaces **8**(44), 30601 (2016)
50. H.D. Lee, T. Feng, L. Yu, D. Mastrogiovanni, A. Wan, T. Gustafsson, E. Garfunkel, Appl. Phys. Lett. **94**(22), 222108 (2009)
51. A. Bouzid, C. Massobrio, J. Chem. Phys. **137**(4), 046101 (2012)
52. A. Bouzid, C. Massobrio, M. Boero, G. Ori, K. Sykina, E. Furet, Phys. Rev. B **92**(13), 134208 (2015)
53. K. Wezka, A. Bouzid, K.J. Pizzey, P.S. Salmon, A. Zeidler, S. Klotz, H.E. Fischer, C.L. Bull, M.G. Tucker, M. Boero et al., Phys. Rev. B **90**(5), 054206 (2014)
54. A. Bouzid, K.J. Pizzey, A. Zeidler, G. Ori, M. Boero, C. Massobrio, S. Klotz, H.E. Fischer, C.L. Bull, P.S. Salmon, Phys. Rev. B **93**(1), 014202 (2016)
55. A. Bouzid, G. Ori, M. Boero, E. Lampin, C. Massobrio, Phys. Rev. B **96**(22), 224204 (2017)
56. D. Colleoni, G. Pourtois, A. Pasquarello, Appl. Phys. Lett. **110**(11), 111602 (2017)
57. D. Colleoni, G. Miceli, A. Pasquarello, Appl. Phys. Lett. **107**(21), 211601 (2015)
58. P. Broqvist, A. Pasquarello, Microelectron. Eng. **84**(9), 2022 (2007)

Chapter 4
Enhancing the Flexibility of First Principles Simulations of Materials via Wavelets

Laura E. Ratcliff and Luigi Genovese

Abstract We illustrate how the properties of a Daubechies wavelet basis set can be exploited to build an effective computational method that enables one to perform electronic structure calculations of systems containing up to many thousands of atoms. This is achieved by implementing a ladder of approaches of different scaling behaviours and decreasing computational complexity. We will explain that such an approach is suitable both for extended systems and for systems with molecular character. We define quantitative indicators that provide guidelines to the end-user about the pertinence of the employed methodology, thereby guaranteeing limited impact on the precision of the result. We provide a quantitative illustration of these concepts to defective systems with an extended character, by presenting the differences in computational walltime and in precision among the various methodological steps of the ladders.

4.1 Introduction

Since the days of their foundations, disciplines like Computational Physics or Quantum Chemistry have had to deal with the problem of the computational reliability of results. A computational result may be considered reliable for various reasons. For example, it may be in agreement with some previous observation or measurement that the employed computational model is susceptible to reproduce. The model itself may in this way be classified as "accurate"; it can then be employed to extract, in silico, other quantities which might be used to predict or interpret experiments. Another aspect of reliability is the concept of "precision", which is more focused on the actual results that a given computational model provides for a specific set

L. E. Ratcliff (✉)
Department of Materials, Imperial College London, London SW7 2AZ, UK
e-mail: laura.ratcliff08@imperial.ac.uk

L. Genovese
Univ. Grenoble Alpes, CEA, INAC-SP2M, L_Sim, 38000 Grenoble, France
e-mail: luigi.genovese@cea.fr

E. V. Levchenko et al. (eds.), *Theory and Simulation in Physics for Materials Applications*, Springer Series in Materials Science 296,
https://doi.org/10.1007/978-3-030-37790-8_4

of quantities, regardless of their potential agreement with experimental data. The aim of a precise approach is to reduce the computational uncertainties of quantities extracted out of a well-defined model, and provide reference results, to which other computer codes employing the same model can be compared.

For a theoretical approach which is based on numerical calculations, for which no analytic reference solution exists, reducing the computational uncertainty is the only possible way to shed light on the predictive power of the model. In other terms, the "accuracy" of a result with respect to experimental data may be reliably quantified only when the computational uncertainty is guaranteed to be significantly lower than the observed discrepancy.

When linking together various length scales, such considerations are not just optional, but rather they become compulsory. Having said that, however, performing a set of production quantum mechanical (QM) simulations with an unnecessarily costly approach would provide a study of poor quality, as the simulation scheme would entangle interactions with different length scales and couplings. In other terms, the dogma "the more complex the simulation the better" is not necessarily true in such situations. Taking these considerations into account allows one to focus on the regions of the system which *require* a high level of theory, providing a better understanding of the fundamental mechanisms and avoiding an unnecessary waste of computational resources. Indeed, such is the motivation behind multiscale QM/QM or QM/MM (molecular mechanics) approaches which combine different levels of theory within a given simulation. It is interesting to discuss such a delicate balance of computational complexity within the framework of large-scale density functional theory (DFT) calculations.

Linear scaling algorithms for Kohn-Sham (KS) DFT [1, 2], developed already some time ago [3, 4], have recently become accessible to a broader community thanks to the introduction of reliable and robust approaches (see e.g. [5] and references therein). This has important consequences for the interpretation and design of first-principles approaches, as the possibility of tackling systems of unconventionally large sizes allows one to address new problems and questions. From a computational point of view, one of the most important characteristics of a DFT code is the set of basis functions used to express the KS orbitals. The domain of applicability of a code is tightly connected to this choice. For example, a non-localized basis set like plane waves is highly suitable for electronic structure calculations of periodic and/or homogeneous systems like crystals or solids, while it is much less efficient in expanding localized information, due to the wider range of components in reciprocal space. For these reasons DFT codes based on plane waves are not convenient for simulating inhomogeneous or isolated systems like molecules, due to the resulting high memory requirements.

An important distinction should be also made between codes which use systematic and non-systematic basis sets. A systematic basis set allows one to calculate the exact solution of the KS equations with arbitrarily high precision by increasing the number of basis functions. In other terms, the numerical precision of the results is related to the number of basis functions used to expand the KS orbitals. With such a basis set it is thus possible to obtain results which are free of errors related to the choice of the basis, eliminating a source of uncertainty. In this context, it is therefore very important to have at hand a computational formalism which is able to provide, at the same time:

- a set of reliable results, in the sense that they can be systematically improved by the end-user, in view of increasing—when needed—the *precision* of the calculations;
- a flexible approach, namely by providing the possibility of explicitly implementing the desired models without having to deal with correction terms and intrinsic approximations;
- an efficient computer program, which enables the optimal use of computational resources, especially in the context of high performance computing;
- the ability to connect together different levels of theory, where various approaches might be linked within a given computational setup.

To this end, in the following we show how the use of a systematic wavelet basis set, as adopted in the BigDFT code [6], facilitates these requirements. This includes the ability to define a computational setup which is adapted to different system sizes, such as a linear scaling approach, as well as the definition of various indicators which allow one to assess whether or not the chosen approach is appropriate given the inherent complexity of the system in question. We focus primarily on the treatment of extended systems, nonetheless one may also apply a similar treatment to (supra)molecular systems.

4.2 Density Functional Theory with Wavelets

The set of basis functions used to express the KS orbitals is of key importance for the nature of the computational operations which have to be performed. In the BigDFT code, the KS wavefunctions are expressed in Daubechies wavelets [7]. The latter is a set of localized, real-space based set of orthogonal functions which allows for a systematic, multi-resolution description. A more complete description of these operations can be found in the BigDFT reference paper [6].

A wavelet basis consists of a family of functions generated from a mother function ϕ and its translations on the points of a uniform grid of spacing h. The most important property of these functions is that they satisfy the so-called refinement equations

$$\phi(x) = \sqrt{2} \sum_{j=1-m}^{m} h_j \, \phi(2x - j) \tag{4.1}$$

$$\psi(x) = \sqrt{2} \sum_{j=1-m}^{m} g_j \, \phi(2x - j)$$

which establishes a relation between the scaling functions on a grid with grid spacing h and another one with spacing $h/2$. h_j and $g_j = (-1)^j h_{-j+1}$ are the elements of a filter that characterizes the wavelet family, and m is the order of the scaling function-wavelet family. The wavelet functions are indicated by the function ψ. All the properties of these functions can be obtained from the relations (4.1). The full basis set can be obtained from all translations by a certain grid spacing h of the mother function centered at the origin. The mother function is localized, with compact support. The maximally symmetric Daubechies scaling function and wavelet of order 16 that are used in this work are shown in Fig. 4.1. The number of basis functions is increased by decreasing the value of h. Thanks to the systematicity of the basis, this will make the numerical description more precise. The degree of smoothness determines the speed with which one converges to the exact result as h is decreased. The degree of smoothness increases as one goes to higher order Daubechies wavelets. This together with the fact that our method is quasi variational gives a convergence rate of h^{14}. Obtaining such a high convergence rate is essential in the context of electronic structure calculations where one needs highly accurate results for basis sets of acceptable size.

The Daubechies wavelets formalism implemented in the BigDFT code has proven to provide results of high precision and reliability on all the systems that can be treated within the pseudopotential (PSP) approximation. Moreover, it has been possible to show that the PSP approximation is of very good quality and can provide the same results as all-electron calculations on a wide variety of ground state quantities, see e.g. [8].

Furthermore, as is natural in an approach which has a direct connection with the real space, the boundary conditions of the calculations can be explicitly specified by the user. Molecular systems as well as systems with 3D periodic boundary conditions can be explicitly simulated, but slabs and wire-like boundary conditions are also accessible. This enables an unbiased treatment of systems that are polarizable (for slabs and wires) or charged (for molecules and clusters).

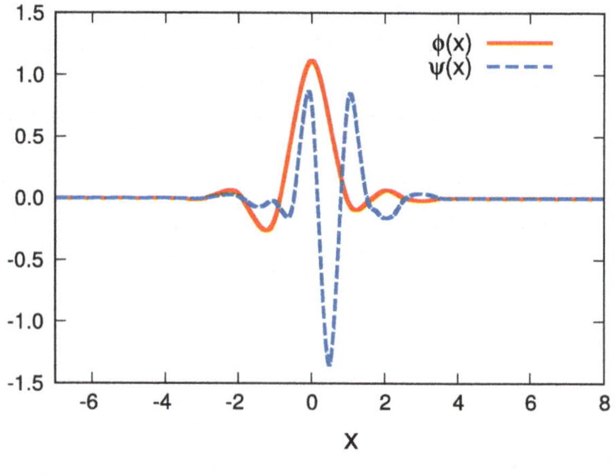

Fig. 4.1 Daubechies scaling function ϕ and wavelet ψ of order 16. Both are different from zero only in the interval from -7 to 8

Such an approach therefore represents an ideal playground for the development of high quality solutions to handle large systems within an efficient and precise algorithm. That is what has been done within the BigDFT developers consortium by implementing a linear scaling algorithm for Kohn-Sham density functional theory.

4.3 Adaptive Localized Orbitals

In order to obtain an algorithm which can scale linearly with respect to the size of the system, one has to abandon the concept of the extended KS orbitals ψ_i and work with other quantities which are strictly localized. To this end one can express the KS orbitals in terms of a set of strictly localized support functions (SFs), $\phi^\alpha(\mathbf{r})$, where $\psi_i(\mathbf{r}) = \sum_\alpha c_i^\alpha \phi^\alpha(\mathbf{r})$. One such quantity is the density matrix

$$\mathbf{F}(\mathbf{r}, \mathbf{r}') = \sum_i f_i \psi_i(\mathbf{r}) \psi_i(\mathbf{r}') = \sum_{\alpha,\beta} |\phi_\alpha\rangle K^{\alpha\beta} \langle\phi_\beta| \tag{4.2}$$

with f_i being the occupation number of orbital i. We also denote with $\mathbf{K}$ the so-called density kernel. One particular approach is to write the density matrix in separable form in terms of the kernel and SFs [9]; an idea that has already been used in various codes such as ONETEP [10], Conquest [11], CP2K [12] and SIESTA [13].

The SFs can also be thought of as a minimal localized basis to directly represent the Kohn-Sham orbitals. In BigDFT, they are atom-centred with a strict user-defined localization radius, beyond which they are set to zero. Consequently, if one neglects elements of the density matrix which are below a given threshold, the number of non-zero elements scales only linearly with respect to the size of the system, thus paving the way towards an algorithm with a reduced complexity. A direct comparison of the results with the cubic scaling algorithm may be realized with a one-off (generalized) diagonalization of the Hamiltonian expressed in the (non-orthogonal) SF basis.

This second "rung" of the "ladder" of computational approaches that can be built thanks to the wavelet properties can be therefore be compared to the original, cubic scaling, approach, where the only approximation is given by the use of PSPs. It is important to introduce the concept of the *crossover point*; above a given number of atoms, it would become computationally more convenient to employ the LS algorithm. The employed figure of merit for the crossover point may be either the walltime or the memory, depending on the user's need and the type of calculations that have to be carried out.

In the LS BigDFT approach, the SFs are expressed in a Daubechies wavelet basis and are defined such as to optimize the usual band structure term of the density-dependent KS Hamiltonian plus a confining operator $\hat{V}_c$ such that

$$\langle\phi_\alpha|\hat{V}_c|\phi_\beta\rangle = \delta_{\alpha\beta}\langle\phi_\alpha|\hat{V}_c^\alpha|\phi_\alpha\rangle \ , \tag{4.3}$$

$$V_c^\alpha(\mathbf{r}) = c_\alpha |\mathbf{r} - \mathbf{R}_\alpha|^4, \tag{4.4}$$

which has the purpose of keeping the SFs confined in their localization regions, centered in the position $\mathbf{R}_\alpha$, while reducing the KS band structure energy. Usually $\mathbf{R}_\alpha$ coincides with the position $\mathbf{R}_a$ of the atom a where ϕ_α is initially centered at the beginning of the SCF optimization procedure. In a typical calculation the strength of the confining potential is reduced by gradually decreasing the value of c_α, until its effect is negligible. Alongside the SF optimization, the density kernel is also optimized using one of three possible methods. These and a detailed description of LS BigDFT are given in [14, 15].

The SFs in BigDFT are therefore optimized in such a manner that they reflect their local environment. Therefore, if two optimized SFs associated with different atoms can be shown to be identical, then one can infer that the local environments of the two atoms are also identical. Such an observation might then lead to the intuition that it is possible to define a type of "fragment approach", by employing SFs from similar environments multiple times. The central idea is to take a group of atoms, for instance an isolated molecule, and fully optimize the SFs. These SFs can then be used as a fixed basis for a system containing several such molecules.

4.4 Reducing the Complexity: Template Support Functions

By employing a fragment approach as described above, the degrees of freedom of a calculation may be reduced further, as there will be no need to optimize the entire set of SFs but only the ones which are associated to the reference ("template") regions. Such considerations are independent of whether the system is composed of molecules or infinitely extended. However, the problem of the *identification* of such template regions will strongly depend on the electronic structure features of the system, which are generally different between molecular and extended systems.

The fragment approach in BigDFT is conceptually straightforward, and can be summarized in three steps:

1. **template calculation**: generate the SF basis $\{\phi_\alpha^T\}$ for each template region T by performing a full LS calculation.
2. **SF replication**: take the SFs from the template calculations and replicate them for the full system of interest.
3. **full calculation**: perform a LS calculation for the full system using the replicated SFs as a fixed basis, optimizing only the density kernel.

Both the first and last steps use the standard machinery of a LS calculation. The second step can be easily automated considering a suitable set of rototranslation transformations $\mathcal{RT}$. Namely, given the range of potential orientations and positions of fragments in a particular target system, it is essential to have a method which is able to rototranslate the template SFs from their template coordinates to the correct location and orientation in the full system of interest. Further details on the definition and application of the fragment approach to (supra)molecular systems are given in [16, 17], while its application to periodic systems is presented in [18]. Irrespective of the details of the system in question, the applicability of such a fragment approach depends on the definition of a set of suitable *indicators* which may quantify the pertinence of a template region.

When a QM system is susceptible to be genuinely separable into fragment states $|\Psi^{\mathfrak{F}}\rangle$, it should be possible to define a projection operator $\hat{W}^{\mathfrak{F}}$ associated with each fragment $\mathfrak{F}$. We may assume that such a projector can be represented in the basis of the SFs by a matrix $\mathbf{W}$, such that $\hat{W}^{\mathfrak{F}}|\Psi\rangle = |\Psi^{\mathfrak{F}}\rangle$, $|\Psi\rangle$ being the ground-state wavefunction. Performing such a fragmentation operation *a posteriori* is a procedure that presents of course some degree of arbitrariness and is susceptible to provide, in the worst case, a partitioning into physically meaningless moieties. The spirit of the fragmentation procedure described in [19] is to provide indicators which help to assess the *physical pertinence* of a given fragmentation.

To this end, we have introduced what we refer to as the *purity indicator*, defined by

$$\Pi_{\mathfrak{F}} = \frac{1}{q}\mathrm{tr}\left(\left(\mathbf{KS}^{\mathfrak{F}}\right)^2 - \mathbf{KS}^{\mathfrak{F}}\right), \tag{4.5}$$

where q is the total number of electrons of the isolated fragment in gas phase and $\mathbf{S}^{\mathfrak{F}} \equiv \mathbf{SR}^{\mathfrak{F}}\mathbf{S}$. We call a fragment *pure* if its projection satisfies the condition $\Pi \simeq 0$. Such a condition, which we emphasize to be non-linear in the projector matrix elements $R^{\mathfrak{F}}$, when fulfilled, enables one to interpret the *fragment*-expectation value

$$\langle \hat{O} \rangle_{\mathfrak{F}} \equiv \mathrm{tr}\left(\hat{F}^{\mathfrak{F}}\hat{O}\right) = \mathrm{tr}\left(\mathbf{KSR}^{\mathfrak{F}}\mathbf{O}\right) \ , \tag{4.6}$$

as a pseudo-observable of the fragment $\mathfrak{F}$.

We have shown that such a projection operator proves to be powerful in the identification of a system's fragments for a system with a molecular electronic structure. In other terms, the approach described so far is well-suited for molecular systems, or more precisely for systems where the fundamental constituents have molecular character. The key question is therefore whether it is possible to define an equivalent approach for extended systems, where it is not possible to define a fragment projection operator which can be used as such.

To this end, we define a quantity which we refer to as the onsite overlap matrix, S^{onsite}. This is defined as

$$S^{\text{onsite}}_{\alpha\beta} \equiv \langle \mathcal{RT}^{\mathbf{R}_\alpha \to \mathbf{R}_0}\phi^{(\ell)}_\alpha | \mathcal{RT}^{\mathbf{R}_\beta \to \mathbf{R}_0}\phi^{(\ell)}_\beta \rangle, \tag{4.7}$$

i.e. as the overlap between two SFs (which may originally be centred on different atoms) if they were to be located on the same site of centre $\mathbf{R}_0$. Where $S^{\text{onsite}}_{\alpha\beta} = 1$, the SFs α and β are identical, and thus have both the same character (s, p_x etc.) and local environment. The onsite overlap is therefore independent of the physical location of the SFs and is instead rather a measure of the similarity between two SFs and correspondingly the chemical environment of the atoms on which they are located. Such a quantity is therefore important where we know that the system may be represented as a replica of the same pattern, with the SFs surrounded by the same chemical environment. In such a case, we refer to the fragments as pseudo-fragments, since, as discussed, they are not truly separable fragments of the system.

With these two different indicators (Π and S^{onsite}) at hand, one can then define a so-called "template" region, which may be constituted either of a system's moiety or by a group of atoms. However, it is evident that such an approach would work especially for the case where there is no internal deformation of a given fragment, i.e. when the transformation from template to system fragment is a rigid rotation. In such cases one can easily find the appropriate transformation. However it is important, both in the molecular and extended system approaches, to also allow for (small) deformations in fragments. We therefore define a cost function J

$$J\left(\mathcal{R}^{T\to S}\right) = \frac{1}{2}\sum_{a=1}^{N} ||\mathbf{R}_a^S - \sum_{b=1}^{N} \mathcal{R}_{ab}^{T\to S}\mathbf{R}_a^T||^2, \tag{4.8}$$

where N is the number of atoms in the fragment and $\mathbf{R}_a^{T(S)}$ are the coordinates of the template (system) fragment. The identification of the optimal transformation from such a cost function is a well known problem [20, 21], and may easily be found using a simple singular value decomposition based approach [22].

We have demonstrated that such a cost function may also be used as a necessary condition for the fragment approach to be pertinent. Stated otherwise, the Π and S^{onsite} indicators might be useful only when there is not too much geometrical difference—quantified by the cost function J—between the template region and the target system.

4.5 Case Study: Graphene

The different methods introduced above are illustrated schematically in Fig. 4.2, where we also put them in context with a multiscale approach, which might be the subject of future work. We show both the key approximation which is introduced for a given method and an indication of approximate applicable lengthscales. In the following, we now show how such methods may be used in practice. Specifically, we are interested in assessing the impact of successive approximations on both the computational cost and the accuracy of a given computational setup. We do so by studying in detail the example of graphene.

4.5.1 Computational Cost: Effect of Dimensionality

As discussed above, when selecting a method to treat a given system, one must choose an approach which is able to model the system in sufficient detail to simulate the properties of interest, while at the same time avoiding a treatment which is

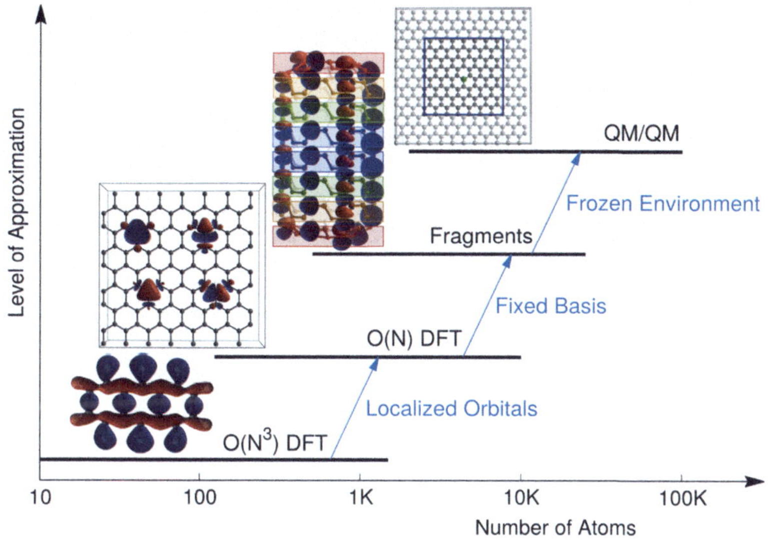

Fig. 4.2 Ladder of various computational methods that can be implemented thanks to the properties of wavelet functions. The base "rung" is provided by the cubic scaling PSP approach, which would limit its range of applicability to systems with less than a thousand atoms. The linear scaling approach, which exploits the locality of the support function basis to express the KS orbitals, is of great utility above this range, not only thanks to its reduced computational cost, but also since it provides a mapping between the KS description and localization regions. Such a concept is fundamental for the conception of the third rung, the fragment approach based on support functions, which are therefore pre-optimized in template regions and used as fixed basis sets to represent the electronic structure of the system. With a similar setup at hand, it is possible to connect together various levels of theory, by embedding the KS QM region into another QM treatment, like a pre-defined KS Hamiltonian or a Tight-Binding treatment. This concept is represented by the fourth rung, which opens the path towards a multi-scale approach for the treatment of materials and assemblies that can go above the nanoscale

unnecessarily costly. To this end, it is important to understand how the cubic scaling, linear scaling and (pseudo-)fragment approaches differ in terms of computational cost. Aside from the size of the system in question, the relative computational cost of the three methods is also affected by other properties such as the band gap (see [23] for considerations concerning the treatment of metallic systems) or the dimensionality of the system. In the following we explore the effects of the latter.

To this end we take two related systems with different dimensionalities, namely a quasi-one-dimensional graphene nanoribbon (Fig. 4.3a) and a fully two-dimensional sheet of graphene (Fig. 4.3b). The importance of the dimensionality of the system may be understood by considering the qualitative differences between the associated SF matrices. Both systems are treated with explicit surface boundary conditions, however as shown in Fig. 4.4, the density kernel of the nanoribbon is more sparse than that of graphene for a given system size, since there are fewer overlapping SFs (assuming the SF localization radii do not differ). As such, we expect to have a lower

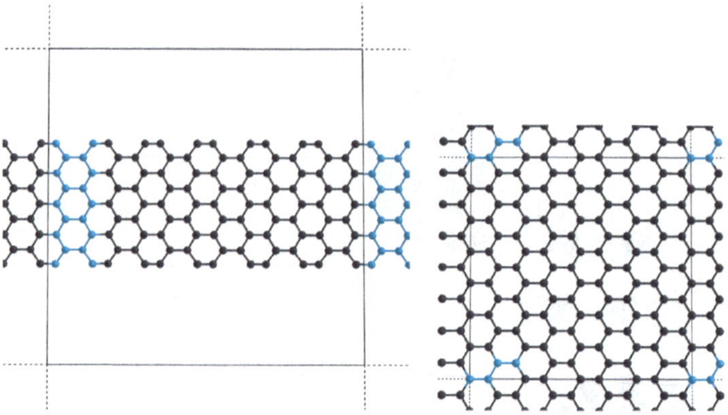

(a) nanoribbon (b) graphene

Fig. 4.3 Depiction of the smallest supercells used for the two systems, where the box represents the size of the cell. The template pseudo-fragment is indicated by the atoms in blue

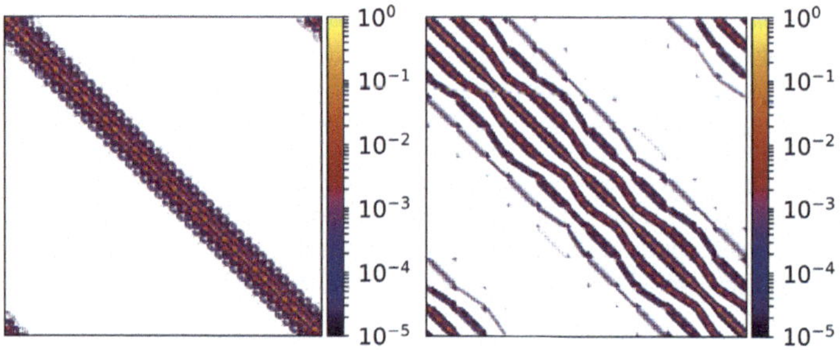

Fig. 4.4 Plots showing the sparsity of the density kernel matrix for 576 atom supercells of the nanoribbon (left) and graphene (right)

crossover point with respect to the cubic scaling approach for the nanoribbon than for graphene.

There are a number of ways to set up a pseudo-fragment approach, however the particular choice should have no impact on either the accuracy or the cost of the calculations. For convenience, we make the choices depicted in Fig. 4.3. We performed calculations for a range of system sizes using a wavelet grid spacing of 0.24 Å and four SFs per atom with localization radii of 3.7 Å. Moderate convergence criteria were used for the LS approach, corresponding to an error of around 10–15 meV/atom in the total energy compared to the cubic scaling approach. We used the PBE functional [24] and a HGH PSP [25] employing non-linear core corrections [8]. The Fermi Operator Expansion approach was used to calculate the density kernel.

As a comparison of the three approaches, we show in Fig. 4.5 the density of states (DoS) for the two systems for each approach for a fixed system size of 576 atoms in each cell. Since the SFs are optimized to minimize the band structure energy, one does not expect that such a basis is able to represent the unoccupied states. However, we

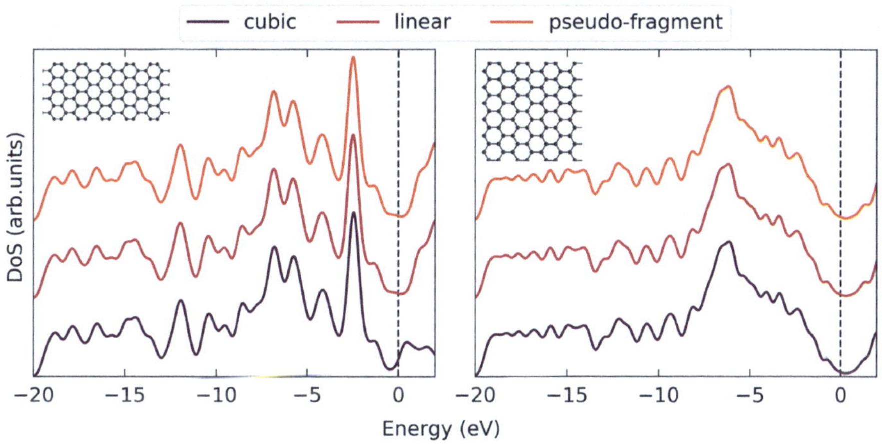

Fig. 4.5 Density of states for 576 atom supercells of the nanoribbon (left) and graphene (right), for the cubic scaling, linear scaling and pseudo-fragment approaches. Gaussian smearing of 0.25 eV has been applied and the curves have been shifted so that the highest occupied molecular orbital (HOMO) is at zero

expect good agreement for the occupied states. It can be seen that the three methods are indeed in excellent agreement up to the Fermi level. In other terms, if one were interested in calculating the DoS of either of these systems, one should choose the least costly method for the given system size, since all approaches are able to capture the complexity of the systems.

In Fig. 4.6 we show the scaling of the three methods for a fixed number of computing cores, where a cubic polynomial has been fitted to the results for the cubic scaling method and a linear function for the remainder of the calculations. There are a number of points to discuss. First, the cubic scaling calculations of the nanoribbon have a higher cost compared to graphene for the same number of atoms. This is due to the larger cell size relative to graphene, since the cost of empty space is non-negligible in the cubic scaling approach. This serves to demonstrate the importance of using appropriate boundary conditions—in this example, the use of explicit wire boundary conditions would be a better choice. A similar effect may be seen on the required memory. On the other hand, the cost of the LS simulations is very similar between systems for a given system size, both in terms of time and memory. Although less severe than in the cubic scaling case, the SFs nonetheless extend into the portion of empty space which is close to the nanoribbon, increasing the computational cost. This effect is however balanced out by the reduced sparsity and resulting lower cost of sparse matrix algebra.

For both systems the use of the pseudo-fragment approach results in a significantly lower prefactor in terms of time. However, the associated memory requirements increase slightly compared to the LS approach. In a standard LS calculation the SFs are reordered and distributed among the MPI tasks in such a manner as to optimize the load balancing. However, for the pseudo-fragment approach the SFs are kept in

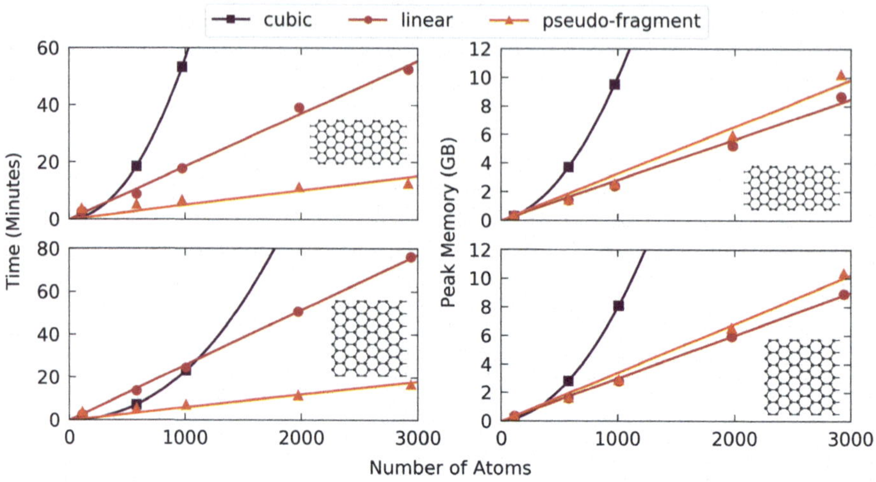

Fig. 4.6 Plot showing the time (left) and memory (right) scaling for the nanoribbon (top) and graphene (bottom), for the cubic scaling, linear scaling and pseudo-fragment approaches. The cost of the template calculation has been included in the total cost for the pseudo-fragment approach. Calculations were performed on 20 nodes on Archer, using 6 MPI tasks and 4 OpenMP threads per node. The memory is the peak memory usage of the root node, as reported by BigDFT

the same order as the input file, in order to simplify the association between a SF and the fragment instance to which it belongs. This can result in a less optimal load balancing, explaining the slightly higher memory costs.

In Table 4.1 we give the various crossover points with respect to the cubic scaling approach. That is, beyond the stated number of atoms it will be cheaper to use the specified approach. The values were calculated using the fitted curves and therefore also take into account the cost of the template calculation. However, we emphasize that the crossover depends not just on the dimensionality and electronic structure (i.e. band gap) of the material considered, but also the given simulation *parameters*. This includes not just the SF localization radii, but also the convergence criteria, the number of iterations of the SF and kernel convergence loops and so on. As such, the values might change by up to a few hundred atoms if a different set of parameters

Table 4.1 Calculated crossover points for the employed parameters, i.e. number of atoms beyond which the specified method is cheaper than the cubic scaling approach. Both time and memory values are given for the linear scaling and pseudo-fragment methods for the two systems considered

Method	Nanoribbon		Graphene	
	Time	Memory	Time	Memory
Linear scaling	329	136	1111	310
Pseudo-fragment	76	191	273	367

were employed. However, since we have used the same parameters between systems, we may nonetheless make some useful comparisons.

It is unsurprising that the crossover point for the nanoribbon is indeed lower than that for graphene for both time and memory. In both cases, the LS approach already uses less memory for a few hundred atoms. Furthermore, the computational savings associated with the pseudo-fragment approach are large enough such that the crossover in time also occurs at less than 300 atoms. In the case of the nanoribbon, the approach is already the cheapest in terms of computer time for less than 100 atoms. Therefore, given a system which contains a significant degree of repetition to permit an appropriate pseudo-fragmentation, the pseudo-fragment approach is expected to be the most efficient approach even for very small system sizes.

4.5.2 Defective Graphene

Having shown above that the pseudo-fragment approach enables accurate calculations of large supercells of pristine graphene at substantially reduced computational cost, we now consider the example of defective graphene. Such a system represents more of a challenge, since there is clearly a region close to the defect which requires a different treatment to atoms far away, which might be treated in the same way as pristine graphene. We have previously shown [18] for the example of a SiC nanotube that it is possible to divide a material into bulk-like and defective regions, using the onsite overlap and cost function indicators described above. In the following we show how such an approach can be generalized to identify the "active" region around a point defect wherein SFs must be adapted to the defect, and an "environment" region where generic SFs from pristine graphene may be used.

If one were able to identify the extent of the active region, then SFs could be generated in a defective supercell which is just large enough to contain the active region, and then reused in a much larger defective supercell with additional bulk-like SFs. Such a concept is illustrated in Fig. 4.7. Before employing such an approach, it is necessary to determine how big the active region should be, in other terms how big does the smaller defective supercell need to be in order to contain all non-bulk-like SFs. In order to answer this question, we take the example of a Si substitutional defect in graphene, and explore the electronic structure and behaviour of the indicators in detail for a fixed supercell size. We take a supercell containing 336 atoms (depicted on the right of Fig. 4.7), since we expect this to be large enough that the atoms furthest from the defect are indeed bulk-like.

4.5.2.1 Computational Setup

Calculations employed a wavelet grid spacing of 0.19 Å, with four SFs per C atom and eight per Si atom, all with localization radii of 4.2 Å. We used the PBE functional [24] and HGH PSPs [25]. For all calculations the direct minimization approach was used

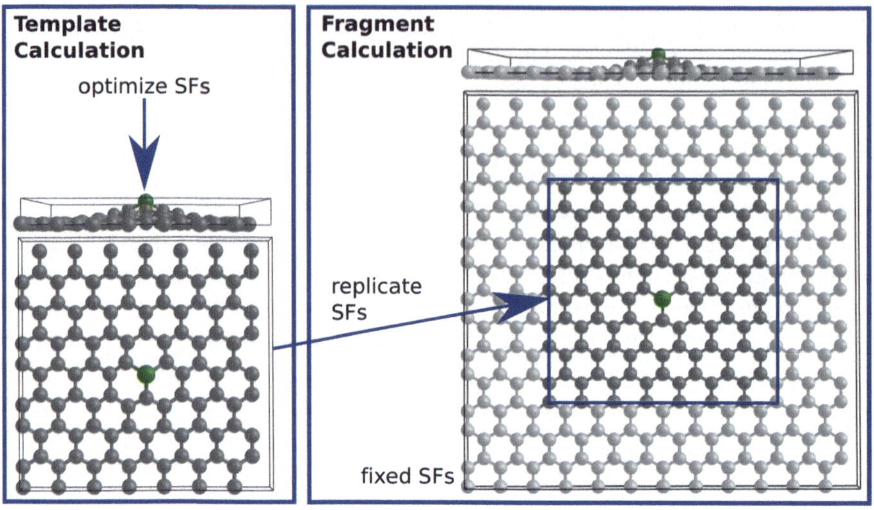

Fig. 4.7 Schematic illustrating the workflow of a pseudo-fragment calculation for graphene containing a Si substitutional defect. The support functions are initially generated for a small defective supercell, then replicated in a large defective supercell and used as a fixed basis. Atoms in the large supercell which are outside the boundaries of the small supercell are considered to be bulk-like. For such atoms, one may choose to use the support functions corresponding to the atom which is furthest from the defect in the small supercell, or directly take optimized support functions from a calculation of pristine graphene

to calculate the density kernel. Tight convergence criteria were applied, such that the difference in total energy with respect to the cubic scaling approach was around 5 meV/atom. The atomic structure was relaxed using the cubic scaling approach, so that the maximum force component was below 0.2 eV/Å. Since the aim is to understand the extent of the active region associated with the defect, rather than a precise calculation of the defect formation energy, a more strict force tolerance was not deemed necessary.

A number of different approaches could be used to divide the system into pseudo-fragments. We choose to treat each atom as a pseudo-fragment. The key question is therefore whether or not a given pseudo-fragments can be considered to be bulk-like (i.e. treated as if it is in pristine graphene), or whether it is sufficiently affected by the presence of the defect to be considered part of the active region. In the latter case, the pseudo-fragment must be associated with SFs which are explicitly optimized to represent its local chemical environment. In order to explore this question, we label each atom according to its distance from the Si atom, d_f, for all atoms which are within 15 Å from the defect. Although some atoms within the system are related by symmetry and could therefore be considered to be different instances of the same pseudo-fragment, we nonetheless consider each atom to be a unique pseudo-fragment. This avoids the introduction of a small but nonetheless non-zero error associated with needing to rotate the SFs between template and system positions.

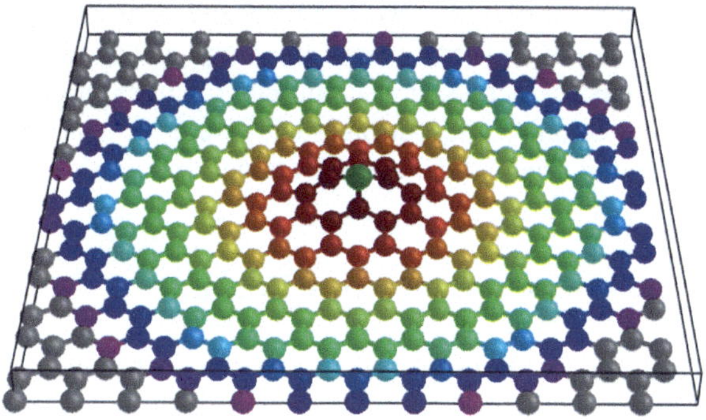

Fig. 4.8 Illustration of the pseudo-fragment setup for the 336 atom defective graphene supercell. Atoms depicted in grey are considered to be bulk-like, while the remainder of the atoms are coloured according to distance d_f from the Si substitutional defect, with each atom considered to be a unique pseudo-fragment

Fig. 4.9 Schematic showing the two types of graphene atoms which are used to specify the atomic pseudo-fragments within their local environment: 'A' (medium grey) and 'B' (dark grey). The distinction is purely one of convenience

The remainder of the atoms are always considered to be bulk-like, as depicted in Fig. 4.8. For convenience, the bulk-like C atoms are divided into two types ('A' and 'B') according to their orientation within the supercell, as illustrated in Fig. 4.9. Such a definition allows one to avoid the need for rotating the SFs associated with atoms in a different relative orientation.

In the first instance, we fully optimize the SFs for all atoms within the supercell. We then perform a series of pseudo-fragment calculations using a fixed basis wherein all atoms up to and including a given d_f are treated with their correctly optimized template SFs, while all atoms beyond d_f are treated using generic bulk-like SFs. For the latter case, the SFs are those corresponding to the atom (of type A or B as appropriate) which is furthest from the defect. A number of values of d_f are used so as to consider a range of cases between two extreme scenarios:

1. $d_f = 0.0$ Å: all C atoms are considered to be bulk-like, i.e. the active region corresponds to the Si atom only.
2. $d_f = 15.0$ Å: all C atoms up to 15.0 Å are considered to be part of the active region and thus treated with SFs which are optimized for their appropriate local chemical environment, with only relatively few atoms considered to be bulk-like.

For each value of d_f, we calculate the total energy and the density of states. By comparing with the results of the template calculation (i.e. fully optimized SFs for all atoms), we are able to quantitatively assess the effect of varying the size of the active region, while putting such results in context with the values of the onsite overlap matrix and the cost function J.

4.5.2.2 Assessing the Active Region

The key quantity of interest for the pseudo-fragment calculations is the total energy. We expect that the larger the active region (i.e. the larger the value of d_f) the smaller the induced error in the total energy. As can be seen from the upper panel of Fig. 4.10, this is indeed the case. Furthermore, apart from a small increase in energy around $d_f = 4.5$ Å, the error in energy otherwise decreases monotonically as d_f is increased. While the error for $d_f = 0$ Å is more than 100 meV/atom, the inclusion of just a few nearest neighbours in the active region is sufficient to reduce the error to around 20 meV/atom. Depending on the accuracy requirements of a given simulation, one might choose the size of the active region accordingly. For example, an active region of radius of around 6 Å would already give an error of less than 10 meV/atom, while in cases where very precise energies are required, a radius of 15 Å gives a negligible error, i.e. less than 1 meV/atom.

Although we choose to focus primarily on the total energy, it is interesting to also consider the convergence of other quantities with respect to the size of the active region. In the first instance, we consider the DoS, which is shown for selected pseudo-fragment setups compared to the full LS result in Fig. 4.11. In order to examine the differences more closely, we apply only a small level of Gaussian smearing. As can be seen, the DoS converges very quickly with increasing active region. Indeed, even for $d_f = 0$ Å, the result is already relatively good, aside from a small additional peak appearing below the HOMO. For $d_f = 3$ Å the DoS is already virtually indistinguishable from the full LS reference. If one is primarily interested in the DoS and does not need a very precise calculation of the total energy, it would therefore be possible to use a rather small active region.

We also consider the convergence of the electronic density. Figure 4.12 shows the absolute error in density relative to the LS approach for select pseudo-fragment approaches, averaged over the y-axis. For reference, the electronic density from the LS calculation averaged over the same axis is also shown. For $d_f = 0$ Å there is a large error in the electronic density around the defect, as might be expected given the unsuitability of bulk-like SFs for the region close to the defect. With increasing d_f the average error in this region quickly decreases, so that by $d_f = 9$ Å there is no particular error in the region close to the defect, but rather only a small error which is distributed across the system.

Although it is interesting to be able to relate the induced error in energy to the size of the active region, it would be more useful in practice if such an error could be predicted ahead of time. That is, if one could already determine how big the active region needs to be without performing a series of calculations. Such an approach

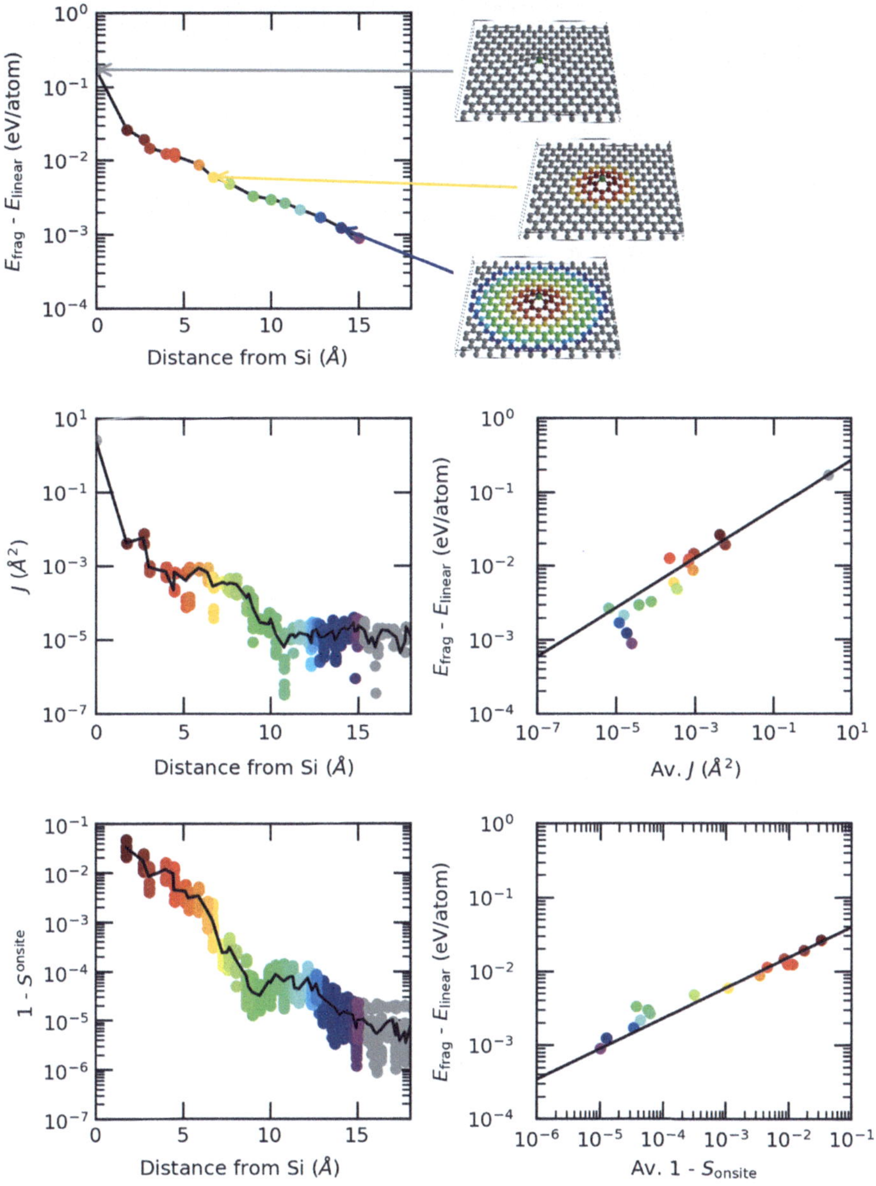

Fig. 4.10 Plots showing various quantities for defective graphene for the different pseudo-fragment calculations described in the text. These include error in total energy with respect to a full linear scaling calculation, where select pseudo-fragment setups are illustrated (top); cost function, J, versus distance from the defect (middle left) and correlation between J and error in energy (middle right); $1 - S^{onsite}$ versus distance from the defect (bottom left) and correlation between $1 - S^{onsite}$ and error in energy (bottom right). The points are coloured according to d_f, as depicted in Fig. 4.8

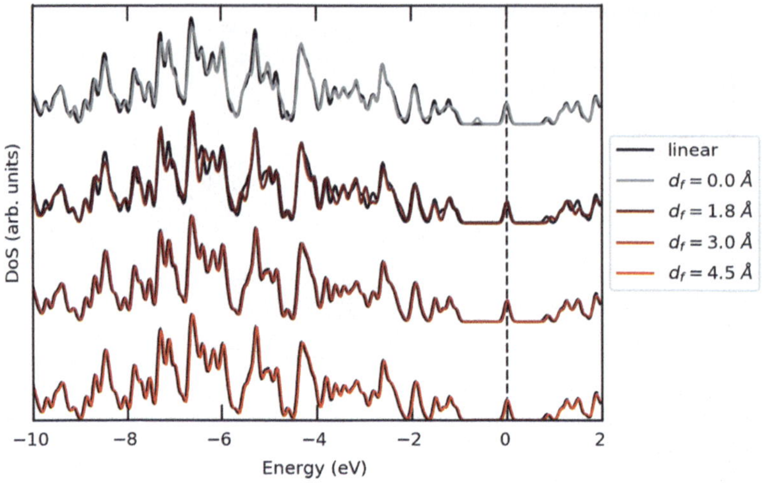

Fig. 4.11 Density of states for defective graphene for different pseudo-fragment setups for which all atoms up to some distance d_f employ correctly optimized support functions, while beyond d_f bulk-like support functions are used. The results are compared to the full linear scaling reference. Gaussian smearing of 0.05 eV has been applied and the curves have been shifted so that the HOMO is at zero

would avoid the need for manually converging the size of the active region for each system, and could potentially be highly useful for the setup of QM/QM or QM/MM calculations wherein an active region treated at a higher level of theory is embedded in an environment treated at a lower level of theory. To this end, both the cost function and onsite overlap matrix introduced above could provide useful indicators for predicting the minimum size required for the active region in order to achieve a given accuracy.

4.5.2.3 Cost Function

We first consider the cost function, J. Such a parameter can be used as a measure of the deviation in the local atomic environment compared to the equivalent pristine material, i.e. the effect of geometrical distortions. Close to the Si atom, the graphene is strongly distorted, so that the material is no longer planar. On the other hand, far from the defect the C atoms are relatively unaffected by the presence of the defect, so that the local geometry is very similar to pristine graphene. We expect to see such behaviour reflected in the value of J.

In order to verify that this is the case, we take the $d_f = 0$ Å setup and calculate the value of J for each atom. We recall that in such a setup, the template pseudo-fragment is the most bulk-like C atom from an equivalent atom, i.e. coming from the atom of type A or B which is furthest from the defect. The value of J is therefore a measure of the deviation between the local atomic structure of a given atom and

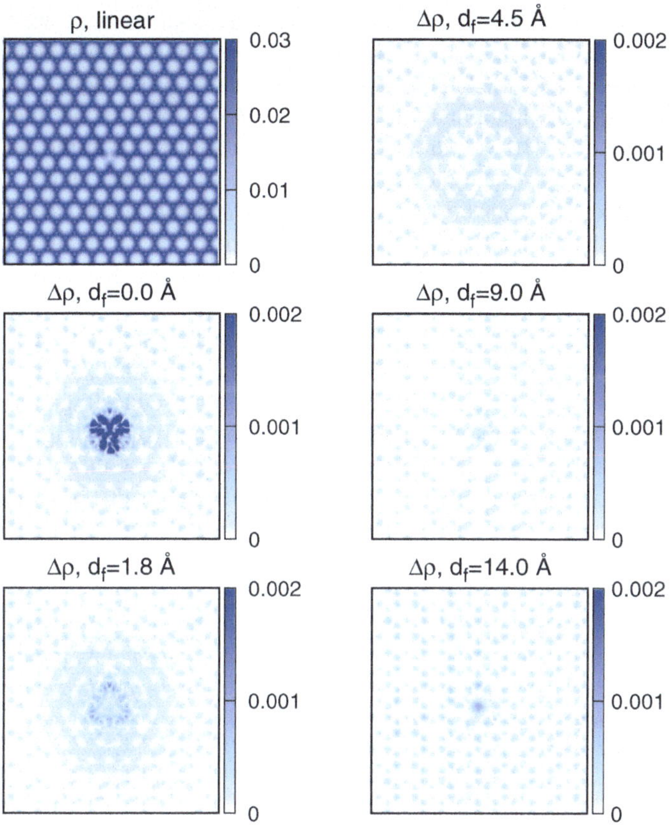

Fig. 4.12 Plot depicting the average electronic density across the y axis, ρ, for the full linear scaling approach, and the absolute error in electronic density relative to the full linear scaling approach averaged across the y axis, $\Delta\rho$, for different pseudo-fragment setups for defective graphene

that of a bulk-like atom. In order to calculate such a quantity correctly, we include the three neighbouring C atoms for both the template and system pseudo-fragment when calculating J for a given pseudo-fragment.

In Fig. 4.10 we plot the value of J for each atom. Since the error in a given pseudo-fragment calculation will be dominated by the atoms just outside the active region, the value of J at a given distance is that coming from the atoms which are at a distance just beyond d_f. For example, the value of J which is plotted at 0.0 Å is that associated with the three C atoms which are bonded to the Si atom. The values are coloured according to distance, while the average value at a given distance is shown in black. As expected, J is very high in the immediate vicinity of the defect, and decreases non-monotonically with distance up to around 10 Å, beyond which point it plateaus at around 10^{-5} Å^{-2}, indicating that the local atomic geometry is hardly affected by the presence of the defect at such distances.

It is reasonable to assume that SFs optimized for a planar structure are not suitable for C atoms in a distorted, non-planar structure, so that atoms corresponding to a high value of J should be included in the active region with appropriately optimized SFs. One can therefore anticipate that a high value of J will correspond to a high error in total energy. In Fig. 4.10 we therefore plot the correlation between the average value of J, as described above, and the error in energy with respect to a LS calculation, ΔE. There is indeed a strong correlation between J and ΔE, as indicated by the line of best fit. This shows that J is indeed a good predictor of the accuracy of a given setup, as we have also shown previously for a one-dimensional system [18].

Although a precise quantitative prediction might require extensive testing across a range of systems to determine whether or not the relationship between J and ΔE varies significantly between systems (given appropriate normalization), one might nonetheless assume that if $J \gtrsim 10^{-3}$ Å^{-2}, then a given pseudo-fragment should be included in the active region. Such a measure is particularly appealing since it relies only on information concerning the atomic structure of a system, i.e. no electronic structure calculations are necessary in order to calculate the value of J. Therefore, it is highly useful even if only used as a qualitative indicator concerning the minimum size of the active region.

Finally, we note that for high values of d_f, the correlation between J and ΔE is less strong, since J has saturated while the energy nonetheless continues to decrease. In such a situation, a measure of the local atomic structure is insufficient, and one must further consider also the effect of the defect on the local *electronic* structure. This may be achieved using the onsite overlap matrix.

4.5.2.4 Onsite Overlap Matrix

Following the above considerations, we consider also the onsite overlap matrix, S^{onsite}, which was calculated for the fully optimized SFs. Since we are interested in comparing with bulk-like SFs, we consider the value of S^{onsite} between each SF and that coming from the same type of atom which is most bulk-like, i.e. the atoms should both be A or both be B. The results are shown in Fig. 4.10, where it can be seen that the SFs generally become more bulk-like as the atoms with which they are associated are further from the defect, as expected. There is a region around 10 Å where the values stop decreasing, before again decreasing and then saturating again around 15 Å.

As with J, there is a strong correlation between $1 - S^{\text{onsite}}$ and ΔE. Unlike J, the trend continues up to $d_f = 15$ Å, with also a stronger overall correlation. In other terms, the variation in SFs versus distance from defect represents the best indicator of the accuracy of a calculation with a given size of active region. This is not surprising, since such an indicator reflects at the same time both variations in local atomic and electronic structure. Although such a quantity requires a full LS-DFT calculation, unlike the calculation of J which may be extracted based purely on geometrical information, the onsite overlap could nonetheless be highly useful in confirming whether or not a given pseudo-fragment setup represents a sufficiently large active region for a given required accuracy. Furthermore, it could be particularly applicable for future developments combining different levels of approximation, where the SFs are optimized on-the-fly for a given active region, with the remainder of the system treated at a lower level of theory.

4.6 Perspective and Conclusions

Thanks to developments in linear scaling approaches to DFT, the scope of first principles calculations has extended to system sizes and materials which were until recently beyond the capabilities of DFT. Furthermore, such approaches bridge the gap between the lengthscales accessible to cubic scaling DFT (of the order of hundreds of atoms) and those which are treated using classical approaches (more than ten thousand atoms). As a result, LS-DFT may be used to test and validate classical approaches, by simulating the system they are designed to treat at a lengthscale which is close to the targeted lengthscale. One may also validate QM/QM or QM/MM models in a similar manner.

In such a context, the ability to choose between methods which are suited to different lengthscales while retaining the same underlying formalism (i.e. basis set) would be invaluable. In other terms, if different approaches can be implemented using a single unifying basis set (and within the same code), then one can compare different methods by adding only one approximation at a time. We have shown how Daubechies wavelets represent a basis set which is ideally suited to such an approach. Since they are a systematic basis set, arbitrary numerical precision may be achieved, while they may also be exploited to directly treat both molecular and extended systems. In the above we have described the setup of cubic scaling, linear scaling and fragment approaches within a wavelet basis set.

Thanks to the implementation within a single framework, we may easily test and compare different approaches. Furthermore, one may even directly transfer results between methods, for example by using the output of a linear scaling calculation as an input guess for a cubic scaling calculation. Such a framework thus also provides an ideal playground for testing new methods and approximations.

In the above, we have demonstrated the practical usage of such concepts through the case study of graphene. We have shown how the dimensionality of the system affects computational cost, where for pristine periodic systems the use of a pseudo-fragment approach is computationally advantageous even for a few hundred atoms. We have also demonstrated the applicability of the different methods to defective graphene. In particular, we have shown how different indicators may be used to predict and quantify the division of a system into active and environment regions. Such indicators could be further exploited in future for the setup of multiscale approaches, avoiding the need for manually converging the size of the active region which is treated using a higher level of theory.

Acknowledgements LER acknowledges an EPSRC Early Career Research Fellowship (EP/P033253/1) and the Thomas Young Centre under grant number TYC-101. We are grateful to the UK Materials and Molecular Modelling Hub for computational resources, which is partially funded by EPSRC (EP/P020194/1). Calculations were also performed on the Imperial College High Performance Computing Service and the ARCHER UK National Supercomputing Service.

References

1. P. Hohenberg, W. Kohn, Phys. Rev. **136**(3B), B864 (1964)
2. W. Kohn, L.J. Sham, Phys. Rev. **140**(4A), A1133 (1965)
3. S. Goedecker, Rev. Mod. Phys. **71**, 1085 (1999)
4. D.R. Bowler, T. Miyazaki, Rep. Prog. Phys. **75**(3), 036503 (2012)
5. L.E. Ratcliff, S. Mohr, G. Huhs, T. Deutsch, M. Masella, L. Genovese, WIREs Comput. Mol. Sci. **7**(1), e1290 (2017)
6. L. Genovese, A. Neelov, S. Goedecker, T. Deutsch, S.A. Ghasemi, A. Willand, D. Caliste, O. Zilberberg, M. Rayson, A. Bergman, R. Schneider, J. Chem. Phys. **129**(1), 014109 (2008)
7. I. Daubechies, *Ten Lectures on Wavelets* (SIAM, 1992)
8. A. Willand, Y.O. Kvashnin, L. Genovese, A. Vázquez-Mayagoitia, A.K. Deb, A. Sadeghi, T. Deutsch, S. Goedecker, J. Chem. Phys. **138**(10), 104109 (2013)
9. E. Hernández, M.J. Gillan, Phys. Rev. B **51**(15), 10157 (1995)
10. C.K. Skylaris, P.D. Haynes, A.A. Mostofi, M.C. Payne, J. Chem. Phys. **122**(8), 84119 (2005)
11. D.R. Bowler, T. Miyazaki, J. Phys.: Condens. Matter **22**(7), 074207 (2010)
12. J. VandeVondele, M. Krack, F. Mohamed, M. Parrinello, T. Chassaing, J. Hutter, Comput. Phys. Commun. **167**(2), 103 (2005)
13. J.M. Soler, E. Artacho, J.D. Gale, A. García, J. Junquera, P. Ordejón, D. Sánchez-Portal, J. Phys.: Condens. Matter **14**(11), 2745 (2002)
14. S. Mohr, L.E. Ratcliff, P. Boulanger, L. Genovese, D. Caliste, T. Deutsch, S. Goedecker, J. Chem. Phys. **140**(20), 204110 (2014)
15. S. Mohr, L.E. Ratcliff, L. Genovese, D. Caliste, P. Boulanger, S. Goedecker, T. Deutsch, Phys. Chem. Chem. Phys. **17**, 31360 (2015)
16. L.E. Ratcliff, L. Genovese, S. Mohr, T. Deutsch, J. Chem. Phys. **142**(23), 234105 (2015)
17. L.E. Ratcliff, L. Grisanti, L. Genovese, T. Deutsch, T. Neumann, D. Danilov, W. Wenzel, D. Beljonne, J. Cornil, J. Chem. Theory Comput. **11**(5), 2077 (2015)
18. L.E. Ratcliff, L. Genovese, J. Phys.: Condens. Matter **31**(28), 285901 (2019)
19. S. Mohr, M. Masella, L.E. Ratcliff, L. Genovese, J. Chem. Theory Comput. **13**(9), 4079 (2017)
20. G. Wahba, SIAM Rev. **7**(3), 409 (1965)
21. W. Kabsch, Acta Crystallogr. A **34**(5), 827 (1978)
22. F.L. Markley, J. Astronaut. Sci. **36**(3), 245 (1988)
23. S. Mohr, M. Eixarch, M. Amsler, M.J. Mantsinen, L. Genovese, J. Nucl. Mater. **15**, 64 (2018)
24. J.P. Perdew, K. Burke, M. Ernzerhof, Phys. Rev. Lett. **77**(18), 3865 (1996)
25. C. Hartwigsen, S. Goedecker, J. Hutter, Phys. Rev. B **58**, 3641 (1998)

Chapter 5
Self-consistent Hybrid Functionals: What We've Learned So Far

Daniel Fritsch

Abstract These are exciting times for computational materials science. We are witnessing a wide-spread availability of high-performance computing facilities, a huge increase in accessible computational resources, and an accompanying development of new exchange-correlation functionals within density functional theory. All this contributes to the establishment of density functional theory as an indispensable tool for materials science investigations in general. Here, we want to highlight some examples utilising a recently developed self-consistent hybrid functional, proposed by Shimazaki and Asai (J Chem Phys 130:164702, 2009 [3]) and Skone et al. (Phys Rev B 89:195112, 2014 [4]), allowing for the calculation of accurate material properties using a fully ab initio procedure. The obtained structural and electronic properties of a range of oxide semiconductors will be critically discussed with respect to experimental results, and pave the way towards open questions in the field.

5.1 Introduction

Density functional theory (DFT), introduced by Hohenberg and Kohn [1] and Kohn and Sham [2], provides us with an indispensable tool to perform predictive materials science investigations into the structural, electronic, and optical properties of a wide range of materials. However, every theoretical result has to be compared to experimental data, and the wisdom accumulated over the last decades show that the agreement of theory and experiment strongly depends on the underlying approximation of the unknown exchange-correlation functional utilised in the calculations. On more formal grounds we can write down the generalised Kohn-Sham (GKS) potential $v_{\mathrm{GKS}}(\mathbf{r}, \mathbf{r}')$ as

$$v_{\mathrm{GKS}}(\mathbf{r}, \mathbf{r}') = v_{\mathrm{H}}(\mathbf{r}) + v_{xc}(\mathbf{r}, \mathbf{r}') + v_{\mathrm{ext}}(\mathbf{r}) . \tag{5.1}$$

D. Fritsch (✉)
Helmholtz-Zentrum Berlin für Materialien und Energie, Hahn-Meitner-Platz 1, 14109 Berlin, Germany
e-mail: daniel.fritsch@helmholtz-berlin.de

E. V. Levchenko et al. (eds.), *Theory and Simulation in Physics for Materials Applications*, Springer Series in Materials Science 296,
https://doi.org/10.1007/978-3-030-37790-8_5

It is expressed as the sum of the Hartree $v_{\mathrm{H}}(\mathbf{r})$, the exchange-correlation $v_{xc}(\mathbf{r}, \mathbf{r}')$, and the external $v_{\mathrm{ext}}(\mathbf{r})$ potential, respectively, where $v_{\mathrm{H}}(\mathbf{r})$ and $v_{\mathrm{ext}}(\mathbf{r})$ are in principle known, and only $v_{xc}(\mathbf{r}, \mathbf{r}')$ requires a more or less sophisticated approximation.

Among others, the particular choice of an exchange-correlation functional may be influenced by the size of the unit cell and their respective number of atoms, the material properties we're interested in, and the available computational resources. With the advent and general availability of local, regional, and national high-performance computing facilities, recent years witnessed a surge in the application of computationally more demanding so-called hybrid functionals, where a particular fraction α of the underlying semilocal exchange potential is replaced by a Hartree-Fock exact-exchange term

$$v_{xc}(\mathbf{r}, \mathbf{r}') = \alpha v_x^{ex}(\mathbf{r}, \mathbf{r}') + (1 - \alpha) v_x(\mathbf{r}, \mathbf{r}') + v_c(\mathbf{r})\,. \tag{5.2}$$

Applying these hybrid functionals to semiconducting materials in particular, it has soon been noticed, that there exists a correlation between the amount α of Hartree-Fock exact-exchange and the particular band gap of a material. Based on more theoretical grounds, i.e. the screening behaviour of the Coulomb interaction, the fraction α of Hartree-Fock exact-exchange can be identified as approximately the inverse of the static dielectric constant ε_∞ of a material

$$\alpha \approx \frac{1}{\varepsilon_\infty}\,. \tag{5.3}$$

In particular (5.3) laid the foundation for the development and utilisation of so-called *dielectric-dependent* hybrid functionals. There are several possible ways to utilise (5.3). The simplest one relies on experimental input, either in the form of an experimentally determined static dielectric constant to be used with (5.3), or by adjusting α directly to reproduce experimentally known band gaps. Somewhat more sophisticated investigations perform a one-off calculation of the static dielectric constant based on a previously relaxed ground state geometry. The most involved way, however, is the one introduced by Shimazaki and Asai [3] and Skone et al. [4], where the amount α of Hartree-Fock exact-exchange is self-consistently determined via a second loop inside the structural geometry optimisation procedure. In this way, no experimental input is required, and the true ab initio character of the calculations is restored. The whole class of *dielectric-dependent* hybrid functionals have been applied to a large range of materials. For a broader overview we refer to [5–12].

5.2 Computational Methods

The calculations of the present work have been performed utilising the Vienna ab initio simulation package (VASP) [13–15] together with the projector-augmented wave (PAW) method [16]. We used the PAW potentials supplied with the VASP package that contributed 6, 10, 14, 12, and 14 valence electrons for the O ($2s^2\ 2p^4$), Ti ($3p^6\ 4s^2\ 3d^2$), Fe ($3p^6\ 4s^2\ 3d^6$), Zn ($3d^{10}\ 4s^2$), and Sn ($5s^2\ 4d^{10}\ 5p^2$) atoms, respectively.

While the focus of the present work is on the application of the self-consistent hybrid functional scPBE0, in a first step calculations based on the PBE0 functional [17] have been performed as well. Cited results from previous investigations also utilised generalised gradient approximation (GGA) functionals in the parametrisations of Perdew et al. [18] and the one revised for solids [19], as well as the hybrid functional HSE06 [20, 21].

Structural relaxations were performed for the various unit cells, with the following numerical parameters in terms of k-point grid, cut-off energy, and convergence criteria for energy and forces on the atoms: ZnO ($8 \times 8 \times 6$, 500 eV, 10^{-5} eV, 10^{-3} eV Å^{-1}), SnO_2 and TiO_2 ($6 \times 6 \times 8$, 500 eV, 10^{-5} eV, 10^{-3} eV Å^{-1}), $ZnFe_2O_4$ ($6 \times 6 \times 6$, 500 eV, 10^{-5} eV, 10^{-3} eV Å^{-1}), amorphous Sn-Ti oxides (Γ-only, 400 eV, 10^{-4} eV, 5×10^{-3} eV Å^{-1}), respectively. For all structural parameters this ensured well-converged results.

5.3 Application 1: Semiconducting Oxides

As a first application we look at the performance of the self-consistent hybrid functional for the structural and electronic properties of bulk semiconducting oxides, namely wurtzite ZnO, and rutile SnO_2 and TiO_2 [10, 32].

ZnO naturally crystallises in the hexagonal wurtzite structure ($P6_3mc$, space group 186). The calculated structural parameters at different levels of theory are summarised in the upper part of Table 5.1. For better visualisation Fig. 5.1 shows the calculated unit cell volumes at different levels of theory in comparison to the experimental unit cell volume. As expected, for plain PBE calculations the calculated unit cell volume is largely overestimated by $\approx 4\%$. This improves already for standard PBE0 calculations, and best agreement with respect to the experimental unit cell volume is observed for the scPBE0 calculations (Fig. 5.1).

The direct Kohn-Sham energy band gap of wurtzite ZnO at PBE level (0.715 eV) is largely underestimated with respect to the experimental band gap of 3.4449 eV [25], thereby showing the well-known "band-gap" problem of plain GGA approaches in DFT calculations. This improves slightly when switching to the PBE0 hybrid functional approach (3.132 eV) and the best agreement with respect to experiment is observed for the scPBE0 calculations (3.425 eV).

SnO_2 and TiO_2 crystallise naturally in the tetragonal rutile structure ($P4_2/mnm$, space group 136). Their respective calculated structural parameters at different levels of theory are given in the middle and lower parts of Table 5.1. As can be seen from the middle parts of Fig. 5.1, SnO_2 follows the same trend as wurtzite ZnO, namely a large overestimation of the plain PBE unit cell volume with respect to the experimental one and better agreement for the plain PBE0 calculations. Again, the best agreement with experiment is observed for the scPBE0 ground state volume. However, while the plain

Table 5.1 Ground state structural parameters for wurtzite ZnO, and rutile SnO_2 and TiO_2, obtained with different approximations for the exchange-correlation potential in comparison to low-temperature experimental data

ZnO	PBE	PBE0	scPBE0	Exp.
a (Å)	3.289	3.258	3.255	3.248 [22]
c (Å)	5.308	5.236	5.230	5.204 [22]
u	0.381	0.381	0.381	0.382 [23]
ε_∞	5.01	3.64	3.58	3.72 [24]
α	–	0.25	0.28	0.27 [24]
E_{gap} (eV)	0.715	3.132	3.425	3.4449 [25]
SnO_2	PBE	PBE0	scPBE0	Exp.
a (Å)	4.834	4.757	4.752	4.737 [26]
c (Å)	3.244	3.193	3.190	3.186 [26]
u	0.307	0.306	0.306	0.307 [26]
ε_∞	4.71	3.76	3.72	3.92 [27]
α	–	0.25	0.27	0.26 [27]
E_{gap} (eV)	0.609	3.590	3.827	3.596 [28]
TiO_2	PBE	PBE0	scPBE0	Exp.
a (Å)	4.649	4.585	4.607	4.587 [29]
c (Å)	2.968	2.948	2.955	2.954 [29]
u	0.305	0.305	0.305	0.305 [29]
ε_∞	7.90	5.86	6.39	7.33 [30]
α	–	0.25	0.16	0.14 [30]
E_{gap} (eV)	1.818	4.152	3.244	3.03 [31]

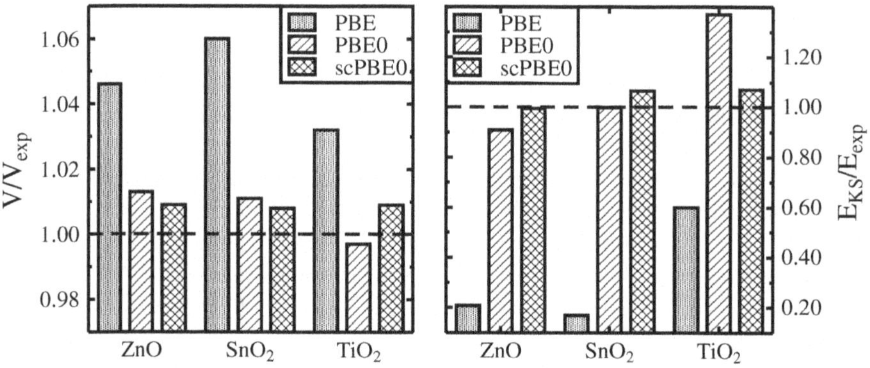

Fig. 5.1 Left: Ground state unit cell volumes V with respect to the experimental volumes V_{exp}. Right: Kohn-Sham energy gaps E_{KS} with respect to the experimental energy gaps E_{exp}. Shown are the results for different levels of theory, namely PBE (dotted), PBE0 (lined), and scPBE0 (squared). Experimental data corresponds to the dashed horizontal lines

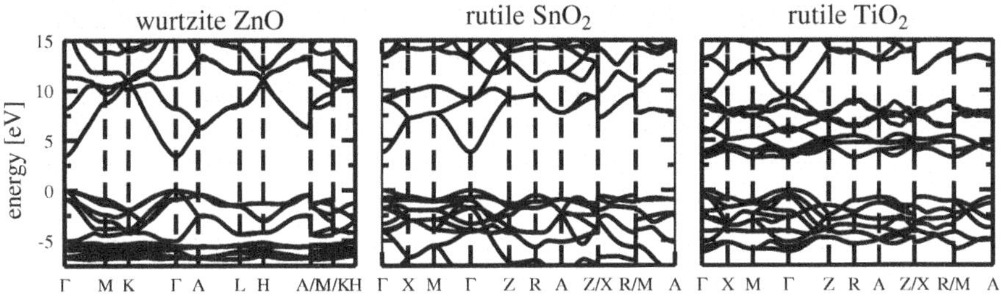

Fig. 5.2 Electronic band structures of wurtzite ZnO (left panel), rutile SnO_2 (middle panel), and rutile TiO_2 (right panel), calculated with the scPBE0 functional. Energies are in electron volt (eV) with the valence band maximum set to zero

PBE ground state volume yields worst agreement with experiment for rutile TiO_2, plain PBE0 gives a slighty smaller ground state volume compared to experiment. The scPBE0 calculations are overestimated with respect to the experimental ground state volume and are slightly worse compared to the plain PBE0 results (Fig. 5.1).

The direct Kohn-Sham energy band gaps of the plain PBE calculations are again strongly underestimated compared to the experimental results. For the hybrid functional calculations, we observe an opposite trend compared to the ground state structural properties. For rutile SnO_2 the plain PBE0 calculations yield the best agreement compared to the experimental band gap, while the scPBE0 is slightly worse. However, for rutile TiO_2, where plain PBE0 calculations yielded best agreement in terms of structural properties, the Kohn-Sham energy band gap is strongly overestimated. The slightly worse results in the scPBE0 structural properties has to be counterbalanced by the best agreement of the Kohn-Sham energy band gap compared to the experimental value.

Based on the relaxed ground state structural and electronic properties (Fig. 5.2) it is in principle possible to calculate the optical properties as well, i.e. the real and imaginary parts of the dielectric function, which subsequently would give access to experimentally measurable quantities such as the absorption coefficient or reflectivity [10, 32].

5.4 Application 2: Ferro(i)magnetic Oxides

As a second application we look at the performance of the self-consistent hybrid functional for bulk ferro(i)magnetic oxides in the spinel structure. In particular, spinel ferrites show a plethora of interactions due to their unique interplay of structural, electronic, and magnetic properties, e.g. $CoFe_2O_4$ and $NiFe_2O_4$ have shown huge potential as building blocks of artificial multiferroic [33, 34] or spintronics devices [35]. However, here we're concerned with $ZnFe_2O_4$ [36] whose magnetic ground state is still under dispute.

While a previous investigation of $ZnFe_2O_4$ was more concerned with the influence of the inversion degree between the two cationic sublattices on the structural, electronic, and optical properties [36], it also showed that for calculations using the hybrid PBE0 functional the ferromagnetic ground state in the normal spinel phase is favoured over the anti-ferromagnetic one, contrary to experimental findings. Applying the scPBE0 functional separately to the ferromagnetic and anti-ferromagnetic spin arrangements of normal $ZnFe_2O_4$ confirms this result, i.e. the ferromagnetic spin arrangement in the normal spinel phase is the ground state in both cases. The lattice constant calculated with the scPBE0 functional of $a = 8.427$ Å is slighty reduced compared to the PBE0 calculations ($a = 8.452$ Å). Figure 5.3 shows the resulting total and projected density of states (DOS) of $ZnFe_2O_4$ (left panel), whereas in the right panels the electronic band structure of the majority- and minority-spin states is shown. The scPBE0 DOS is nearly indistinguishable from the PBE0 calculations [36], and the electronic band gap is again between the majority-spin valence band and the minority-spin conduction band. However, due to the increased fraction α of Hartree-Fock exact-exchange, the band gap increases from 3.13 eV (PBE0) to 4.28 eV (scPBE0).

The ferromagnetic spin-arrangement in the normal spinel phase of $ZnFe_2O_4$ has recently been confirmed by Ulpe et al. [37] who applied the scPBE0 functional to a range of ternary transition metal oxides. These investigations show that the scPBE0 functional can not only be applied to investigate the structural and electronic properties of semiconductors, but is capable to go beyond and yield information on magnetic properties as well.

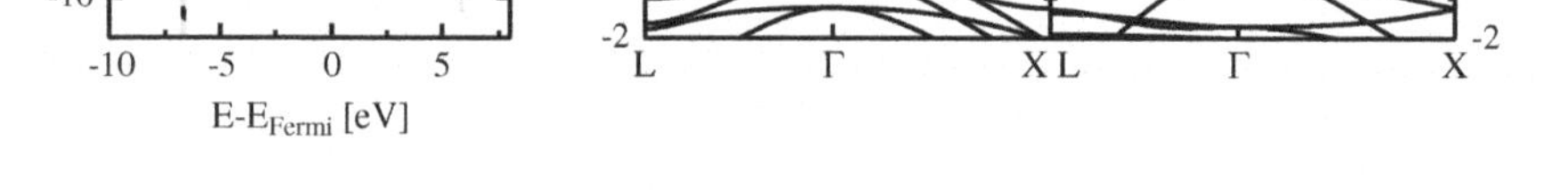

Fig. 5.3 Left panel: Total and projected density of states per formula unit for $ZnFe_2O_4$. The octahedral O_h (tetrahedral T_d) states of Fe (Zn) are shown as solid (dashed) lines, and the shaded grey area depicts the total density of states. Minority spin projections are shown using negative values. Right panels: Electronic band structure for $ZnFe_2O_4$, where the majority-spin (minority-spin) bands are shown in the left (right) panel. The zero energy is set to the valence band maximum in all panels

5.5 Application 3: Amorphous Oxides

As a third application we look at the performance of the self-consistent hybrid functional for the electronic and optical properties of transparent amorphous semiconducting oxides which got into focus for applications ranging from displays to solar cells [38]. Here, we're concerned with the particular example of amorphous $(SnO_2)_{1-x}(TiO_2)_x$ ($0 \leq x \leq 1$) oxides (Sn-Ti oxides) [32], which have been the focus of a previous investigation. Since a full ab initio molecular dynamics generation of amorphous structure models based on the scPBE0 hybrid functional is out of scope of even nowadays computational resources we rely on the previously generated amorphous structure models for the whole range of $(SnO_2)_{1-x}(TiO_2)_x$ ($0 \leq x \leq 1$) based on the PBEsol functional [19] and applying a permutation operation. Reference [32] shows in detail that this combined approach yields reliable amorphous structure models which can then be utilised to analyse different structure models for a given composition x of $(SnO_2)_{1-x}(TiO_2)_x$ and moreover, the subsequent influence of different structure models on the optical properties (dielectric functions, optical gaps) can be scrutinised. While the previous investigation showed that plain PBEsol optical gaps are severely underestimated with respect to the experimental values [32], even hybrid functional calculations based on the HSE06 functional [20, 21] showed still some discrepancies.

Here, we follow a slightly different approach. In general, amorphous materials are unique in the sense that on the one hand they lack the long-range order of crystalline materials, but on the other hand a short range order is still present. This can be seen from analysing the radial distribution functions, which measure the statistical distribution of pairwise atomic distances [32]. Since the local atomic arrangement between crystalline phases and the generated amorphous structure models for SnO_2 and TiO_2 are quite similar, here we transfer the previously determined fractions α of Hartree-Fock exact-exchange to the respective amorphous phases, and interpolate linearly for the intermediate concentrations x of $(SnO_2)_{1-x}(TiO_2)_x$.

Figure 5.4 summarises the results. While the upper panels show the calculated real and imaginary dielectric functions utilising the scPBE0 hybrid functional (solid lines), compared to the previously reported results from PBEsol calculations (dashed lines) [32], the lower panels show the absorption coefficients suitably scaled for a subsequent Tauc-plot analysis to extract the optical band gaps. Similarly to our earlier investigation, different amorphous structure models in the intermediate range yield optical band gaps with an error margin of 0.1 eV. Moreover, the extracted optical band gaps of amorphous SnO_2 (4.2 eV) and TiO_2 (3.4 eV) are in favourable agreement with the respective experimental values of around 4.0 eV [38] and 3.2–3.5 eV [39]. Moreover, we significantly improve over the previously reported values of 2.9 eV (SnO_2) and 3.8 eV (TiO_2) determined using the hybrid HSE06 functional [32].

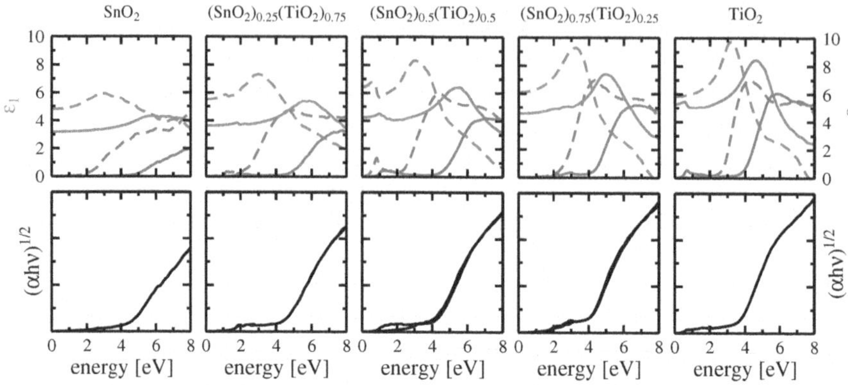

Fig. 5.4 Upper panels: Real (light) and imaginary (dark) parts of the dielectric functions for one particular amorphous structure model for $(SnO_2)_{1-x}(TiO_2)_x$ $(0 \leq x \leq 1)$ calculated with the PBEsol (dashed lines) and the scPBE0 hybrid functional (solid lines), respectively. Lower panels: Tauc plots ($\sqrt{\alpha h\nu}$ vs. energy) based on the scPBE0 calculated absorption coefficients for various amorphous structure models showing the optical gaps

5.6 Summary and Outlook

In summary, we presented an up-to-date compilation on what we've learned so far about the application and performance of a new self-consistent hybrid functional, recently introduced by Shimazaki and Asai [3] and Skone et al. [4]. Applications included technologically important oxide semiconductors, and lesser known examples such as magnetic and amorphous oxides. In all cases, the knowledge gain in terms of structural, electronic, magnetic, and optical properties, outweighs the additional computational effort put into this method. Future research directions will focus on the detailed calculation of the static dielectric constant of the materials, the deep underlying connection to many-body perturbation theory, and possibly the further improvement of exchange-correlation functionals to be applied in computational materials science.

Acknowledgements This work made use of computational resources provided by the North-German Supercomputing Alliance (HLRN, www.hlrn.de) and the ARCHER UK National Supercomputing Service (www.archer.ac.uk) via the membership of the UKs HPC Materials Chemistry Consortium, funded by EPSRC (EP/L000202).

References

1. P. Hohenberg, W. Kohn, Phys. Rev. **136**, B864 (1964)
2. W. Kohn, L.J. Sham, Phys. Rev. **140**, A1133 (1965)
3. T. Shimazaki, T. Nakajima, J. Chem. Phys. **130**, 164702 (2009)
4. J.H. Skone, M. Govoni, G. Galli, Phys. Rev. B **89**, 195112 (2014)

5. M.A.L. Marques, J. Vidal, M.J.T. Oliveira, L. Reining, S. Botti, Phys. Rev. B **83**, 035119 (2011)
6. J.C. Conesa, J. Phys. Chem. C **116**, 18884 (2012)
7. D. Koller, P. Blaha, F. Tran, J. Phys.: Condens. Matter **25**, 435503 (2013)
8. E. Menéndez-Proupin, P. Palacios, P. Wahnón, J.C. Conesa, Phys. Rev. B **90**, 045207 (2014)
9. M. Gerosa, C.E. Bottani, L. Caramella, G. Onida, C. Di Valentin, G. Pacchioni, Phys. Rev. B **91**, 155201 (2015)
10. D. Fritsch, B.J. Morgan, A. Walsh, Nanoscale Res. Lett. **12**, 19 (2017)
11. A. Erba, J. Phys.: Condens. Matter **29**, 314001 (2017)
12. J. He, C. Franchini, J. Phys.: Condens. Matter **29**, 454004 (2017)
13. G. Kresse, J. Hafner, Phys. Rev. B **47**, 558 (1993)
14. G. Kresse, J. Hafner, Phys. Rev. B **49**, 14251 (1994)
15. G. Kresse, J. Furthmüller, Comp. Mat. Sci. **6**, 15 (1996)
16. P.E. Blöchl, Phys. Rev. B **50**, 17953 (1994)
17. C. Adamo, V. Barone, J. Chem. Phys. **110**, 6158 (1999)
18. J.P. Perdew, K. Burke, M. Ernzerhof, Phys. Rev. Lett. **77**, 3865 (1996)
19. J.P. Perdew, A. Ruzsinszky, G.I. Csonka, O.A. Vydrov, G.E. Scuseria, L.A. Constantin, X. Zhou, K. Burke, Phys. Rev. Lett. **100**, 136406 (2008)
20. J. Heyd, G.E. Scuseria, M. Ernzerhof, J. Chem. Phys. **118**, 8207 (2003)
21. J. Heyd, G.E. Scuseria, M. Ernzerhof, J. Chem. Phys. **124**, 219906 (2006)
22. R.R. Reeber, J. Appl. Phys. **41**, 5063 (1970)
23. H. Schulz, K.H. Thiemann, Solid State Commun. **32**, 783 (1979)
24. E.C. Heltemes, H.L. Swinney, J. Appl. Phys. **38**, 2387 (1967)
25. W.Y. Liang, A.D. Yoffe, Phys. Rev. Lett. **20**, 59 (1968)
26. J. Haines, J.M. Léger, Phys. Rev. B **55**, 11144 (1996)
27. R. Summitt, J. Appl. Phys. **39**, 3762 (1968)
28. K. Reimann, M. Steube, Solid State Commun. **105**, 649 (1998)
29. J.K. Burdett, T. Hughbanks, G.J. Miller, J.W. Richardson Jr., J.V. Smith, J. Am. Chem. Soc. **109**, 3639 (1987)
30. M. Dou, C. Persson, J. Appl. Phys. **113**, 083703 (2013)
31. A. Amtout, R. Leonelli, Phys. Rev. B **51**, 6842 (1995)
32. D. Fritsch, Phys. Status Solidi A **215**, 1800071 (2018)
33. D. Fritsch, C. Ederer, Phys. Rev. B **82**, 104117 (2010)
34. D. Fritsch, C. Ederer, Phys. Rev. B **86**, 014406 (2012)
35. N. Caffrey, D. Fritsch, T. Archer, S. Sanvito, C. Ederer, Phys. Rev. B **87**, 024419 (2013)
36. D. Fritsch, J. Phys.: Condens. Matter **30**, 095502 (2018)
37. A.C. Ulpe, K.C.L. Bauerfeind, T. Bredow, ACS Omega **4**, 4138 (2019)
38. M.J. Wahila, Z.W. Lebens-Higgins, K.T. Butler, D. Fritsch, R.E. Treharne, R.G. Palgrave, J.C. Woicik, B.J. Morgan, A. Walsh, L.F.J. Piper, APL Mater. **7**, 022509 (2019)
39. M. Landmann, T. Köhler, S. Köppen, E. Rauls, T. Frauenheim, W.G. Schmidt, Phys. Rev. B **86**, 064201 (2012)

Chapter 6
Simulation of the Phonon Drag of Point Defects in a Harmonic Crystal

I. L. Bataronov, V. A. Yuryev, E. V. Levchenko, M. V. Yuryeva and N. A. Yuyukin

Abstract In the framework of quantum physical kinetics methods, an expression is formulated for the force acting on a point defect from the side of the nonequilibrium phonon subsystem under the conditions of phonon heat transfer. Based on the solution of the problem of phonon scattering by a mass defect of a harmonic crystal, the expression for the transition t-matrix is written in the framework of the Lifshitz method in the continuum approximation. Using the representation of the t-matrix in terms of the Green's function of an ideal crystal, expressions for the partial coefficients of phonon drag in a cubic crystal are obtained. Based on the pole representation of the Green's function, an expression is obtained for the partial coefficients in terms of the phonon dispersion law in the crystal. To execute a computational experiment, the Born–Karman phonon spectrum model was used, modified taking into account the conditions for the phonon spectrum at the boundary of the Brillouin cubic zone and the presence of sections of the "negative" dispersion. The size of these sections is characterized by the parameter ε. A method is proposed for calculating surface integrals over isofrequency surfaces based on the Samarsky method. For various values of the parameter, the integrals are calculated and the partial drag coefficients are calculated. An analysis of the drag coefficient of point defects by the flow of thermal phonons showed that, with the parameter values $\varepsilon > 0.27$, the drag force is directed against the heat flux. The asymmetry of the drag coefficient with respect to the sign of the defect power due to the effect of resonant phonon scattering is established.

I. L. Bataronov · V. A. Yuryev (✉) · M. V. Yuryeva · N. A. Yuyukin
Voronezh State Technical University, 14 Moskovsky Prospect, Voronezh, Russia
e-mail: yururvva@gmail.com

E. V. Levchenko
Priority Research Centre for Computer-Assisted Research Mathematics and Its Applications (CARMA), The University of Newcastle, Callaghan, NSW 2308, Australia

E. V. Levchenko et al. (eds.), *Theory and Simulation in Physics for Materials Applications*, Springer Series in Materials Science 296,
https://doi.org/10.1007/978-3-030-37790-8_6

6.1 Introduction

When considering the influence of the temperature field on the distribution and motion of defects in the crystal structure (coherent twin boundaries, dislocations, point defects), usually only the effect of thermoelastic stresses and thermodynamic driving forces [1–4] is taken into account. Along with this, one should take into account the nonequilibrium character of the phonon distribution in the presence of a temperature gradient. Under the conditions of the phonon heat flux, the effect of "drag" of defects takes place, first indicated for impurity atoms in liquid metals by Fiks [5] and considered in simplified model approaches [6, 7]. This effect is similar to the well-known "phonon wind" effect in the dynamic dragging of moving dislocations [8]. The action of the thermal phonon flux can lead to a qualitative change in the pattern of redistribution of point defects in a nonuniformly heated crystal [9]. Therefore, it is necessary to develop more accurate methods for calculating the force of phonon drag. This force consists of several effects. One of them is the scattering of phonons on the deformation potential of a defect [8]. This effect is associated with the anharmonic properties of a real crystal. In order to avoid taking it into account, in this paper we consider a harmonic crystal.

For processes related to scattering by point defects, it may turn out to be a significant deviation of the phonon dispersion law from linear for large values of the wave number. Quantitative accounting of the phonon dispersion law at present can only be performed on the basis of a complex numerical calculation of [10], therefore, it is of interest to develop an approach that uses model forms of the dispersion law. One of such models is the Born–Karman model [11]. However, in such a model, the density of phonon states becomes infinite on a Debye sphere which is modeling the boundary of the Brillouin zone. Therefore, in the present work, the model was modified by passing to the cubic Brillouin zone, which allowed us to exclude the indicated nonphysical feature.

6.2 The Heat Drag Force of a Point Defect

We will proceed from the general expression for the force acting on a point defect from the side of the nonequilibrium phonon subsystem of the crystal in the harmonic approximation [12]

$$\vec{F} = \frac{\hbar}{2(2\pi)^3}\sum_{\alpha'}\sum_{\alpha}\iint \mathrm{d}^3k\mathrm{d}^3k'\left(\vec{k}'-\vec{k}\right)\left[\delta N_{\alpha'}\left(\vec{k}'\right)-\delta N_{\alpha}\left(\vec{k}\right)\right]W_{\alpha'\alpha}\left(\vec{k}',\vec{k}\right). \quad (6.1)$$

Here $W_{\alpha'\alpha}(\vec{k}',\vec{k})\mathrm{d}^3k'$, is the probability of a phonon to transfer per unit time from a state characterized by a wave vector $\vec{k}$ and polarization α to an element d^3k' in the neighbourhood of the state $(\alpha',\vec{k}')$ as a result of scattering by a defect to which a momentum is transmitted in a single scattering event $\hbar(\vec{k}-\vec{k}')$, $\delta N_\alpha(\vec{k},\vec{r})$ is a nonequilibrium addition to the locally equilibrium distribution function of phonons $N_0(\omega_\alpha(\vec{k}))$, depending on their frequency $\omega_\alpha(\vec{k})$.

We use this result to calculate the force $\vec{F}$ in the presence of a temperature gradient. An analysis of the thermal conductivity of solids shows [13–15] that results consistent with experiment can be obtained in the approximation of a frequency-dependent relaxation time $\tau(\omega_\alpha)$, within which the solution of the linearized Boltzmann equation for a nonequilibrium addition to the phonon distribution function has the form

$$\delta N = \tau(\omega_\alpha)\frac{\omega_\alpha}{T}\frac{\partial N_{0\alpha}}{\partial\omega_\alpha}\vec{u}^{(\alpha)}\nabla T. \tag{6.2}$$

Here $\vec{u}^{(\alpha)}$ is the group velocity of the phonon with polarization α.

Using formula (6.2), we rewrite expression (6.1) in the form

where

$$F_i = \frac{\nabla_j T}{T}\int d\omega\frac{\partial N_0}{\partial\omega}\tau(\omega)\beta'_{ij}(\omega)c_s^2, \tag{6.3}$$

$$\begin{aligned}\beta'_{ij}(\omega) = \frac{\hbar}{2\hbar(2\pi)^5 c_s^2}\sum_{\alpha',\alpha}\iint d^3k d^3k'\left(\vec{k}'_i-\vec{k}_i\right)\left(u_j^{(\alpha')}-u_j^{(\alpha')}\right)\\ \cdot\,\delta\left(\omega-\omega_\alpha\left(\vec{k}\right)\right)\delta\left(\omega-\omega_{\alpha'}\left(\vec{k}'\right)\right)\left|t_{\alpha'\alpha}\left(\vec{k}',\vec{k}\right)\right|^2,\end{aligned} \tag{6.4}$$

c_s is the average phonon velocity in the crystal, and the transition probability $W_{\alpha'\alpha}(\vec{k},\vec{k}')$ is expressed through the transition matrix [16].

In expression (6.4), we can go from integration over d^3k to integration over frequency ω and over the isofrequency surface S. Then the integrals over ω with δ-functions are calculated directly. As a result, we get

$$\beta'_{ij} = \frac{\omega}{2\hbar(2\pi)^5 c_s^2}\sum_{\alpha'}\sum_{\alpha}\iint\frac{dSdS'}{u^{(\alpha)}u^{(\alpha')}}\left(k'_i-k_i\right)\left(u_j^{(\alpha')}-u_j^{(\alpha)}\right)\left|t_{\alpha'\alpha}\left(\vec{k}',\vec{k}\right)\right|^2. \tag{6.5}$$

In the last formula, integration is carried out over isofrequency surfaces $\omega_\alpha(\vec{k}) = \omega$ and $\omega_{\alpha'}(\vec{k}') = \omega$ in $\vec{k}$ space.

According to the theory of thermal conductivity [14], the thermal diffusivity can be represented as

$$\chi = \frac{1}{3}c_s^2\bar{\tau}_T,$$

where $\bar{\tau}_T$ there is a temperature-dependent average relaxation time. Then formula (6.3) can be written in a convenient form

$$\vec{F} = -3\chi\hat{B}'\frac{\nabla T}{T}, \tag{6.7}$$

Here, $\hat{B}'$ the phonon drag coefficient tensor is determined by the formula

$$B'_{ij} = -\int_0^{\omega_m} d\omega \frac{\partial N_0}{\partial \omega}\frac{\tau(\omega)}{\bar{\tau}_T}\beta'_{ij}(\omega), \tag{6.8}$$

where ω_m is the maximum phonon frequency.

6.3 The Problem of Phonon Scattering on a Defect with a Different Mass

To determine the tensor β', it is necessary to find the transition matrix. The problem of phonon scattering on a point defect with a local perturbation of the mass and force constants was solved in general by Lifshitz [17] and was considered in [18–20] for various models of defects. However, the practical use of the results obtained to calculate the drag force actually requires the use of complex numerical calculations [20] and uses model ideas about atomic interaction in a crystal. Therefore, to obtain an analytical result, we use the continuum approximation to solve the scattering problem using the Lifshitz method.

To avoid insignificant complications of the problem, we consider a simple cubic lattice with one atom per cell. For simplicity, we assume that a defect differs from crystal atoms only in mass. The difference in elastic constants can also be taken into account [12], but the magnitude of this difference is difficult to determine for a real crystal. We place at the point $\vec{r} = 0$ a point defect whose mass M differs from the atoms m of the matrix by Δm ($\Delta m = M - m$ for the substitution defect and $\Delta m = M$ for the interstitial defect).

Based on the general solution of the scattering problem for the transition matrix [12], for the model under consideration we obtain the expression

$$\left|t_{\alpha'\alpha}\left(\vec{k}',\vec{k}\right)\right|^2 = \tilde{G}_{ij}(\omega)\tilde{G}^*_{kl}(\omega)\left[\varphi_i^{(\alpha')}\left(\vec{k}'\right)\right]^*\varphi_k^{(\alpha')}\left(\vec{k}'\right)\varphi_j^{(\alpha)}\left(\vec{k}\right)\left[\varphi_l^{(\alpha)}\left(\vec{k}\right)\right]^*. \tag{6.9}$$

where the notation is used

$$\varphi_i^{(\alpha)}\left(\vec{k}\right) = \sqrt{\frac{\hbar}{2\rho\omega_\alpha(\vec{k})}}e_i^{(\alpha)}\left(\vec{k}\right);\; M(\omega) = \Delta m\omega^2. \tag{6.10}$$

The tensor in $G^{-1}(\omega)$ (6.5) is expressed in terms of the inverse tensor

$$\tilde{G}_{ij}^{-1}(\omega) = g_{ij}(\omega) - \frac{1}{M(\omega)}\delta_{ij}, \tag{6.11}$$

where is the function used

$$g_{ij}(\omega) = \frac{2}{\hbar}\sum_{\alpha}\int \frac{\varphi_i^{(\alpha)}\left(\vec{k}\right)\left[\varphi_j^{(\alpha)}\left(\vec{k}\right)\right]^{*}\omega_{\alpha}}{(\omega + io)^2 - \omega_{\alpha}^2}\frac{\mathrm{d}^3k}{(2\pi)^2}. \tag{6.12}$$

We express it through the Green function $G_{ij}(\vec{k},\omega)$ of an ideal crystal. We use the representation [21] for this function

$$G_{ij}(\vec{k},\omega) = \sum_{\alpha}\frac{e_i^{(\alpha)}(\vec{k})\left[e_j^{(\alpha)}(\vec{k})\right]^{*}}{\rho\left[\omega_{\alpha}^2 - (\omega + io)^2\right]}. \tag{6.13}$$

Then, according to (6.12), taking into account notation (6.10), we have

$$g_{ij}(\omega) = \int \frac{d^3k}{(2\pi)^3}\sum_{\alpha}\frac{1}{\rho}\frac{e_i^{(\alpha)}\left[e_j^{(\alpha)}\right]^{*}}{\omega_{\alpha}^2 - (\omega + io)^2} = \int \frac{d^3k}{(2\pi)^3}G_{ij}. \tag{6.14}$$

As a result, formulas (6.9), (6.11–6.14) fully express the transition matrix through the Fourier transform of the Green's function (6.13) of an ideal crystal and thereby solve the problem.

6.4 Partial Phonon Drag Coefficients

We use the obtained results to record the drag entrainment coefficient. Preliminarily transform expression (6.5), opening the brackets in it

$$(k_i' - k_i)(u_j' - u_j) = k_i'u_j' + k_iu_j - k_iu_j' - k_i'u_j$$

and presenting it as the sum of four integrals. Note that the integrands in the last two integrals are odd in the variables k_i, k_i', and the corresponding integrals taken in symmetric limits are equal to zero. The first integral reduces to the second. As a result, we obtain the formula

$$\beta'_{ij} = \frac{\omega}{\hbar(2\pi)^5 c_s^2} \sum_{\alpha,\alpha'} \int \mathrm{d}^3k \mathrm{d}^3k' \delta(\omega - \omega_{\alpha\vec{k}}) \delta(\omega - \omega_{\alpha'\vec{k}'}) k_i u_j^{(\alpha)} \left| t_{\alpha'\alpha}\left(\vec{k}', \vec{k}\right) \right|^2. \quad (6.15)$$

Next, we use the Sokhotsky theorem

$$\mathrm{Im} \frac{2\omega_\alpha}{\omega_\alpha^2 - (\omega + i0)^2} = \pi\delta(\omega - \omega_\alpha); \quad \omega > 0, \quad (6.16)$$

Given the formula (6.16) in the expression (6.12), we obtain the equality

$$\sum_{\alpha'} \int \frac{\mathrm{d}^3k'}{(2\pi)^3} \left[\varphi_i^{(\alpha')}\left(\vec{k}'\right)\right]^* \varphi_j^{(\alpha')}\left(\vec{k}'\right) \delta\left(\omega - \omega_{\alpha'\vec{k}'}\right) = \frac{\hbar}{\pi} \mathrm{Im} g_{ij}(\omega). \quad (6.17)$$

As a result, taking into account (6.9) and (6.17), expression (6.15) takes the form

$$\begin{aligned} \beta'_{ij} = & \frac{\omega}{4\pi^3 c_s^2} G_{km}(\omega) G^*_{\mathrm{ln}}(\omega) \mathrm{Im} g_{kl}(\omega) \\ & \cdot \sum_\alpha \int \mathrm{d}^3k\, k_i u_j^{(\alpha)} \delta(\omega - \omega_\alpha) \varphi_m^{(\alpha)}\left(\vec{k}\right) \left[\varphi_n^{(\alpha)}\left(\vec{k}\right)\right]^*. \end{aligned} \quad (6.18)$$

We will take into account further that in a cubic crystal the tensor g_{ij} is spherical: $g_{ij}(\omega) = g(\omega)\delta_{ij}$. By virtue of (6.11), this will also be the tensor $\tilde{G}_{ij}$, so we immediately find:

$$\tilde{G}_{ij} = \frac{M(\omega)}{g(\omega)M(\omega) - 1} \delta_{ij}. \quad (6.19)$$

Moreover, according to (6.14), we have

$$g(\omega) = \frac{1}{3} g_{ii}(\omega) = \frac{1}{3} \int \frac{\mathrm{d}^3k}{(2\pi)^3} G_{ii}. \quad (6.20)$$

As a result, using formulas (6.19) and (6.20), expression (6.18) takes the form

$$\begin{aligned} \beta'_{ij} = & \frac{\omega}{4\pi^3 c_s^2} \frac{M^2(\omega)}{|1 - M(\omega)g(\omega)|^2} \mathrm{Im} g(\omega) \\ & \sum_\alpha \int \mathrm{d}^3k\, k_i u_j^{(\alpha)} \delta(\omega - \omega_\alpha) \varphi_m^{(\alpha)}\left(\vec{k}\right) \left[\varphi_m^{(\alpha)}\left(\vec{k}\right)\right]^*. \end{aligned} \quad (6.21)$$

Using formulas (6.10) and taking into account the normalization of polarization vectors, we finally obtain from formula (6.21):

$$\beta'_{ij} = \frac{\hbar}{8\pi^3 \rho c_s^2} \frac{M^2(\omega)}{|1 - M(\omega) g(\omega)|^2} \mathrm{Im} g(\omega) \cdot \sum_\alpha \int \mathrm{d}^3 k k_i u_j^{(\alpha)} \delta(\omega - \omega_\alpha). \tag{6.22}$$

In a cubic crystal, the integral standing here is also a spherical tensor, therefore, we can rewrite expression (6.22) in the form:

$$\begin{aligned} \beta'_{ij} &= \beta' \delta_{ij}, \\ \beta' &= \frac{\hbar}{24\pi^3 \rho c_s^2} \frac{M^2(\omega)}{|1 - M(\omega) g(\omega)|^2} \mathrm{Im} g(\omega) \cdot \sum_\alpha \int \mathrm{d}^3 k \left(k_i u_i^{(\alpha)}\right) \delta(\omega - \omega_\alpha). \end{aligned} \tag{6.23}$$

The resulting formula, taking into account (6.20), expresses the partial coefficient of phonon drag through the Green function of an ideal harmonic crystal.

6.5 Pole Representation for Phonon Drag Coefficient

We now find the pole representation of the resulting formula. According to (6.13), taking into account the normalization of polarization vectors, we have

$$G_{ii}\left(\vec{k}, \omega\right) = \sum_\alpha \frac{e_i^{(\alpha)}\left(\vec{k}\right) \left[e_i^{(\alpha)}\left(\vec{k}\right)\right]^*}{\rho \left[\omega_\alpha^2 - (\omega + io)^2\right]} = \frac{1}{\rho} \sum_\alpha \frac{1}{\omega_\alpha^2 - (\omega + io)^2}. \tag{6.24}$$

Then the integral (6.20) is written in the form:

$$g(\omega) = \frac{1}{3\rho} \sum_\alpha \int \frac{\mathrm{d}^3 k}{(2\pi)^3} \frac{1}{\omega_\alpha^2 - (\omega + io)^2}. \tag{6.25}$$

This integral is singular, so we transform it. We consider a separate term from (6.25) and proceed in it to integration over isofrequency surfaces and over frequency. Given that the Jacobian of the transition is $1/u_\alpha$, where u_α is a modulus of the group velocity of the phonon, we obtain

$$\int \frac{\mathrm{d}^3 k}{(2\pi)^3} \frac{1}{\omega_\alpha^2 - (\omega + io)^2} = \frac{1}{(2\pi)^3} \int_0^{\omega_{m\alpha}} \frac{\mathrm{d}\omega'}{\omega'^2 - (\omega + io)^2} \int_{\omega_\alpha = \omega'} \frac{\mathrm{d}S}{u_\alpha}. \tag{6.26}$$

The internal integral standing here is a continuous function of frequency. Denote this integral by

$$\sigma_\alpha(\omega) = \int\limits_{\omega_\alpha=\omega} \frac{dS}{u_\alpha}. \tag{6.27}$$

Then the integral in (6.26) can be transformed by the method of selection of singularities (we omit the polarization index for simplicity):

$$\int\limits_0^{\omega_m} \frac{\sigma(\omega')\mathrm{d}\omega'}{\omega'^2 - (\omega + io)^2} = \int\limits_0^{\omega_m} \frac{\sigma(\omega)\mathrm{d}\omega'}{\omega'^2 - (\omega + io)^2} + \int\limits_0^{\omega_m} \frac{(\sigma(\omega') - \sigma(\omega))\mathrm{d}\omega'}{\omega'^2 - \omega^2}$$

The first of the integrals on the right-hand side is calculated directly, and in the second there is no singularity:

$$\int\limits_0^{\omega_m} \frac{\sigma(\omega')\mathrm{d}\omega'}{\omega'^2 - \omega^2} = \frac{\sigma(\omega)}{2\omega} \ln \frac{\omega - \omega_m}{\omega + \omega_m} + \int\limits_0^{\omega_m} \frac{(\sigma(\omega') - \sigma(\omega))\mathrm{d}\omega'}{\omega'^2 - \omega^2}.$$

It should be noted that the logarithm standing here is understood as a function of a complex variable, therefore, when $\omega < \omega_{m\alpha}$ it has an imaginary part equal to π. The rest of the written expression is real. We denote

$$\sigma'_\alpha(\omega) = 2\omega \int\limits_0^{\omega_{m\alpha}} \frac{\sigma_\alpha(\omega') - \sigma_\alpha(\omega)}{\omega'^2 - \omega^2} \mathrm{d}\omega'. \tag{6.28}$$

Then we finally have

$$g(\omega) = \frac{1}{6\rho(2\pi)^3\omega} \sum_\alpha \left(\sigma_\alpha(\omega) \ln \frac{\omega - \omega_{m\alpha}}{\omega + \omega_{m\alpha}} + \sigma'_\alpha(\omega) \right). \tag{6.29}$$

In particular, it follows that

$$\mathrm{Im} g(\omega) = \frac{\pi}{6\rho(2\pi)^3\omega} \sum_\alpha \sigma_\alpha(\omega)\theta(\omega_{m\alpha} - \omega). \tag{6.30}$$

Here $\theta(x)$ is the Heaviside function.

Finally, we consider the integral in expression (6.23). We transform it similarly to (6.26)

$$\int \mathrm{d}^3k\left(k_i u_i^{(\alpha)}\right)\delta(\omega-\omega_\alpha)=\int\limits_0^{\omega_{m\alpha}}\delta(\omega-\omega')\mathrm{d}\omega'\int\limits_{\omega_\alpha=\omega'}\left(k_i u_i^{(\alpha)}\right)\frac{\mathrm{d}S}{u_\alpha}$$
$$=\theta(\omega_{m\alpha}-\omega)\int\limits_{\omega_\alpha=\omega}\left(k_i u_i^{(\alpha)}\right)\frac{\mathrm{d}S}{u_\alpha}$$

The integral standing here has the dimension of volume in phase space. Denote it as

$$\Omega_\alpha(\omega)=\int\limits_{\omega_\alpha=\omega}\left(k_i u_i^{(\alpha)}\right)\frac{\mathrm{d}S}{u_\alpha}. \tag{6.31}$$

As a result, expression (6.23) finally takes the form

$$\beta'=\frac{\hbar V_a^2}{36(2\pi)^5 c_s^2}\frac{(\Delta m/m)^2\omega^3}{\left|1-\frac{\Delta m}{m}\omega\frac{V_a}{6(2\pi)^3}\sum\limits_\alpha\left(\sigma_\alpha(\omega)\ln\frac{\omega-\omega_{m\alpha}}{\omega+\omega_{m\alpha}}+\sigma'_\alpha(\omega)\right)\right|^2}\sum_\alpha\sigma_\alpha(\omega)\theta_\alpha$$
$$\cdot\sum_\alpha\Omega_\alpha(\omega)\theta_\alpha, \tag{6.32}$$

where V_a is the atomic volume, and is indicated $\theta_\alpha=\theta(\omega_{m\alpha}-\omega)$. Thus, the calculation of the phonon drag coefficient was reduced to the calculation of surface integrals (6.27), (6.31), and the usual eigen integral (6.28). These integrals are determined only by the phonon spectrum of the crystal.

6.6 Computational Experiment

To calculate the integrals, we use the modified Born–Karman model with a model spectrum of the form

$$\omega_\alpha^2\left(\vec{k}\right)=c_\alpha^2k_0^2\left[\psi^2(k_x/k_0)+\psi^2\left(k_y/k_0\right)+\psi^2(k_z/k_0)\right], \tag{6.33}$$

where the function $\psi(x)$ satisfies the condition $\psi'(1)=0$, $2k_0$ is the length of the edge of the cubic Brillouin zone. Then, on the entire boundary of the Brillouin zone, the necessary condition will automatically be fulfilled: $\partial\omega_\alpha(\vec{k})/\partial n=0$.

We set the function $\psi(x)$ in the form consistent with the calculation of the phonon spectrum in the atomic model with interaction with the nearest neighbors:

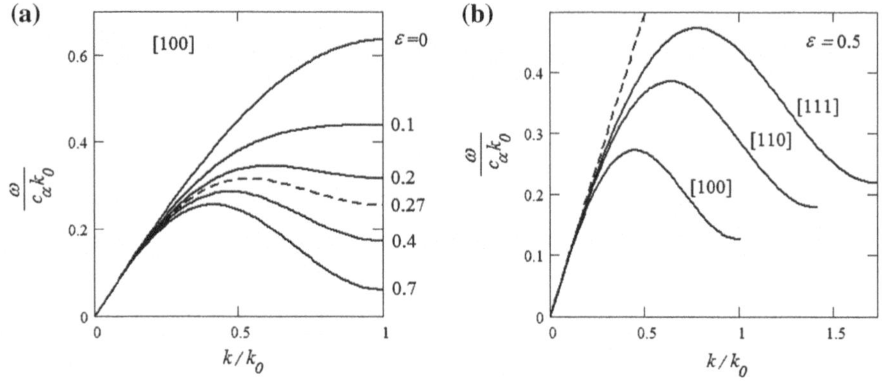

Fig. 6.1 The form of the model law of phonon dispersion for various values of the parameter ε (**a**) and different directions in the crystal (**b**). The dashed line on **b** shows the dispersion law in the Debye model

$$\psi(x) = \frac{2}{\pi(1+3\varepsilon)}\left(\sin\left(\frac{\pi x}{2}\right) + \varepsilon\sin\left(\frac{3\pi x}{2}\right)\right). \tag{6.34}$$

Here the parameter $\varepsilon(\varepsilon \in [0, 1])$ determines the size of the "negative" dispersion plot (Fig. 6.1). With an increase in ε, the length of this section increases. The model value of ε can be determined by the limiting phonon frequency at the boundary of the Brillouin zone in the [100] direction:

$$\varepsilon = \frac{2c_\alpha k_0 - \pi\omega_\alpha(k_0, 0, 0)}{2c_\alpha k_0 + 3\pi\omega_\alpha(k_0, 0, 0)}.$$

We calculate the integrals by the Samarsky method. To do this, we write the identity

$$\int\limits_{\omega_\alpha=\omega} f\left(\vec{k}\right)\frac{dS}{u_\alpha} = \int d^3k f\left(\vec{k}\right)\delta(\omega-\omega_\alpha),\ \omega<\omega_{m\alpha}$$

Here we replace the δ-function by its approximation δ_h with the blur parameter h. Then we get approximately

$$\int\limits_{\omega_\alpha=\omega} f\left(\vec{k}\right)\frac{dS}{u_\alpha} \approx \int d^3k f\left(\vec{k}\right)\delta_h(\omega-\omega_\alpha). \tag{6.35}$$

As a result, the surface integral is reduced to the volume integral. The advantage of this method is that there is no need to find the explicit coordinates of the

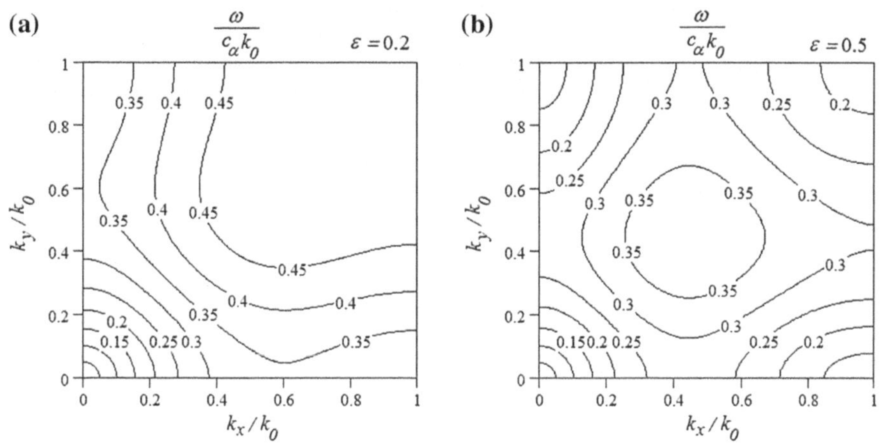

Fig. 6.2 Traces of isofrequency surfaces in the (001) plane for two values of the parameter ε

isofrequency surface. For the phonon spectrum, this presents a significant difficulty in view of the multiplicity and non-closeness of the isofrequency surface (Fig. 6.2).

We introduce dimensionless normalized variables:

$$\xi = \frac{k}{k_0},\ \zeta_\alpha = \frac{\omega}{c_\alpha k_0},\ \eta_\alpha\left(\vec{\xi}\right) = \frac{\omega_\alpha\left(\vec{k}\right)}{c_\alpha k_0}. \tag{6.36}$$

Then the integrals (6.27), (6.28) and (6.31) using the formula (6.35) and taking into account the expression (6.33) are transformed to a unit cube:

$$\sigma_\alpha(\omega) = \frac{8k_0^2}{c_\alpha} J_1(\zeta_\alpha),\ J_1(\zeta) = \int_0^1\int_0^1\int_0^1 \delta_h(\zeta - \eta_\alpha)\mathrm{d}\xi_1\mathrm{d}\xi_2\mathrm{d}\xi_3. \tag{6.37}$$

$$\Omega_\alpha(\omega) = \frac{8k_0^3}{\zeta_\alpha} J_2(\zeta_\alpha),\ J_2(\zeta) = \int_0^1\int_0^1\int_0^1 \delta_h(\zeta - \eta_\alpha)\sum_{i=1}^{3} \xi_i\psi_\alpha(\xi_i)\psi'_\alpha(\xi_i)\mathrm{d}\xi_1\mathrm{d}\xi_2\mathrm{d}\xi_3. \tag{6.38}$$

$$\sigma'_\alpha(\omega) = \frac{2k_0^2\zeta_\alpha}{c_\alpha} J_3(\zeta_\alpha),\ J_3(\zeta) = \int_0^{\eta_{m\alpha}} \frac{J_1(\zeta') - J_1(\zeta)}{\zeta'^2 - \zeta^2}\mathrm{d}\zeta'. \tag{6.39}$$

Here, in the integrals (6.37), (6.38), the cubic symmetry of the integrand is taken into account. Further consideration of this symmetry allows one to reduce the integration volume by a factor of six:

$$\int_0^1\int_0^1\int_0^1 f\left(\vec{\xi}\right)d\xi_1 d\xi_2 d\xi_3 = 6\int_0^1 d\xi_1 \int_0^{\xi_1} d\xi_2 \int_0^{\xi_2} f\left(\vec{\xi}\right)d\xi_3$$

By virtue of the adopted model of the dispersion law (6.33), (6.34), the expressions for J_i depend on polarization only through the parameter ε, which can be different for different polarizations. For simplicity, we accept it the same for all polarizations. Then the functions $\eta_\alpha(\vec{\xi})$ and $J_i(\zeta)$ will be universal, and the dependence on polarization is completely transferred to the argument ζ_α in formulas (6.37)–(6.39).

To calculate the integral (6.39), the maximum frequency of the phonon spectrum is required. According to formulas (6.33), (6.34), this frequency is achieved in the [111] direction (Fig. 6.1). As a result, we find

$$\eta_m = \begin{cases} \frac{2\sqrt{3}}{\pi}\frac{1-\varepsilon}{1+3\varepsilon}, & \varepsilon \le \frac{1}{9} \\ \frac{6\varepsilon+2}{3\pi\sqrt{\varepsilon(1+3\varepsilon)}}, & \varepsilon > \frac{1}{9} \end{cases}.$$

As a function, we choose a Gaussian function that is analytical and well localized:

$$\delta_h(x) = \frac{1}{\sqrt{2\pi}h}\exp\left(-\frac{x^2}{2h^2}\right).$$

The results of calculating the integrals (6.37), (6.38) are presented in Figs. 6.3 and 6.4.

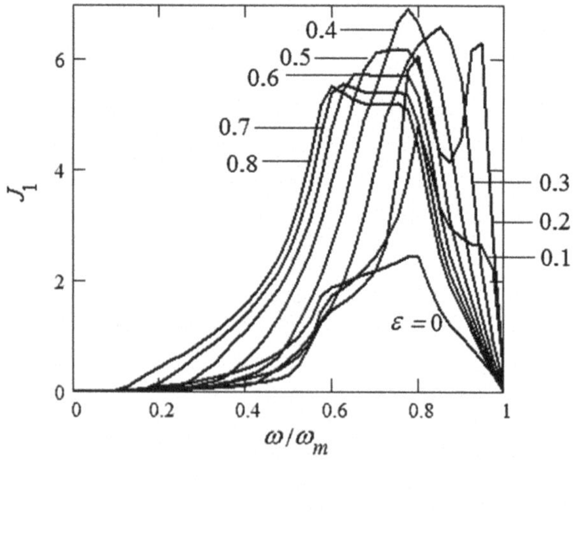

Fig. 6.3 Graphs of calculated values of the function J_1, presented depending on the frequency assigned to the maximum phonon frequency of a given polarization

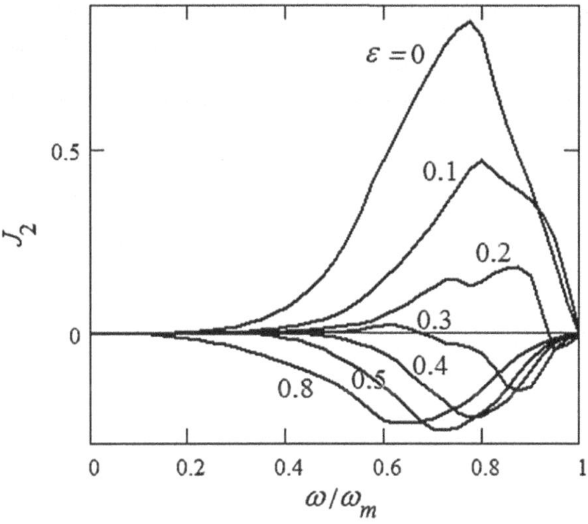

Fig. 6.4 Graphs of calculated values of the function, presented depending on the frequency assigned to the maximum phonon frequency of a given polarization

The function J_1 is proportional to the density of phonon states of a given polarization, and its form is consistent with theoretical ideas about the nature of the frequency dependence of the density of states [11, 15].

The function J_2 is proportional to the momentum transmitted by the defect in one scattering event. It is characteristic that with an increase in the parameter ε this function changes sign, and at values $\varepsilon > 0.3$ becomes completely negative.

The calculation of the integral (6.39) was carried out on the basis of the values of the function J_1 obtained in the previous calculation. Unlike J_1 and J_2, this function is not equal to zero at frequencies exceeding the limiting phonon frequency, therefore, the calculation was carried out up to double the maximum frequency. The calculation results are presented in Fig. 6.5.

In accordance with definition (6.27), the first term in parentheses in expression (6.29) vanishes at $\omega > \omega_m$, and the logarithmic dependence is transferred to the second term, as follows from definition (6.28). The nature of this dependence is manifested in the "tails" of the graphs in Fig. 6.5.

Let us further consider formula (6.32). We substitute expressions (6.37), (6.38), (6.39) into it and take into account the relation between the atomic volume and the volume of the first Brillouin zone: $V_a k_0^3 = \pi^3$. After simple transformations, we obtain the following formula:

$$\beta' = \frac{\hbar\pi k_0^2}{18}\frac{(\omega/c_s k_0)^2 \sum_\alpha \frac{J_1(\zeta_\alpha)}{c_\alpha}\theta_\alpha \cdot \sum_\alpha c_\alpha J_2(\zeta_\alpha)\theta_\alpha}{\left|1-\frac{1}{24}\frac{\Delta m}{m}\sum_\alpha\left(\frac{4\omega}{c_\alpha k_0}J_1(\zeta_\alpha)\ln\frac{\omega-\omega_{m\alpha}}{\omega+\omega_{m\alpha}}+\frac{\omega^2}{(c_\alpha k_0)^2}J_3(\zeta_\alpha)\right)\right|^2}\left(\frac{\Delta m}{m}\right)^2, \quad (6.40)$$

For the calculation by formula (6.40), the specification of the long-wavelength phonon velocities is required. In the Born–Karman model, they are accepted in the

Fig. 6.5 Graphs of calculated values of the function J_3, presented depending on the frequency assigned to the maximum phonon frequency of a given polarization

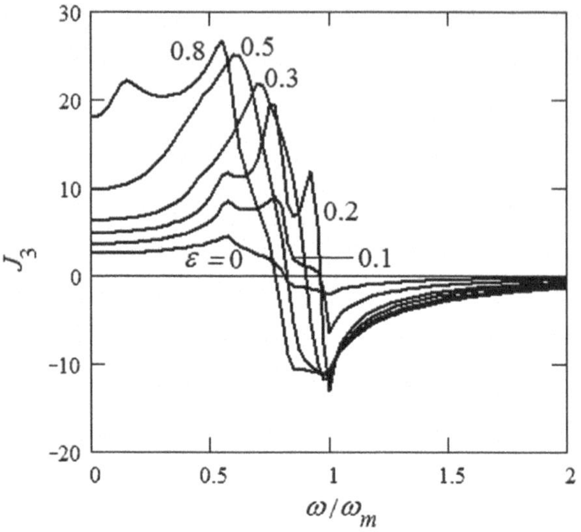

isotropic approximation, respectively, c_t and c_l are the propagation velocities of the longitudinal and transverse vibrations of the lattice. We denote $\gamma = c_l/c_t$.

We also turn to the general normalized frequency $z = \omega/\omega_m$, where ω_m is the maximum phonon frequency of all polarizations. As a result, expression (6.40) is converted to the form:

$$\beta' = \beta'_0 f(z),$$

$$f(z) = \frac{z^2[2\gamma J_1(\gamma z)\theta(1-\gamma z) + J_1(z)\theta(1-z)]\left[\frac{2}{\gamma}J_2(\gamma z)\theta(1-\gamma z) + J_2(z)\theta(1-z)\right]}{\left|1 - \frac{1}{24}\frac{\Delta m}{m}\left[8\eta_m\gamma z J_1(\gamma z)\ln\frac{\gamma z-1}{\gamma z+1} + 4\eta_m z J_1(z)\ln\frac{z-1}{z+1} + 2(\eta_m\gamma z)^2 J_3(\gamma z) + (\eta_m z)^2 J_3(z)\right]\right|^2} \tag{6.41}$$

where

$$\beta'_0 = \frac{\hbar\pi\omega_m^2}{18c_s^2}\left(\frac{\Delta m}{m}\right)^2. \tag{6.42}$$

For further calculations we will use a typical value $\gamma = 2$. The dependences calculated according to the results of previous calculations and formula (6.41) $\beta'(\omega)$ are shown in Fig. 6.6. The dashed line corresponds to the value at which the power of the defect $\Delta m = 0$. According to formula (6.43), the coefficient $\beta'(\omega)$ is also equal to zero. However, for a small value of the defect power ($|\Delta m| \ll m$), the denominator in formula (6.41) can be considered equal to unity, and the dependence on power will be entirely determined by formula (6.42). In this case, the value of the function $f(z)$ is considered to be independent of the power of the defect and is represented by the dashed line in Fig. 6.6.

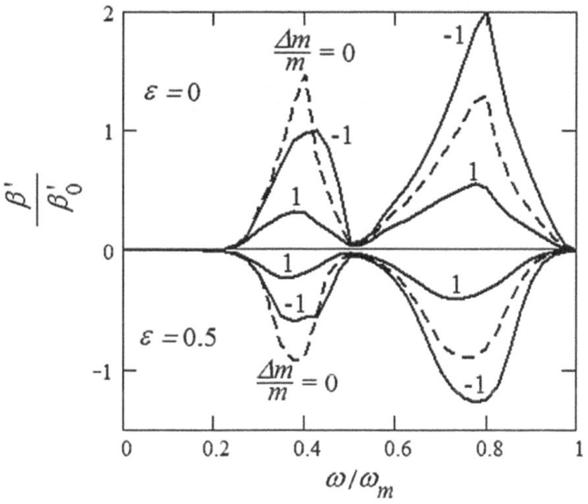

Fig. 6.6 The calculated coefficient β' versus frequency for two parameter values ε and three point defect power values $\Delta m/m$

On the curves of Fig. 6.6, two extrema are distinguished, corresponding to the scattering of transverse and longitudinal phonons. It is seen that the partial drag coefficients from longitudinal phonons are much larger than from transverse ones. However, when phonon state occupation numbers are taken into account (see formula (6.8)), this relation changes to the opposite. Also in Fig. 6.6 asymmetry is noted with respect to the sign of power: with a negative sign, the drag effect increases, with a positive sign it decreases.

6.7 Results and Discussion

We apply the obtained results to the analysis of the coefficient of phonon drag of a point defect by a heat flux (6.8).

In the temperature range in which the diffusion mobility of point defects is manifested significantly and for which it makes sense to speak of phonon drag, the equilibrium distribution function has the classical form [1, 11]:

$$N_0(\omega) = \frac{k_B T}{\hbar\omega}, \tag{6.43}$$

where k_B is the Boltzmann constant. At frequencies approaching to ω_m that make the main contribution to the integral (6.8), and at a concentration of defects that is sufficiently small to neglect the processes of combined phonon-impurity phonon scattering, the relaxation time $\tau(\omega)$ can be assumed to be independent of the frequency [13–15]. Then formula (6.8) with the use of (6.41), (6.42) and taking into account (6.43) is transformed to the form:

$$B' = B'_0 \cdot \int_0^1 f(z) \frac{dz}{z^2}, \tag{6.44}$$

where denote

$$B'_0 = \frac{\pi}{18} \left(\frac{\Delta m}{m} \right)^2 \frac{k_B T \omega_m}{c_s^2}. \tag{6.45}$$

The calculation results according to formula (6.44) based on previous calculations and for various parameters are presented in Fig. 6.7.

As follows from the results of Fig. 6.7a, regardless of the power of the defect with a change in the parameter ε, the drag coefficient changes sign at $\varepsilon \approx 0.27$, the corresponding dispersion law is shown in Fig. 6.1a with dashed line. In crystals for which the phonon dispersion law has a parameter ε greater than this threshold value, the drag force is directed against the heat flux. This effect is manifested even with a small "negative" dispersion, since such a part of the dispersion takes place at frequencies adjacent to the maximum. It is with such frequencies that phonons make a decisive contribution to the drag force (Fig. 6.6.).

The dependence of the drag coefficient on the power of a point defect (Fig. 6.7b) is asymmetric in character of the power sign. For defects with a negative power, that is, impurity atoms with a mass, the smaller the mass of the atoms of the crystal (in particular, vacancies), the effect is greater than for power-positive defects. This effect is associated with the influence of resonant phonon scattering on a defect that has its own quasilocal vibrational states. Such states, remote from the boundary of the phonon spectrum, appear only at a negative defect power [19] and lead to a

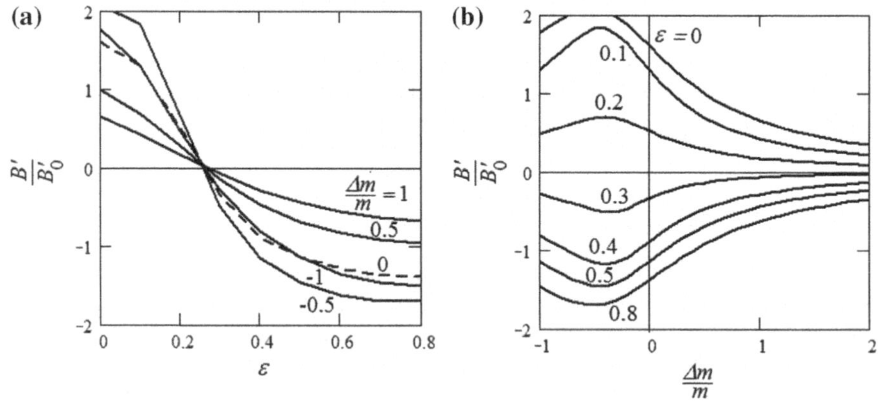

Fig. 6.7 Dependence of the drag coefficient on the magnitude of the negative portion of the phonon dispersion law (**a**) and on the power of the defect (**b**)

decrease in the real part of the expression in the denominator of formula (6.32) and, accordingly, to an increase in the partial drag coefficient.

The considered drag effect should be taken into account in combination with forces of a different nature, including thermodynamic forces, when determining the real redistribution of impurity atoms in a temperature gradient field [9].

6.8 Conclusions

The possibility of different directions of the drag force in the crystal, depending on the nature of the phonon spectrum, was indicated earlier [5]. In the present work, an attempt is made to quantify the relationship between the magnitude and direction of the drag force with the degree of nonlinearity of the phonon dispersion law. We did not take into account the role of optical phonons in the formation of drag force, since their consideration in the framework of the continuum model presents certain difficulties. Also, the anharmonicity of the crystal can have a certain effect on the drag force [8]. This effect is associated with the size effect of a point defect and does not occur in the case of a mass defect.

References

1. Ch. Kittel, *Introduction to Solid State Physics*, 8th edn. (Wiley, New York, 2005)
2. H.G. Van Bueren, *Imperfections in Crystals* (North-Holland Publishing Company, Amsterdam, 1960)
3. Y.E. Geguzin, M.E. Krivoglaz, *Motion of Macroscopic Inclusions in Solids* (Mettallurgiya, Moscow, 1971)
4. B.Y. Lyubov, *Kinetic Theory of Phase Transformations* (Mettallurgiya, Moscow, 1969)
5. V.B. Fiks, On the mechanism of thermodiffusion in liquids. Sov Phys Sol Stat **3**, 994 (1961)
6. G. Schottky, A theory of thermal diffusion based on lattice dynamics of a linear chain. Phys. Stat. Sol. **8**, 357 (1965)
7. P.P. Kuzmenko, On dragging of diffusing ions with phonons in metals. Ukr. J. Phys. **15**, 1982 (1970)
8. V.I. Al'shits, V.L. Indenbom, Dynamic dragging of dislocations. Sov. Phys. Usp. **18**, 1 (1975)
9. M.V. Yur'yeva, I.L. Bataronov, A.M. Roshchupkin, V.A. Yur'yev, Influence of electric current on impurity diffusion in bicrystal. Bull. Russ. Acad. Sci. Phys. **61**(2), 77 (1995)
10. S.V. Vonsovsky, M.I. Katsnelson, *Quantum Physics of Solids* (Nauka, Moscow, 1983)
11. J.M. Ziman, *Electrons and Phonons. The Theory of Transport Phenomena in Solids* (Clarendon Press, Oxford, 1960)
12. I.L. Bataronov, A.M. Roshchupkin, M.V. Yur'yeva, On dragging of defects in crystals with heat flow. Bull. Russ. Acad. Sci. Phys. **61**(5), 927 (1997)
13. V.S. Oskotski, I.A. Smirnov, *Defects in Crystals and Thermal Conductivity* (Nauka, Leningrad, 1972)
14. R. Berman, *Thermal Conduction in Solids* (Clarendon Press, Oxford, 1976)
15. V.L. Gurevich, *Kinetics of Phonon Systems* (Nauka, Moscow, 1980)
16. A.G. Sitenko, *The Theory of Scattering* (Visha Shkola, Kiev, 1975)
17. I.M. Lifshits, Scattering of short elastic waves in crystal lattice. JETF **18**(3), 293 (1948)

18. I.M. Lifshitz, Some problems of the dynamic theory of non-ideal crystall lattices. Nuovo Cim. **3**(4 Suppl), 716 (1956)
19. I.M. Lifshits, A.M. Kosevich, The dynamic of a crystal lattice with defects. Rep. Prog. Phys. **29**(1), 217 (1966)
20. G. Leibfried, N. Breuer, *Point Defects in Metals. 1. Introduction to the Theory* (Springer-Verlag, Berlin, Heidelberg, New York 1978)
21. A.M. Kosevich, *Physical Mechanics of Real Crystals* (Naukova Dumka, Kiev, 1981)

Part II
Recent Advances in Molecular Dynamics and Monte Carlo Simulations of Transport Properties of Materials

Chapter 7
Diffusion Kinetics in Binary Liquid Alloys with Ordering and Demixing Tendencies

Andreas Kromik, E. V. Levchenko and Alexander V. Evteev

Abstract Theoretical relationship between collective and tracer diffusion coefficients has been derived and tested for different types of binary melts: (i) with an ordering tendency (case study on Ni–Al and Ni–Zr melts) and (ii) with a demixing tendency (case study on Cu–Ag melts). The obtained relationship explicitly demonstrates microscopic cross-correlation effects in the kinetics of collective diffusion. Our approach incorporates molecular dynamics calculations, modelling and statistical mechanical analysis based on fundamental concepts of the fluctuation-dissipation theorem, generalized Langevin equation and Mori-Zwanzig formalism. We also applied the developed theory to interpret recent available experimental data as well as our molecular dynamics data of diffusion kinetics in different types of binary melts: with chemical ordering and contrarily with demixing tendency.

7.1 Introduction

The alloys of the studied systems (Ni–Al, Ni–Zr, Cu–Ag) attracted attention thanks to their unique physical, chemical and electrical properties [1–8]. Since the studied alloys have significant engineering importance, it is necessary to have an excellent understanding of mass transport properties of these alloys within the liquid state. In particular, the knowledge of diffusion coefficients in liquid alloys is essential for controlling of the crystalline microstructure during solidification process in

---

Alexander V. Evteev: Deceased 8 December 2018

A. Kromik · A. V. Evteev
School of Engineering, Priority Research Centre for Computer-Assisted Research Mathematics and Its Applications (CARMA), The University of Newcastle, Callaghan, NSW 2308, Australia

E. V. Levchenko (✉) · A. V. Evteev
School of Mathematical and Physical Sciences, CARMA, The University of Newcastle, Callaghan, NSW 2308, Australia
e-mail: elena.levchenko@newcastle.edu.au

E. V. Levchenko et al. (eds.), *Theory and Simulation in Physics for Materials Applications*, Springer Series in Materials Science 296,
https://doi.org/10.1007/978-3-030-37790-8_7

experiment [9] as well as for essential parameters input in phase field modelling [9, 10]. However, the current capacity of existing experimental diffusion database of liquid alloys is very limited [11–18] because of technical difficulties and high costs associated with the experimental measurements of the diffusion properties in the melts. Hence, establishing theoretical relation between different mass transport coefficients is important for a quantitative prediction of materials properties as some of the unknown coefficients can be expressed via others, which are reliably measurable in the experiment [19].

Dynamical and transport properties of many-body system can be expressed in terms of the time-correlation functions of appropriate physical variables. For example, the frequency-dependent thermal conductivity and diffusion coefficient of a many-body system are the one-sided Fourier transforms of the time-correlation functions of the heat and mass currents, respectively. Therefore, for understanding, predicting, controlling dynamical, and transport properties of many-body systems, it is important to develop methods of calculation, modelling and analysis of this time-correlation functions. Our group developed an innovative approach based on fundamental concepts of the Brownian motion [20, 21], the fluctuation-dissipation theorem [20, 21], generalized Langevin equation [20, 21] and Mori-Zwanzig formalism [22–25]. These concepts state a general relationship between response of a given system to an external disturbance and the internal fluctuations of the system in the absence of the disturbance. In this chapter, we will focus on mass transport properties in different types of binary melts: with chemical ordering (case study on Ni–Al and Ni–Zr melts) and with phase separation tendency (case study on Cu–Ag melts).

In 1948 Darken [26] introduced an equation for a binary system, which expressed the interdiffusion coefficient, D_c, via two self-diffusion coefficients, D_1 and D_2:

$$D_c = \Phi(c_2 D_1 + c_1 D_2), \tag{7.1}$$

where c_1 and c_2 are the atomic fractions of species 1 and 2, Φ is the thermodynamic factor. Note $c_1 = \frac{N_1}{N}, c_2 = \frac{N_2}{N}$, where N_1 is number of atoms of species 1, N_2 is number of atoms of species 2, N is the total number of atoms in the system, i.e. $N = N_1 + N_2$.

Equation (7.1) is known as Darken equation. Both D_c and Φ are in principal accessible in experiment. The thermodynamic factor Φ is the normalized second derivative of the molar Gibbs free energy G/N with respect to the composition $c_1(c_2)$ at constant temperature T and pressure P:

$$\Phi = \frac{c_1 c_2}{k_B T}\left(\frac{\partial^2 (G/N)}{\partial c_1^2}\right)_{T,P} = \frac{c_1 c_2}{k_B T}\left(\frac{\partial^2 (G/N)}{\partial c_2^2}\right)_{T,P}, \tag{7.2}$$

where k_B is the Boltzmann constant.

In 1961 Manning [27] included a certain correction factor of the microscopic kinetic origin, S, into the (7.1):

$$D_c = \Phi S(c_2 D_1 + c_1 D_2), \tag{7.3}$$

Equation (7.3) is known as the Darken-Manning equation [28].

On the other hand, the Onsager phenomenological transport equation can be written as [29, 30]

$$\boldsymbol{J}_c = L_{cc}\boldsymbol{X}'_c. \tag{7.4}$$

where $\boldsymbol{J}_c$ is the interdiffusion flux, L_{cc} is Onsager phenomenological coefficient for mass transport and $\boldsymbol{X}'_c$ conjugated thermodynamic force related to interdiffusion flux. Interdiffusion flux $\boldsymbol{J}_c$ links with interdiffusion coefficients D_c by Fick's law:

$$\boldsymbol{J}_c = c_2\boldsymbol{J}_1 - c_1\boldsymbol{J}_2 = -\frac{N}{V}D_c\nabla c_1 = \frac{N}{V}D_c\nabla c_2 = -\frac{N}{V}D_c(c_2\nabla c_1 - c_1\nabla c_2), \tag{7.5}$$

where V is the volume of the system.

Furthermore, according to the Onsager formalism of the thermodynamics of irreversible processes for an isotropic binary melt [21, 22], the ratio D_c/Φ is equal to the renormalized phenomenological Onsager coefficient $\tilde{L}_{cc}$ [23]:

$$\frac{D_c}{\Phi} = \frac{Vk_{\mathrm{B}}T}{Nc_1c_2}L_{cc} = \tilde{L}_{cc}. \tag{7.6}$$

Now for correction factor we have:

$$S = \frac{D_c}{\Phi(c_2D_1 + c_1D_2)} = \frac{Vk_{\mathrm{B}}TL_{cc}}{Nc_1c_2(c_2D_1 + c_1D_2)} = \frac{\tilde{L}_{cc}}{c_2D_1 + c_1D_2} \tag{7.7}$$

Hence, the correction factor S characterises cross-correlation effects in collective diffusion process. Available experimental [17] and simulation data [12, 19, 31] for the correction factor suggested that it should be approximately less than unity for alloys with ordering tendency. However, a detailed formula for the correction factor S involving a cross correlation term has not been obtained until now.

In this chapter, we will derive an expression relating collective and tracer diffusion coefficients. We are aiming to reveal explicitly the cross-correlation effects in collective diffusion. We will analyse the obtained relation for different types of binary melts: with chemical ordering (case study on Ni–Al and Ni–Zr melts) and with phase separation tendency (case study on Cu–Ag melts). Finally, we apply our theoretical findings for interpretation of recent available experimental data and our molecular dynamics (MD) data of diffusion kinetics.

7.2 Theoretical Treatment

7.2.1 *Generalized Langevin Equations for the Velocities and Integral-Differential Equations for the Velocity Autocorrelation Functions*

The generalised Langevin equation plays a key role in the theoretical development through use of the projection-operator formalism of statistical mechanics introduced by Zwanzig and Mori [22–25, 32]. First, consider an isotropic binary liquid alloy consisting of N_1 atoms with mass m_1 and N_2 atoms having mass m_2 in a fixed volume V at thermal equilibrium.

By the Mori-Zwanzig formalism [22–25, 32] the total force $\boldsymbol{f}_{\alpha i}(t)$ acting on the tagged particle can be decomposed on the sum of systematic and random terms, the first and the second summands on the RHS of (7.8) respectively:

$$m_\alpha \frac{\mathrm{d}\boldsymbol{v}_{\alpha i}(t)}{\mathrm{d}t} = \boldsymbol{f}_{\alpha i}(t) = -m_\alpha \int\limits_0^t K_\alpha(t-t')\boldsymbol{v}_{\alpha i}(t')\mathrm{d}t' + \boldsymbol{R}_{\alpha i}(t). \tag{7.8}$$

This is a generalized Langevin equation for the velocities $\boldsymbol{v}_{\alpha i}(t)$ for a binary system, where $\alpha = 1, 2(\alpha = 1 : i \in [1, \ldots, N_1], \alpha = 2 : i \in [1, \ldots, N_2])$ are tagged atoms of species 1 and 2; $K_\alpha(t)$ is the memory kernel for the evolution of the systematic/frictional forces acting on the tagged atom α at time t. The memory kernel is proportional but opposite to the velocities of the tagged atoms at times prior to time $t, 0 \le t' \le t$.

$\boldsymbol{R}_{\alpha i}(t)$ is the random force. A random force possesses three properties [22–25, 32]:

(1) It vanishes in the mean $\langle \boldsymbol{R}_{\alpha i}(t) \rangle = 0$ (notation $\langle \cdots \rangle$ means the statistical time average at thermal equilibrium),
(2) $\boldsymbol{f}_{\alpha i}(0) = \boldsymbol{R}_{\alpha i}(0)$, i.e. initially, the total and random forces are the same (see (7.8)).
(3) Total and random forces evolve differently with time and relation $\langle \boldsymbol{R}_{\alpha i}(t)\boldsymbol{v}_{\alpha i}(0) \rangle = 0$ is preserved for all times t. In other words: random force is not correlated with the initial velocity $\boldsymbol{v}_{\alpha i}(0)$.

Hence, if we multiply (7.8) by $\boldsymbol{v}_{\alpha i}(0)$ and take the thermal average, using the above mentioned third property of random force we will get the integral-differential equation for the velocity autocorrelation functions, which often called a satellite equation for Langevin generalised (7.8):

$$\frac{\mathrm{d}\varepsilon_\alpha(t)}{\mathrm{d}t} = -\int\limits_0^t K_\alpha(t-t')\varepsilon_\alpha(t')\mathrm{d}t', \tag{7.9}$$

where $\varepsilon_\alpha(t) = \left\langle \boldsymbol{v}_{\alpha i}(t)\boldsymbol{v}_{\alpha i}(0)/\boldsymbol{v}_{\alpha i}^2\right\rangle = m_\alpha \langle \boldsymbol{v}_{\alpha i}(t)\boldsymbol{v}_{\alpha i}(0)\rangle/3k_BT$ are the normalized velocity autocorrelation functions. The squared static velocities, also called thermal velocities, are equal to $\left\langle \boldsymbol{v}_{\alpha i}^2\right\rangle = 3k_BT/m_\alpha$, so that $\varepsilon_\alpha(0) = 1$.

7.2.2 *Properties of the Correlation Functions of Dynamical Variables in Equilibrium*

We should note, that throughout the chapter, above and below, we implicitly employ useful properties of the correlation functions of dynamical variables in equilibrium. Namely, in equilibrium the derivative of the time correlation function of dynamical variables $Z(t)$ and $Y(t)$ with respect to the initial time t_0 must be zero, namely: $\frac{\mathrm{d}}{\mathrm{d}t_0}\langle Z(t+t_0)Y(t_0)\rangle = 0$. So, applying the product rule the LHS of the above expression we get:

$$\begin{aligned}&\left\langle \frac{\mathrm{d}}{\mathrm{d}t_0}(Z(t+t_0))Y(t_0)\right\rangle + \left\langle Z(t+t_0)\frac{\mathrm{d}}{\mathrm{d}t_0}(Y(t_0))\right\rangle \\ &\quad = \left\langle \dot{Z}(t+t_0)Y(t_0)\right\rangle + \left\langle Z(t+t_0)\dot{Y}(t_0)\right\rangle = 0.\end{aligned} \tag{7.10}$$

And in case of autocorrelation function, i.e. $Z(t) = Y(t)$, in we get $\langle \dot{Z}(t)Z(0)\rangle = -\langle Z(t)\dot{Z}(0)\rangle$, $\langle \dot{Z}(t)Z(t)\rangle = \langle \dot{Z}(0)Z(0)(0)\rangle = \left\langle \dot{Z}Z\right\rangle = 0$, and for the second derivative $\langle \ddot{Z}(t)Z(0)\rangle = -\langle \dot{Z}(t)\dot{Z}(0)\rangle$, where $\dot{Z} = \frac{\mathrm{d}z}{\mathrm{d}t}$ and $\ddot{Z} = \frac{\mathrm{d}^2z}{\mathrm{d}t^2}$.

Using these properties and random force properties, we can now express the memory kernels as autocorrelation of the random forces [22–25, 32]. Indeed, multiplying (7.8) by $\boldsymbol{R}_{\alpha i}(0)$, or by $\boldsymbol{f}_{\alpha i}(0)$, which is the same by the above mentioned second property of the random force, and taking the thermal average and using (7.9), we will get:

$$K_\alpha(t) = \langle \boldsymbol{R}_{\alpha i}(t)\boldsymbol{R}_{\alpha i}(0)\rangle/m_\alpha^2\left\langle \boldsymbol{v}_{\alpha i}^2\right\rangle \tag{7.11}$$

or

$$K_\alpha(t) = \langle \boldsymbol{R}_{\alpha i}(t)\boldsymbol{R}_{\alpha i}(0)\rangle/3m_\alpha k_BT, \tag{7.12}$$

using expression for squared static velocities $\left\langle \boldsymbol{v}_{\alpha i}^2\right\rangle = 3k_BT/m_\alpha$.

In addition, it can be seen from (7.9) that functions $\varepsilon_\alpha(t)$ and $K_\alpha(t)$ are even. It is interesting to highlight that at zero time, the memory kernels represent a square of Einstein frequency, $K_\alpha(0) = \Omega_\alpha^2$, a frequency, at which a tagged atom of species α would vibrate on average if they were experiencing small oscillations in the potential wells generated by the neighbouring atoms when retained at their mean equilibrium positions around the tagged atom. Indeed, $K_\alpha(0) = \frac{\langle \dot{\boldsymbol{R}}^2_{\alpha i}\rangle}{3m_\alpha k_B T} = \frac{m_\alpha \langle \dot{\boldsymbol{v}}^2_{\alpha i}\rangle}{3k_B T}$, where $\dot{\boldsymbol{v}}_{\alpha i} = \frac{d\boldsymbol{v}_{\alpha i}(t)}{dt} = -m_\alpha^{-1}\nabla_{\alpha i}U$ is acceleration and U is potential energy.

7.2.3 Total Force Decomposition

Total force decomposition involves two steps [24]:

(1) The total force is represented of the sum: $\boldsymbol{f}_{\alpha i}(t) = \boldsymbol{f}^{(\mathrm{I})}_{\alpha i}(\boldsymbol{v}_{\alpha i}(t')) + \boldsymbol{f}^{(\mathrm{II})}_{\alpha i}(t)$, where the first summand $\boldsymbol{f}^{(\mathrm{I})}_{\alpha i}(\boldsymbol{v}_{\alpha i}(t'))$ depends on the past history of velocity $\boldsymbol{v}_{\alpha i}$ since $0 \le t' \le t$ while the second summand $\boldsymbol{f}^{(\mathrm{II})}_{\alpha i}(t)$ describes contribution which depends explicitly on dynamics of the other atoms at time t.

(2) The functional $\boldsymbol{f}^{(\mathrm{I})}_{\alpha i}(\boldsymbol{v}_{\alpha i}(t'))$ can be further decomposed into linear and non-linear terms: $\boldsymbol{f}^{(\mathrm{I})}_{\alpha i}(\boldsymbol{v}_{\alpha i}(t')) = \boldsymbol{f}^{(\mathrm{I,\,linear})}_{\alpha i}(\boldsymbol{v}_{\alpha i}(t')) + \boldsymbol{f}^{(\mathrm{I,\,non-linear})}_{\alpha i}(\boldsymbol{v}_{\alpha i}(t'))$. The linear term of the expansion is the systematic force of the generalized Langevin equation, i.e. the first summand in the (7.8) (integral term of (7.8)). While the sum of the non-linear term together with $\boldsymbol{f}^{(\mathrm{II})}_{\alpha i}(t)$ uniquely defines the random force of the generalized Langevin equation, namely the random force is equal to $\boldsymbol{R}_{\alpha i}(t) = \boldsymbol{f}^{(\mathrm{I,\,non-linear})}_{\alpha i}(\boldsymbol{v}_{\alpha i}(t'), 0 \le t' \le t) + \boldsymbol{f}^{(\mathrm{II})}_{\alpha i}(t)$.

Consequently, the total force can be decomposed as:

$$\begin{aligned} \boldsymbol{f}_{\alpha i}(t) = -m_\alpha \int_0^t K_\alpha(t-t')\boldsymbol{v}_{\alpha i}(t')\mathrm{d}t' \\ + \boldsymbol{f}^{(\mathrm{I,\,non-linear})}_{\alpha i}(\boldsymbol{v}_{\alpha i}(t')) + \boldsymbol{f}^{(\mathrm{II})}_{\alpha i}(t) \end{aligned} \qquad (7.13)$$

7.2.4 Generalized Langevin Equations and Its Satellite Equations for the Interdiffusion Flux

As stated by the Mori-Zwanzig formalism [22–25, 32], we can write equations analogous to (7.8) and (7.9) for any arbitrary chosen dynamical variable and its autocorrelation function, as we did for velocity. Our choice for further theoretical treatment in this chapter will be the interdiffusion flux $\boldsymbol{J}_c(t)$, due to its invariance to the choice of reference frame [31, 33, 34] as we will demonstrate below. Therefore, the microscopic expression for the interdiffusion flux $\boldsymbol{J}_c(t)$ of a binary liquid alloy at thermal equilibrium in fixed volume V is equal to [31, 33, 34]:

$$\boldsymbol{J}_c(t) = \frac{N}{V}c_1c_2\bar{\mathbf{v}}_c(t) = \frac{N}{V}c_1c_2[\bar{\mathbf{v}}_1(t) - \bar{\mathbf{v}}_2(t)] = c_2\boldsymbol{J}_1(t) - c_1\boldsymbol{J}_2(t), \tag{7.14}$$

where $\boldsymbol{J}_1(t)$ and $\boldsymbol{J}_2(t)$ are the fluxes of species 1 and 2 and $\bar{\mathbf{v}}_1(t)$ and $\bar{\mathbf{v}}_2(t)$ are the mean velocities of species 1 and 2, $\bar{\mathbf{v}}_c(t) \equiv \bar{\mathbf{v}}_{12}(t)$.

Now from (7.14) we can see that the interdiffusion flux $\boldsymbol{J}_c(t)$ is indeed does not depend on the choice of reference frame because it defines the fluxes of components $\boldsymbol{J}_1(t)$ and $\boldsymbol{J}_2(t)$ relative to each other. In the zero-momentum reference frame, we have $c_1m_1\bar{\mathbf{v}}_1(t) + c_2m_2\bar{\mathbf{v}}_2(t) = 0$ or $m_1\boldsymbol{J}_1(t) + m_2\boldsymbol{J}_2(t) = 0$, so that:

$$\boldsymbol{J}_c(t) = \frac{N}{V}c_1c_2\bar{\mathbf{v}}_c(t) = \frac{m}{m_2}\boldsymbol{J}_1(t) = -\frac{m}{m_1}\boldsymbol{J}_2(t) = \frac{m}{m_2V}\sum_{i=1}^{N_1}\boldsymbol{v}_{1i}(t) = -\frac{m}{m_1V}\sum_{i=1}^{N_2}\boldsymbol{v}_{2i}(t), \tag{7.15}$$

where $m = c_1m_1 + c_2m_2$ is the total system mass per atom. Here, mass per atom carried by flux $\boldsymbol{J}_1$ is m_1 and mass per atom carried by flux $\boldsymbol{J}_2$ is m_2. Then, it follows from (7.15) that the effective mass per atom carried by interdiffusion flux $\boldsymbol{J}_c$ is equal to $m_c = m_1m_2/m$. In other words, at thermal equilibrium the interdiffusion flux is produced by an effective system element containing Nc_1c_2 identical particles of mass m_c in the given volume V with the mean velocity equal to zero and the thermal velocity $(3k_\mathrm{B}T/m_c)^{1/2}$ from the Maxwell–Boltzmann distribution. Furthermore, on a basis of (7.15) we can obtain $3Nc_1c_2k_\mathrm{B}T/m_\mathrm{c} = V^2\langle \boldsymbol{J}_c^2\rangle$ [31]. Hence, we can readily derive from an explicit analytical expression of the form $V^2\langle \boldsymbol{J}_c^2\rangle = 3Nc_1c_2k_\mathrm{B}T/m_\mathrm{c}$. This is analogous to the expression for the squared static velocities $\langle \boldsymbol{v}_{\alpha i}^2\rangle$ derived in Sect. 7.2.1 above and confirmed by our MD simulations [31].

Therefore, the generalized Langevin equation for the interdiffusion flux $\boldsymbol{J}_c(t)$ and the equation for its normalized autocorrelation function $\varepsilon_c(t) = m_cV^2\langle \boldsymbol{J}_c(t)\boldsymbol{J}_c(0)\rangle / 3Nc_1c_2k_\mathrm{B}T$ $(\varepsilon_c(0) = 1)$ can be presented as:

$$m_cV\frac{\mathrm{d}\boldsymbol{J}_c(t)}{\mathrm{d}t} = \boldsymbol{F}_c(t) = -m_cV\int_0^t K_c(t-t')\boldsymbol{J}_c(t')\mathrm{d}t' + \boldsymbol{R}_c(t), \tag{7.16}$$

$$\frac{\mathrm{d}\varepsilon_c(t)}{\mathrm{d}t} = -\int_0^t K_\mathrm{c}(t-t')\varepsilon_c(t')\mathrm{d}t'. \tag{7.17}$$

In these equations, $\boldsymbol{F}_c(t)$ and $\boldsymbol{R}_c(t)$ are the total and random forces, respectively, acting on the mass m_c. Consequently, $\boldsymbol{R}_c(t)$ has random force properties listed in Sect. 7.2.1 and similar to the properties of $\boldsymbol{R}_{\alpha i}(t)$, namely:

(i) $\langle \boldsymbol{R}_c(t)\rangle = 0$,
(ii) $\boldsymbol{F}_c(0) = \boldsymbol{R}_c(0)$
(iii) $\langle \boldsymbol{R}_c(t)\boldsymbol{J}_c(0)\rangle = 0$, and
(iv) $K_c(t) = \langle \boldsymbol{R}_c(t)\boldsymbol{R}_c(0)\rangle / m_c^2V^2\langle \boldsymbol{J}_c^2\rangle = \langle \boldsymbol{R}_c(t)\boldsymbol{R}_c(0)\rangle / 3Nc_1c_2m_\mathrm{c}k_\mathrm{B}T$.

The memory kernel $K_c(t)$ describes the evolution of the systematic force acting on the mass m_c. As we can see from (7.17) both autocorrelation functions $\varepsilon_c(t)$ and $K_c(t)$ are even functions of time. Furthermore, at initial time the memory kernel can be expressed via corresponding effective Einstein frequency Ω_c: $K_c(0) = \Omega_c^2 = \frac{\langle \boldsymbol{R}_c^2 \rangle}{3 m_c k_B T} = \frac{m_c V^2 \langle \dot{\boldsymbol{J}}_c^2 \rangle}{3 N c_1 c_2 k_B T}$, where $\dot{\boldsymbol{J}}_c = \frac{\mathrm{d}\boldsymbol{J}_c(t)}{\mathrm{d}t} = (-1)^{\alpha} m_c^{-1} \sum_{i=1}^{N_\alpha} \nabla_{\alpha i} U$.

7.2.5 Alternative Expression for Interdiffusion Flux via Single-Particle Memory Kernels and Random Forces

Alternatively combining Langevin equation for velocity (7.8) and interdiffusion flux definition (7.14) and (7.15) we will arrive to more significant expression for $\boldsymbol{J}_c(t)$, because it includes the single-particle memory kernels, $K_1(t)$ and $K_2(t)$, and random forces, $\boldsymbol{R}_{1i}(t)$ and $\boldsymbol{R}_{2j}(t)$:

$$m_c V \frac{\mathrm{d}\boldsymbol{J}_c(t)}{\mathrm{d}t} = \boldsymbol{F}_c(t)$$
$$= -m_c V \int_0^t \left[c_2 \frac{m_2}{m} K_1(t-t') + c_1 \frac{m_1}{m} K_2(t-t') \right] \boldsymbol{J}_c(t') \mathrm{d}t' + \boldsymbol{R}_{12}(t). \tag{7.18}$$

where force $\boldsymbol{R}_{12}(t)$ represents the contribution to the total force associated with interdiffusion flux $\boldsymbol{F}_c(t)$ due to the difference in the average random accelerations of atoms of species 1 and 2, $\frac{\bar{\boldsymbol{R}}_1(t)}{m_1}$ and $\frac{\bar{\boldsymbol{R}}_2(t)}{m_2}$ respectively (7.19), so that:

$$\boldsymbol{R}_{12}(t) = N c_1 c_2 m_c \left(\frac{\bar{\boldsymbol{R}}_1(t)}{m_1} - \frac{\bar{\boldsymbol{R}}_2(t)}{m_2} \right), \tag{7.19}$$

where

$$\bar{\boldsymbol{R}}_1(t) = \frac{1}{N_1} \sum_{i=1}^{N_1} \boldsymbol{R}_{1i}(t), \; \bar{\boldsymbol{R}}_2(t) = \frac{1}{N_2} \sum_{j=1}^{N_2} \boldsymbol{R}_{2j}(t) \tag{7.20}$$

are the average random forces acting on atoms of species 1 and 2 at time t, respectively. It follows from (7.18) that $\boldsymbol{F}_c(0) = \boldsymbol{R}_{12}(0) = \boldsymbol{R}_c(0)$ and $\langle \boldsymbol{R}_{12}(t) \rangle = 0$. It should be highlighted that $\boldsymbol{R}_{12}(t)$ is not truly a random force by definition, but the difference between the random forces.

Since (7.16) and (7.18) both describe the total force acting on the mass carried by the interdiffusion flux $\boldsymbol{J}_c$ we can equate them and divide by the effective mass m_c:

$$V\int_0^t \left[c_2\frac{m_2}{m}K_1(t-t')+c_1\frac{m_1}{m}K_2(t-t')-K_c(t-t')\right]\boldsymbol{J}_c(t')\mathrm{d}t' = \frac{\boldsymbol{R}_{12}(t)-\boldsymbol{R}_c(t)}{m_c}. \tag{7.21}$$

Now, after multiplying both sides of (7.21) by $\boldsymbol{J}_c(0)$ and taking the thermal average, we arrive to:

$$\int_0^t \left[c_2\frac{m_2}{m}K_1(t-t')+c_1\frac{m_1}{m}K_2(t-t')-K_c(t-t')\right]\varepsilon_c(t')\mathrm{d}t' = \frac{P_{12}(t)}{k_\mathrm{B}T}, \tag{7.22}$$

where

$$P_{12}(t) = \frac{V}{3Nc_1c_2}\langle \boldsymbol{R}_{12}(t)\boldsymbol{J}_c(0)\rangle. \tag{7.23}$$

corresponds to the correlations between fluctuations of $\boldsymbol{R}_{12}(t)$ and $\boldsymbol{J}_c(0)$ at $t > 0$.

7.2.6 Frequency-Dependent Diffusion Coefficients in a Binary Liquid Alloy

Now, taking the one-sided Fourier transforms of (7.9), (7.17) and (7.22) and combining them, in frequency domain we will get:

$$\tilde{L}_{cc}(\omega) = \frac{m^2D_1(\omega)D_2(\omega)}{c_1m_1^2D_1(\omega)+c_2m_2^2D_2(\omega)}\left[1+\frac{W_{12}(\omega)}{k_\mathrm{B}T}\right]. \tag{7.24}$$

where

$$D_1(\omega) = \frac{k_\mathrm{B}T}{m_1}\int_0^\infty \varepsilon_1(t)\mathrm{e}^{-i\omega t}\mathrm{d}t, \tag{7.25}$$

$$D_2(\omega) = \frac{k_\mathrm{B}T}{m_2}\int_0^\infty \varepsilon_2(t)\mathrm{e}^{-i\omega t}\mathrm{d}t, \tag{7.26}$$

$$\tilde{L}_{cc}(\omega) = \frac{k_\mathrm{B}T}{m_c}\int_0^\infty \varepsilon_c(t)\mathrm{e}^{-i\omega t}\mathrm{d}t, \tag{7.27}$$

are the frequency-dependent diffusion coefficients in a binary liquid alloy represented via related velocity autocorrelation functions consistent with the basic theorem for linear responses [21–25, 32, 35].

$$W_{12}(\omega) = \int_0^\infty P_{12}(t) e^{-i\omega t} dt = \frac{V}{3Nc_1c_2} \int_0^\infty \langle \boldsymbol{R}_{12}(t) \boldsymbol{J}_c(0) \rangle e^{-i\omega t} dt. \quad (7.28)$$

7.2.7 Correlations Between Fluctuations of $\boldsymbol{R}_{12}(t)$ and $\boldsymbol{J}_c(0)$

In short time limit $t \to 0$ the value of $P_{12}(t)$ can be approximated via the first derivative of $P_{12}(t)$ evaluated $t = 0$ [36], as

$$\frac{\dot{P}_{12}(0)}{k_B T} = \frac{1}{k_B T} \frac{dP_{12}(t)}{dt}\bigg|_{t=0} = c_2 \frac{m_2}{m} \Omega_1^2 + c_1 \frac{m_1}{m} \Omega_2^2 - \Omega_c^2, \quad (7.29)$$

Due to common MD simulation configurations of interatomic interactions, in the short time limit $t \to 0$, $P_{12}(t)$ can be estimated as [36]

$$\frac{P_{12}(t)}{k_B T} \approx \sigma_{12} \Omega_c^2 t, \quad (7.30)$$

where

$$\sigma_{12} = \frac{c_2 m_2 \Omega_1^2 + c_1 m_1 \Omega_2^2}{m \Omega_c^2} - 1 \quad (7.31)$$

is the dimensionless factor which defines the initial sign of $P_{12}(t)$ as the correlation between $\boldsymbol{R}_{12}(t)$ and $\boldsymbol{J}_c(0)$ starts to develop with time. As a result, $\sigma_{12} < 0$ and $\sigma_{12} > 0$ indicate that $\boldsymbol{R}_{12}(t)$ tends to initially create with $\boldsymbol{J}_c(0)$ obtuse and acute angles, respectively as it is shown in Fig. 7.1 representing the initial instant of time defined by σ_{12}.

The behaviour $\sigma_{12} < 0$ and $W_{12} < 0$ is expected for a binary melt exhibiting chemical ordering in the normal liquid state. Therefore an the initial angle in short time limit $t \to 0$ for $\sigma_{12} < 0$ (obtuse angle) between $\boldsymbol{R}_{12}(t)$ and $\boldsymbol{J}_c(0)$ should remain obtuse, ensuring a negative value of W_{12} which is expected for these type of melts accounting for their mixing tendency [36]. However, during the transition between the normal liquid state and the undercooled liquid state the angle between $\boldsymbol{R}_{12}(t)$ and $\boldsymbol{J}_c(0)$ may principally transform from and obtuse angle into an acute angle. This can be caused by the short-range atomic ordering which becomes unfavourable in the undercooled liquid state of the binary melt. This change advocates behaviour of the undercooled binary mixing melt towards a more thermodynamically stable state. Consequently, we can observe $W_{12} > 0$ in the undercooled liquid state of a

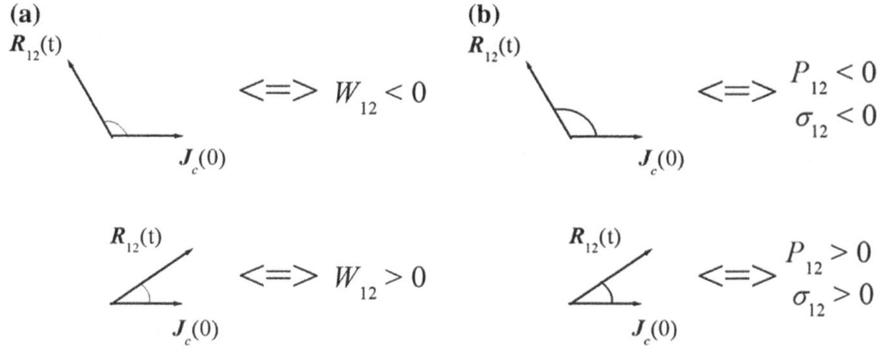

Fig. 7.1 Evaluation of correlation between fluctuations of $\boldsymbol{R}_{12}(t)$ and $\boldsymbol{J}_c(0)$ as **a** the time average over correlation length W_{12} and **b** in short time limit $t \to 0$ for P_{12} (develops over time) and its initial instant σ_{12}

binary mixing melt exhibiting chemical ordering. Therefore, we may conclude that $\sigma_{12} < 0$ and $W_{12} < 0$ can be considered as necessary conditions for a binary mixing melt exhibiting chemical ordering to be in thermodynamic equilibrium. In addition, a relative change in W_{12} can be used to characterize variation in dynamical stability of atomic ordering in a binary mixing melt upon undercooling.

7.3 Results and Discussion

7.3.1 Frequency-Dependent Mass Transport Coefficients in the Hydrodynamic Limit

The most practical case for frequency-dependent mass transport coefficients is when $\omega \to 0$, i.e. $t \to \infty$, so called case of the hydrodynamic limit. In other words, we consequently leave out the frequency dependence in the notation. In this case (7.24) for the Onsager coefficient $\tilde{L}_{cc}$ becomes

$$\tilde{L}_{cc} = \frac{m^2 D_1 D_2}{c_1 m_1^2 D_1 + c_2 m_2^2 D_2} \left[1 + \frac{W_{12}}{k_B T} \right] \tag{7.32}$$

and can be written with a sequence of three nested equations:

$$\tilde{L}_{cc} = S(c_2 D_1 + c_1 D_2), \tag{7.33}$$

where

$$S = S_0\left(1 + \frac{W_{12}}{k_B T}\right) \tag{7.34}$$

with

$$S_0 = \frac{m^2 D_1 D_2}{m^2 D_1 D_2 + c_1 c_2 (m_1 D_1 - m_2 D_2)^2}. \tag{7.35}$$

From the obtained equations for pure components, i.e. for $c_1 = 0$ or $c_2 = 0$, we get $S_0 = 1$ as follows from (7.35) and $W_{12} = 0$ and $S = 1$, since $\boldsymbol{R}_{12} = 0$, as follows from (7.15), (7.19), (7.23) and (7.28). Also from (7.33) on the limit when $c_1 \to 0$, then $\tilde{L}_{cc} \to D_1$ and if $c_2 \to 0$, then $\tilde{L}_{cc} \to D_2$. Moreover, in these cases the interdiffusion coefficient $D_c \to \tilde{L}_{cc}$, because $\Phi \to 1$ (7.6). Hence, within mentioned limits the coefficient of interdiffusion is approaching the coefficient of self-diffusion of the minority species, so that $D_c \to \tilde{L}_{cc} \to D_1$ when $c_1 \to 0$ and $D_c \to \tilde{L}_{cc} \to D_2$ when $c_2 \to 0$.

7.3.2 Decomposition of the Correction Factor

In addition, further investigation of (7.32)–(7.35) provides new insight into a relation of self and collective diffusion in binary melts. It follows from (7.28) that $W_{12} \equiv W_{12}(0)$ is associated to the average amount of generated–dissipated energy owed to the correspondences between fluctuations of $\boldsymbol{R}_{12}$ and $\boldsymbol{J}_c$. Therefore in thermal equilibrium, the absolute value of W_{12} cannot surpass the characteristic thermal energy, $k_B T$, by equipartition law [36], namely: $|W_{12}| \leq k_B T$.

Next, it follows from (7.35) that for all binary alloys: $0 < S_0 \leq 1$. This means that for the correction factor S: $0 \leq S \leq 2S_0$ or simply, $0 \leq S \leq 2$. Consequently, the correction factor S appears to break down into the product of two factors, see (7.34). In binary melts for which a greater tracer diffusion coefficient is carried by the species with greater mass, it can become significant, i.e. substantially less than unity.

The first factor in (7.34) S_0 can be also be written in form of the ratio of the self-diffusion coefficients D_1/D_2, the ratio of the atomic masses m_1/m_2, and the composition c_1 (or c_2). Indeed, if we divide numerator and denominator of (7.35) by $m^2 D_1 D_2$ and rearrange we will get:

$$
\begin{aligned}
S_0 &= \frac{1}{1+\frac{c_1 c_2}{m^2 D_1 D_2}(m_1 D_1 - m_2 D_2)^2} \\
&= \frac{1}{1+\frac{c_1 c_2}{m^2 D_1 D_2}\left(m_1^2 D_1^2 - 2m_1 D_1 m_2 D_2 + m_2^2 D_2^2\right)} \\
&= \frac{1}{1+c_1 c_2\left(\frac{m_1^2}{m^2}\frac{D_1}{D_2} - \frac{2m_1 m_2}{m^2} + \frac{m_2^2}{m^2}\frac{D_2}{D_1}\right)}
\end{aligned}
\tag{7.36}
$$

where $m = m_1 c_1 + m_2 c_2$.

Meanwhile, the second factor in the (7.34) $(1 + W_{12}/k_B T)$ is responsible for a collective energy generation-dissipation effect as a result of the correlations between fluctuations of the interdiffusion flux $\boldsymbol{J}_c$ and the force $\boldsymbol{R}_{12}$, introduced by the average random accelerations $\bar{\boldsymbol{R}}_\alpha$ as stated by (7.19)–(7.20).

For binary mixing melts showing tendency of ordering (Ni–Al, Ni–Zr) the angle between vectors $\boldsymbol{R}_{12}(t)$ and $\boldsymbol{J}_c(0)$ should be obtuse as predicted in Sect. 7.2.7. In other words, during fluctuations the force $\boldsymbol{R}_{12}$ effectively supresses the interdiffusion flux's deviation from equilibrium. Accordingly, in this case W_{12} is negative as well as the enthalpy of formation. According to (7.34), when $W_{12} < 0$, we get $S < 1$.

On the other hand, for binary melts with demixing tendency (Cu–Ag), it is predicted that the angle between $\boldsymbol{R}_{12}(t)$ and $\boldsymbol{J}_c(0)$, should be acute. In this case during fluctuations $\boldsymbol{R}_{12}$ effectively stimulates interdiffusion flux deviation from equilibrium. Consequently, in this case we can predict that W_{12} is positive. For $W_{12} > 0$, $S > 1$ according to (7.34).

Hence, in general we can conclude W_{12} is connected with the free energy of formation of binary melts. Furthermore, W_{12} can be used to introduce the concept of random of a binary liquid random alloy, where $W_{12} = 0$, hence $S = S_0 \leq 1$.

7.3.3 Composition Dependence of S, S_0 and $W_{12}/k_B T$ for Ni–Al, Ni–Zr and Cu–Ag Systems: Molecular Dynamics, Theoretical Predictions and Experimental Data

For illustration, we show in Figs. 7.2, 7.3 and 7.4 the composition dependencies of S, S_0 and $W_{12}/k_B T$ for the different types of systems, namely: with an ordering tendency (a) Ni–Al and (b) Ni–Zr and with a demixing tendency (c) Cu–Ag, respectively.

The melts are MD models of their corresponding systems, calculated at various compositions and temperatures range from 1000–2200 K using the embedded-atom method (EAM) potential developed in [37]. For calculation of S and S_0 based on (7.33) and (7.35), respectively, we made use of our MD data for $\tilde{L}_{cc}$, D_1 and D_2.

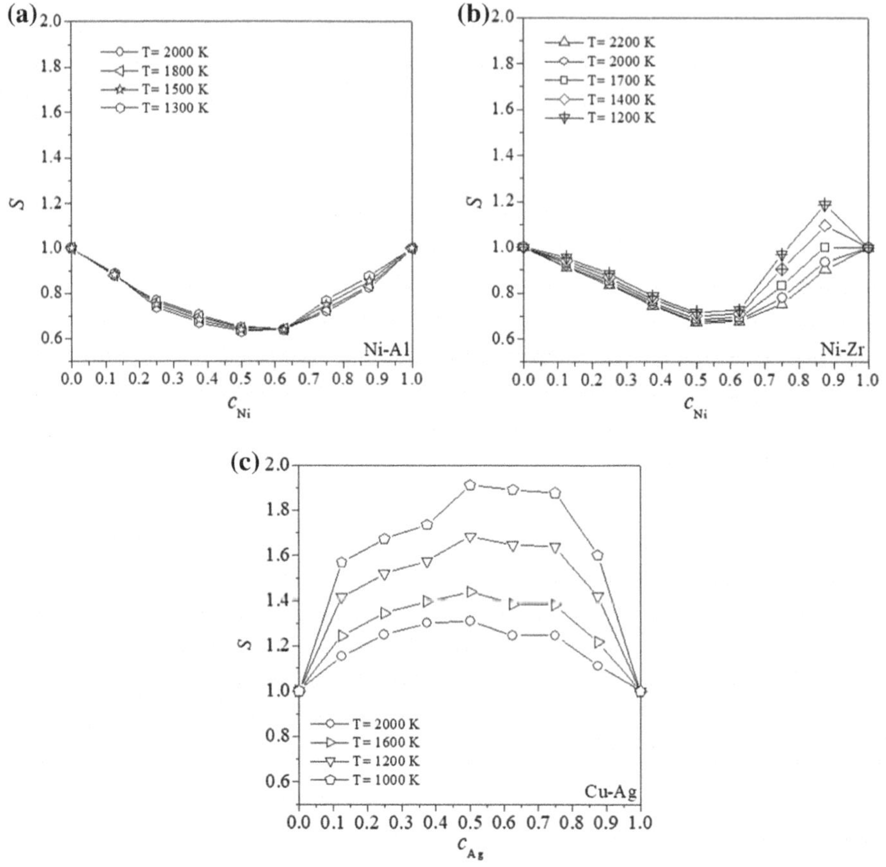

Fig. 7.2 Composition dependence of the correction factor $S = \frac{\tilde{L}_{cc}}{c_2 D_1 + c_1 D_2}$ for **a** Ni–Al melts, **b** Ni–Zr melts and **c** Cu–Ag melts at various temperatures

Meanwhile, $W_{12}/k_B T$ was calculated based on (7.34) using the results for S and S_0, i.e.: $\frac{W_{12}}{k_B T} = \frac{S}{S_0} - 1$.

The Onsager coefficient $\tilde{L}_{cc}$ was evaluated within the framework of the Green-Kubo formalism [21, 32, 35, 38] by means of the time integral of $\varepsilon_c(t)$ (see (7.27) at $\omega = 0$), namely $\tilde{L}_{cc} = \frac{k_B T}{m_c} \lim_{t \to \infty} \int_0^t \frac{\langle \boldsymbol{J}_c(t') \boldsymbol{J}_c(0) \rangle}{\langle \boldsymbol{J}_c^2 \rangle} \mathrm{d}t'$.

The self-diffusion coefficients of Ni and Al atoms [39], Ni and Zr atoms in [36] and Cu and Ag atoms were calculated, according to the well-known Einstein relation [21, 32, 35], as: $D_\alpha = \lim_{t \to \infty} \frac{\left\langle \frac{1}{N_\alpha} \sum_{i=1}^{N_\alpha} \Delta \boldsymbol{r}_{\alpha i}^2(t) \right\rangle}{6t}$, where $\Delta \boldsymbol{r}_{\alpha i}(t) = \boldsymbol{r}_{\alpha i}(t) - \boldsymbol{r}_{\alpha i}(0)$ represent the time-displacements of single atoms of species 1 ($i = 1 \ldots N_1$) and 2 ($j = 1 \ldots N_2$), respectively (we assume here that $\mathrm{Ni, Cu} \equiv 1$ and $\mathrm{Al, Zr, Ag} \equiv 2$).

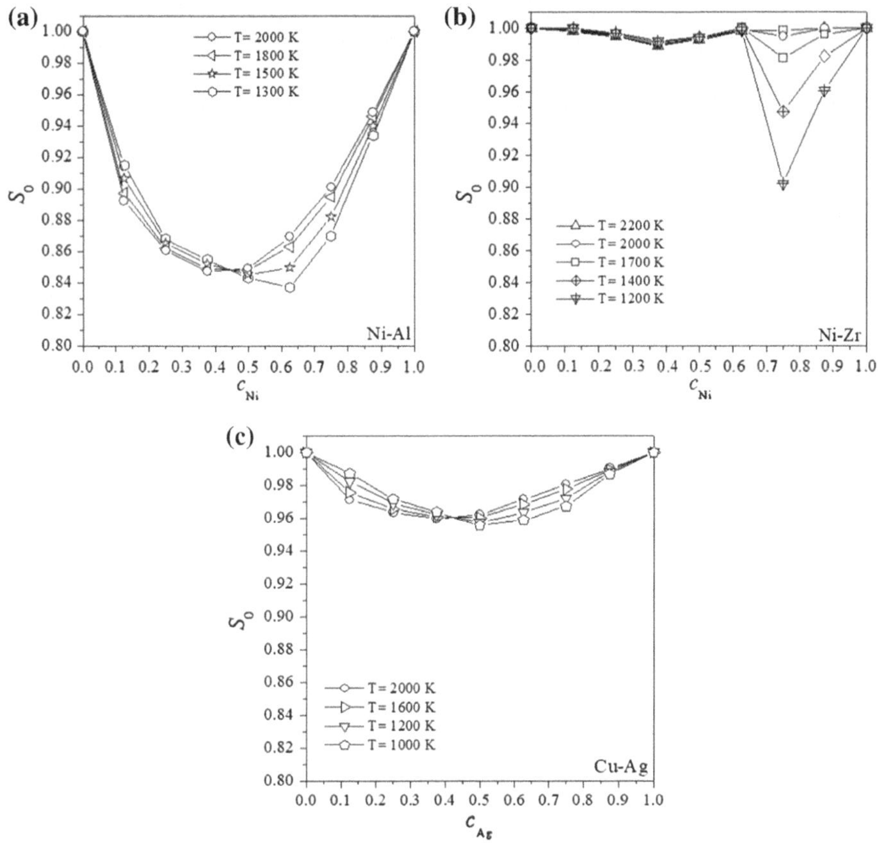

Fig. 7.3 Composition dependence of $S_0 = \frac{m^2 D_1 D_2}{m^2 D_1 D_2 + c_1 c_2 (m_1 D_1 - m_2 D_2)^2}$ for **a** Ni–Al melts, **b** Ni–Zr melts and **c** Cu–Ag melts at various temperatures

We took into consideration seven alloy compositions in Ni–Al system for example: $Ni_{87.5}Al_{12.5}$, $Ni_{75}Al_{25}$, $Ni_{62.5}Al_{37.5}$, $Ni_{50}Al_{50}$, $Ni_{37.5}Al_{62.5}$, $Ni_{25}Al_{75}$ and $Ni_{12.5}Al_{87.5}$. We also considered seven alloy compositions across the Ni–Zr and Cu–Ag systems correspondingly in the similar manner. Calculations were performed in 3D simulation cells with periodic boundary conditions in all three directions using microcanonical ensemble dynamics at zero pressure. The simulation cells consist of about 4000 atoms (for different models the number of atoms varied slightly within the range 4000–4394). For numerical integration of the equations of motion we apply the Verlet algorithm in the velocity form [40] with a time step Δt = 1.5 fs. To ensure the zero-momentum reference frame, we conserved the total momentum of the model systems at a zero value. To obtain the statistical time average at thermal equilibrium of the autocorrelation function of the

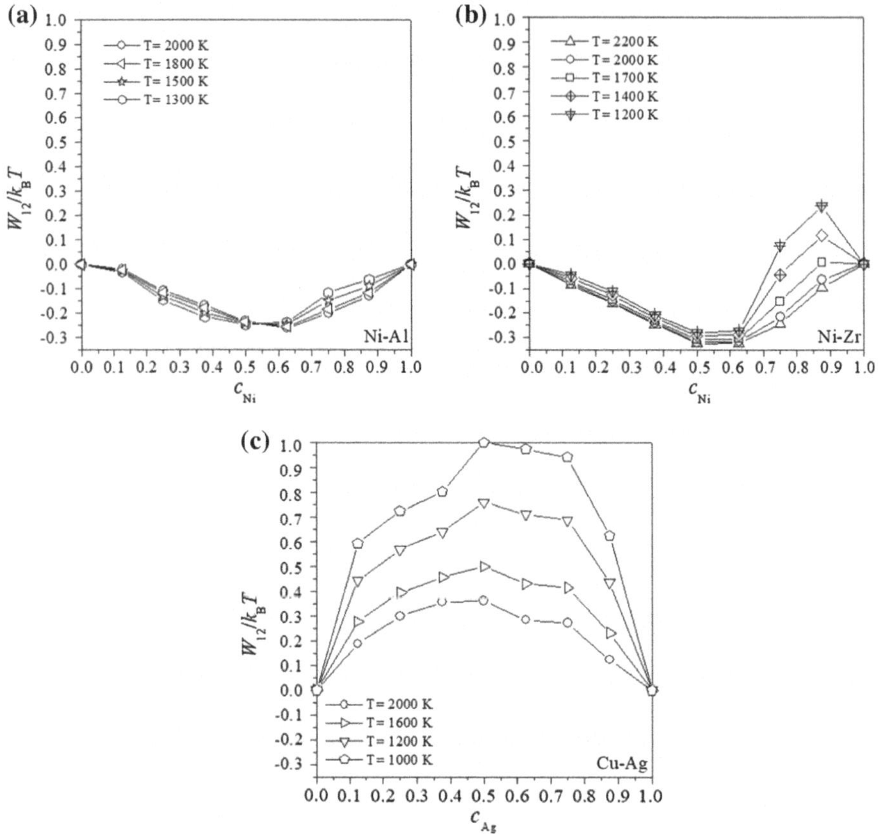

Fig. 7.4 Composition dependence of $\frac{W_{12}}{k_B T} = \frac{S}{S_0} - 1$ for **a** Ni–Al melts, **b** Ni–Zr melts and **c** Cu–Ag melts at various temperatures

interdiffusion flux and the mean-squared displacements of atoms, we use about 1.8×10^7 time origins.

As it is proposed by theoretical treatment above for a binary mixing melt with tendency of ordering, the correction factor S is found to be less than unity $(W_{12} < 0)$ across the whole composition range of the MD models of Ni–Al binary melts (see Fig. 7.2a). The correction factor S of Ni–Zr binary melts (see Fig. 7.2b) shows analogous trends to those of Ni–Al melts at high temperatures i.e. the normal liquid state. In agreement with the approach that, upon increasing concentration of the minority species toward the equi-atomic composition, the kinetics of collective diffusion in mixing binary liquid alloys slow down, minima of S in the composition dependence of Ni–Al and Ni–Zr melts are found to be of similar position and depth.

Confirming our theoretical analysis, the composition dependence of the correction factor S of Cu–Ag binary melts (see Fig. 7.2c), serving as an example for a binary system with demixing tendency, clearly shows a reversed behaviour of the graph, compared to Figs. 7.2a, b. The value of $S \geq 1$ over the whole composition range, reaching its maximum of 1.91 for the composition $Cu_{50}Ag_{50}$ at a temperature of 1000 K. The maximum sort of plateaus towards the Ag-rich side of the Cu–Ag alloys ranging from $Cu_{50}Ag_{50}$ to $Cu_{12.5}Ag_{87.5}$.

The composition dependence of S_0 of Ni–Zr melts (see Fig. 7.3b), despite also showcasing a system with mixing tendency, looks inarguably different to the Ni–Al melt (see Fig. 7.3a). In the case of Ni and Zr, the larger self-diffusion coefficients is carried by the lighter Ni species ($m_{\mathrm{Ni}}/m_{\mathrm{Zr}} \approx 58.69/91.22 \approx 0.64$). This is in contrast to the situation of the Ni–Al melt, where the larger self-diffusion coefficients is also carried by the Ni species, which in this configuration is the heavier ($m_{\mathrm{Ni}}/m_{\mathrm{Al}} \approx 58.69/26.98 \approx 2.18$). The ratio of the masses of Ni and Zr ($m_{\mathrm{Ni}}/m_{\mathrm{Zr}} \approx 0.64$) effectively damp the single-particle kinetic effect due to the ratio $D_{\mathrm{Ni}}/D_{\mathrm{Zr}}$ above unity (varying between 1.2 and 1.8 at high temperatures [36]) resulting in $S_0 \approx 1$ covering the whole composition range. In the case of the configuration of Ni–Al melts, the mass ratio $m_{\mathrm{Ni}}/m_{\mathrm{Al}} \approx 2.18$ magnifies the single-particle kinetic effects due to an increase of the ratio $D_{\mathrm{Ni}}/D_{\mathrm{Al}}$ above unity even further, resulting in a reduction of S_0 below unity at all compositions (see Fig. 7.3a). The minimum of S_0 is reached with a value of 0.85 for the models of Ni–Al melts in the normal liquid state in the vicinity of equi-atomic composition. For the Ni-rich models in Ni–Zr melts (see Fig. 7.3b) a considerable drop of S_0 can be seen upon undercooling. This observation is the result of the increasing value of the ratio of the self-diffusion coefficients ($D_{\mathrm{Ni}}/D_{\mathrm{Zr}}$) for the model of $Ni_{87.5}Zr_{12.5}$ melt at 1200 K [36]. This increase overcomes the damping effect of the mass ratio $m_{\mathrm{Ni}}/m_{\mathrm{Zr}} \approx 0.64$ and therefore producing a sizable drop of S_0 as can be seen on Fig. 7.3b.

According to Fig. 7.4a it can be seen that for Al-rich alloy compositions with $c_{\mathrm{Ni}} \lesssim 0.125$, the term $|W_{12}|/k_{\mathrm{B}}T \lesssim 0.03$, meaning that the correction factor S is less than 3% smaller than S_0 (see Figs. 7.2a and 7.3a respectively). Thus, an Al-rich liquid alloy (with $c_{\mathrm{Ni}} \lesssim 0.125$) of the model system can be considered as a good approximation for the suggested concept of a binary liquid random alloy. Consequently, by assuming that a comparable condition $S \approx S_0$ is satisfied for real Al-rich liquid alloys of Ni–Al binary system, recent experimental data on the correction factor in Al-rich Ni–Al melts [17] can be analysed by making use of (7.33) and (7.35).

In the models of Ni–Zr melts the trend of the composition dependence of S follows, shifted up by one, the trend of the composition dependence of $W_{12}/k_{\mathrm{B}}T$. This is in contrast to the models of Ni–Al melts, for which the composition dependence of S_0 and $(1 + W_{12}/k_{\mathrm{B}}T)$ as contribution of S are found to be similar. During undercooling of the Ni–Zr melts the effect of S_0 becomes notable. The main difference however, upon undercooling the sign of $W_{12}/k_{\mathrm{B}}T$ becomes positive for the Ni-rich and Zr-rich Ni–Zr melts (see Fig. 7.4b). The strongest increase of $W_{12}/k_{\mathrm{B}}T$ upon undercooling is observed in the models of Ni-rich Ni–Zr melts (see Fig. 7.4b), where $W_{12}/k_{\mathrm{B}}T$ reaches 0.48 and 0.08, respectively, in the models of $Ni_{87.5}Zr_{12.5}$ and $Ni_{12.5}Zr_{87.5}$ melts at 1200 K just before the onset of their crystallization at 1150 K. This behaviour is interpreted as an expression of emerging heterogeneity in atomic dynamics of melt due to entering the energy landscape controlled regime upon undercooling derived from a correlated behaviour of W_{12}/k_BT and the ratio D_{Ni}/D_{Zr}.

Though, a noteworthy decoupling of the self-diffusion coefficients of Ni and Zr in the models of Ni-rich Ni–Zr melts upon undercooling marks in a decrease of factor S_0 (see Fig. 7.3b) that accounts for the contribution into the correction factor S due to single-particle kinetic effects. At the same time, the contribution into S due to a collective energy generation-dissipation effect accounted by W_{12}/k_BT is amplified upon transition to diffusion behaviour approaching the energy landscape controlled regime. The variation of the latter contribution significantly controls the variation of the former contribution causing substantial increase in the correction factor S above unity for the models of Ni-rich Ni–Zr melts upon undercooling.

As for a binary system with de-mixing tendency, according to the results of our MD simulations of Cu-Ag melts with the EAM potential developed in [41], the behaviour of $W_{12} > 0$ over the whole composition range (see Fig. 7.4c) as in contrast to $W_{12} < 0$ for systems with mixing tendency like the melts of Ni–Al and Ni–Zr, Fig. 7.1a, b. Moreover, the value of W_{12}/k_BT nears practically unity in the vicinity of the eutectic point of the model system with demixing tendency. Consequently, the correction factor S in the vicinity of the eutectic point is expected to be slightly less than two in real Cu–Ag melts with a reasonable approximation that $D_{Cu}/D_{Ag} \sim 1$ (recall that $m_{Cu}/m_{Ag} \approx 63.55/107.87 \approx 0.59$).

To the best of our knowledge, the shown experimental data are the only available data on the correction factor in binary melts available in literature at the present time. The self-diffusion coefficients of Al, D_{Al} in Ni–Al cannot be measured [12, 13, 17] because of the very low coherent scattering cross section of Al, as well as lack of suitable isotopes. Therefore it was first assumed in [17] that $D_{Al} \approx D_{Ni}$. Next, with this assumption a quantity $S_1 = \tilde{L}_{cc}/D_1$ (we use our notation) was assessed on account of experimental measurements [17]. In fact, as it can be seen from (7.33), $S_1 = S$ if $D_1 = D_2$. Though, (7.33) and (7.35) propose (assuming $S \approx S_0$) that both $S_1 = S(c_2 + c_1 D_2/D_1)$ and S may differ noticeably when the ratio D_1/D_2 reaches a value above unity (note that for Ni–Al system: $m_1/m_2 \approx 58.71/26.98 \approx 2.18$), even already for sufficiently small values of c_1 (Al-rich alloy compositions). In Fig. 7.5, alongside experimental data for S_1 (at 1173 K) [17], we present the predicted composition dependence based on (7.33) and (7.35) in Al-rich Ni–Al melts for: (i) S at fixed $D_1/D_2 = 1$, and (ii) S and S_1 at fixed $D_1/D_2 = 2$. One can clearly see in Fig. 7.5, a good agreement of the predicted results and the experimental results for the composition dependence of S_1 predicted based on our theoretical consideration at fixed $D_1/D_2 = 2$. Now also taking into account our main supposition $S \approx S_0$ for Al-rich Ni–Al melts, on the basis of the results of MD simulations of Ni–Al melts, we should indicate that $D_1/D_2 \sim 2$, the ratio of the tracer diffusion coefficients in real Al-rich Ni–Al melts.

We can see from Fig. 7.4c that at 50% of Ag concentration W_{12} reaches its theoretical upper limit, namely it can not exceed k_BT. In other words, at about eutectic composition W_{12} pushes system to decomposition. Indeed, on Fig. 7.6 we

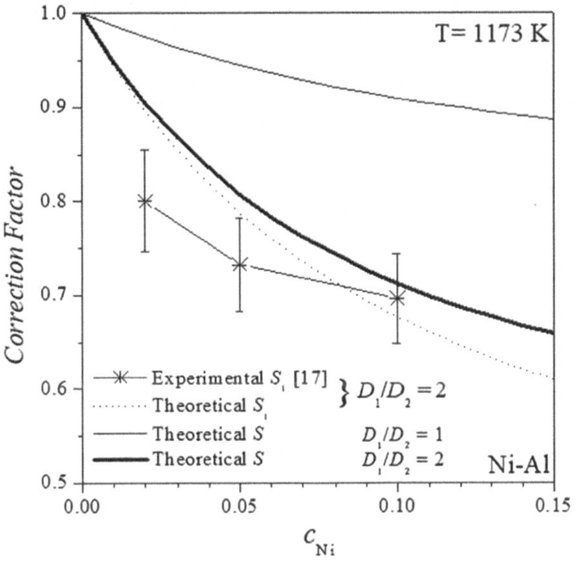

Fig. 7.5 Composition dependence of the correction factor in Al-rich Ni–Al liquid alloy. The symbols and the dashed line represent, respectively, experimental (at 1173 K) [17] and theoretical (at fixed $D_1/D_2 = 2$) data for $S_1 = \tilde{L}_{cc}/D_1$. The thin and thick solid lines represent the theoretical data for S at fixed $D_1/D_2 = 1$ and $D_1/D_2 = 2$, respectively

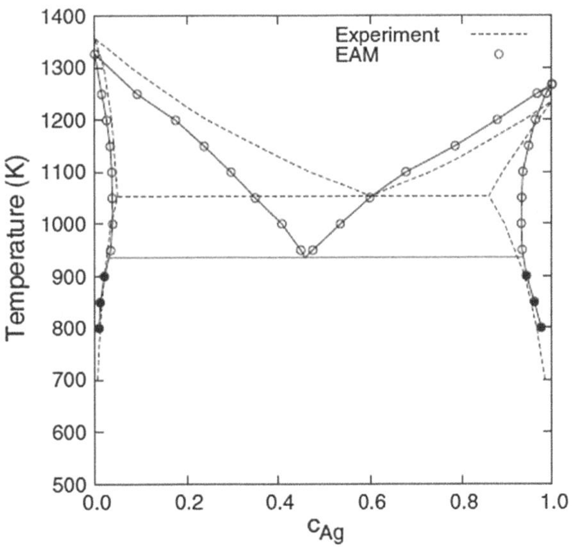

Fig. 7.6 Phase diagram of the Cu–Ag system. Filled symbols representing calculations using a thermodynamic model; open symbols where attained from the solid-melt interface simulations [41]

can see the phase separation at both EAM calculations and experimental data for Cu–Ag melt [41].

Comprehensively, the presented theoretical treatment is a significant step to help analysing the relationship between the kinetic part of single-particle and collective diffusion dynamics in a binary melt which is not far from its equilibrium state. However, we remind that the term $W_{12}/k_B T$ is expected to deviate from its equilibrium range in the vicinity of a phase transition in the liquid state, as in cases of liquid-liquid demixing in addition to upon undercooling below equilibrium melting temperature. Additionally, the self-diffusion coefficients contributing to S_0 may give a temperature dependence to $\tilde{L}_{cc}$ that is even more complex in the undercooled liquid state. The self-diffusion coefficients in this state are rapidly reduced compared to higher temperatures. For future work it would therefore be compelling to study the temperature dependence of relative magnitude of these two contributions, $W_{12}/k_B T$ and S_0, into $\tilde{L}_{cc}$ in the undercooled liquid states of different binary alloys.

7.3.4 Analogy with the Kirkwood-Buff Solution Theory

Finally, we recall that the Kirkwood-Buff solution theory [42] gives a significant link between the thermodynamic factor Φ and the partial pair distribution functions $g_{\alpha\beta}(r)$ of species α and β ($\alpha, \beta = 1, 2$) in a binary melt, as:

$$\Phi = \frac{c_1 c_2}{k_B T}\left(\frac{\partial^2 (G/N)}{\partial c_\alpha^2}\right)_{T,P} = \frac{c_\alpha}{k_B T}\left(\frac{\partial \mu_\alpha}{\partial c_\alpha}\right)_{T,P} = \frac{1}{1 + c_1 c_2 (I_{11} + I_{22} - 2I_{12})}, \tag{7.37}$$

where μ_α denotes the chemical potential of species α, and

$$I_{\alpha\beta} = 4\pi \frac{N}{V} \int_0^\infty \left[g_{\alpha\beta}(r) - 1\right] r^2 \mathrm{d}r \tag{7.38}$$

denotes the so-called Kirkwood-Buff integrals. Physically, $\frac{N}{V} c_\beta g_{\alpha\beta}(r)$ is the average number density of atoms of species β at a distance r from an atom of species α fixed at the centre. Now, by taking (7.6), (7.33)–(7.35) and (7.37) into consideration, then in a binary melt the interdiffusion coefficient, D_c, can now be expressed as:

$$\begin{aligned} D_c &= \Phi \tilde{L}_{cc} \\ &= \frac{\left(1 + \frac{W_{12}}{k_B T}\right)(c_2 D_1 + c_1 D_2)}{\left[1 + c_1 c_2 (I_{11} + I_{22} - 2I_{12})\right]\left[1 + c_1 c_2 \frac{m_1^2 D_1^2 + m_2^2 D_2^2 - 2 m_1 D_1 m_2 D_2}{m^2 D_1 D_2}\right]}. \end{aligned} \tag{7.39}$$

By comparing the algebraic structures of the Kirkwood-Buff expression for the thermodynamic factor Φ and our expression for S_0 a close affinity between them can be seen. Actually, the Kirkwood-Buff expression for the thermodynamic factor Φ gives quantification of the thermodynamic effect caused by deviation from random mixing on collective diffusion in terms of local structuring or rather chemical ordering as $I_{11} + I_{22} - 2I_{12}$ (see (7.38)). Where local structuring describes the relative preference of a species to be surrounded by another species. The decay of the excess or deficiency of atoms of species β around atoms of species α, where the distance is expressed through space integrals of the partial pair correlation functions $g_{\alpha\beta}(r) - 1$ are relevant in regards to the thermodynamic factor Φ. In a similar manner quantifies our expression for S_0 the kinetic effect on collective diffusion in terms of the relative difference in the decay of single-particle dynamics of different species with time phrased via the time integrals of the velocity autocorrelation functions of different species, as $\frac{m_1^2 D_1^2 + m_2^2 D_2^2 - 2m_1 D_1 m_2 D_2}{m^2 D_1 D_2}$ (see (7.25) and (7.26)). Consequently, $S_0 = \left[1 + c_1 c_2 \frac{m_1^2 D_1^2 + m_2^2 D_2^2 - 2m_1 D_1 m_2 D_2}{m^2 D_1 D_2}\right]^{-1}$ expresses the kinetic factor of the interdiffusion coefficient given by (7.39). Eventually, in binary melts, $\frac{W_{12}}{k_B T}$ accounts for cross-coupling between thermodynamic and collective kinetic effects in the interdiffusion processes. Indeed, it captures, according to (7.28), the decay peculiarities of both space and time dependent correlations in dynamics of species in a binary melt. Nevertheless, further and extensive study of the properties of the factor $\frac{W_{12}}{k_B T}$ are needed to propose any more quantitative claims.

7.4 Conclusion

Analysis of diffusion kinetics in binary melts with mixing tendency as well as demixing tendency were carried out in the framework of the Mori-Zwanzig formalism. For the first time, the Onsager coefficient for mass transport has been related to the two self-diffusion coefficients of species in binary melt through an analytical expression. The expression serves as an exceptional alternative option to the well-known Darken equation. We presented the derived expression that includes the correction factor S to the Darken equation.

Additionally, the correction factor S appears to break down into the product of two other factors as: $S = S_0(1 + W_{12}/k_B T)$. The first factor $S_0 \leq 1$ is expressed in terms of the ratio of the tracer diffusion coefficients D_1/D_2, the ratio of the atomic masses m_1/m_2, and the alloy composition. The second factor $(1 + W_{12}/k_B T)$ is related to a collective energy generation-dissipation effect (term $W_{12}/k_B T$) due to the correlations between fluctuations of the interdiffusion flux and the force caused by the difference in the average random accelerations of atoms of different species. Evaluation of the recognizable different behaviour of the correction factor S for melts in regards to their tendency of mixing shows that for binary mixing melts exhibiting mixing tendency (i.e. Ni–Al and Ni–Zr melts) the correction factor should typically be $S < S_0$ $(W_{12} < 0)$, while for binary melts where precursors of liquid–liquid demixing are important (such as Cu–Ag melts) the correction factor should be $S > S_0$ $(W_{12} > 0)$. In the case of thermal equilibrium it was pointed out that for the correction factor $0 \leq S \leq 2S_0$, due to a constraining effect of the energy of thermal fluctuations $(|W_{12}| \leq k_B T)$.

In addition, the extensive results of this study can be utilized to establish a concept of a binary liquid random alloy for which $W_{12} = 0$, so that the correction factor $S = S_0$. A systematic comparison of results of binary systems with mixing tendency i.e. Ni–Al melts and Ni–Zr melts, as well as demixing tendency i.e. Cu–Ag, which has shown opposite behaviour of the correction factor S and its contributions of S_0 and W_{12}/k_BT. Concluded with $S < S_0$ $(W_{12} < 0)$ for binary mixing melts with chemical ordering and $S > S_0$ $(W_{12} > 0)$ for binary demixing melts.

Lastly, this study and the related theory significantly supports analysis for better, more in-depth interpretation of recent experimental data [17] on the correction factor to the Darken equation in Al-rich Ni–Al melts.

Acknowledgements This research was supported by the Australian Research Council through its Discovery Project Grants Scheme and Priority Research Center for Computer-Assisted Research Mathematics and Its Applications at the University of Newcastle.

References

1. S.W. Basuki, A. Bartsch, F. Yang, K. Ratzke, A. Meyer, F. Faupel, Decoupling of component diffusion in a glass-forming $Zr_{46.75}$ $Ti_{8.25}$ $Cu_{7.5}$ Ni_{10} $Be_{27.5}$ melt far above the liquidus temperature. Phys. Rev. Lett. **113**(16), 165901 (2014). https://doi.org/10.1103/physrevlett.113.165901
2. K. Georgarakis, A.R. Yavari, M. Aljerf, D.V. Louzguine-Luzgin, M. Stoica, G. Vaughan, A. Inoue, On the atomic structure of Zr–Ni and Zr–Ni–Al metallic glasses. J. Appl. Phys. **108**(2) (2010). https://doi.org/10.1063/1.3446131
3. R. Kobold, W.W. Kuang, H. Wang, W. Hornfeck, M. Kolbe, D.M. Herlach, Dendrite growth velocity in the undercooled melt of glass forming Ni_{50} Zr_{50} compound. Philos. Mag. Lett. **97** (6), 249–256 (2017). https://doi.org/10.1080/09500839.2017.1330561
4. T. Voigtmann, A. Meyer, D. Holland-Moritz, S. Stüber, Hansen, T., Unruh, T.: Atomic diffusion mechanisms in a binary metallic melt. EPL (Europhys Lett) **82**(6) (2008). https://doi.org/10.1209/0295-5075/82/66001
5. S.K. Yadav, S. Lamichhane, L.N. Jha, N.P. Adhikari, D. Adhikari, Mixing behaviour of Ni–Al melt at 1873 K. Phys. Chem. Liq. **54**(3), 370–383 (2016). https://doi.org/10.1080/00319104.2015.1095640
6. O. Bystrenko, V. Kartuzov, Interface structure and contact melting in AgCu eutectic. A molecular dynamics study. Mater. Res. Express **4**(12), 126503 (2017). https://doi.org/10.1088/2053-1591/aa9b63
7. V. Olufemi Ogunrinde, Y. Odusote, A.I. Popoola, Correlation between thermodynamic and surface properties of liquid Ag–Cu and Cd–Ga binary alloys. J. Sci. Technol. Res. **1**, 1–13 (2019)
8. T. Böhme, Mechanical and thermodynamic materials Properties derived by semi-empirical atomic potentials with special focus on Ag, Cu, and the binary alloy Ag–Cu, in *New Achievements in Continuum Mechanics and Thermodynamics: A Tribute to Wolfgang H. Müller*, ed. by B.E. Abali, H. Altenbach, F. dell'Isola, V.A. Eremeyev, A. Öchsner (Springer International Publishing, Cham, 2019), pp. 51–70
9. W.J. Boettinger, J.A. Warren, C. Beckermann, A. Karma, Phase-field simulation of solidification. Ann. Rev. Mater. Res. **32**(1), 163–194 (2002). https://doi.org/10.1146/annurev.matsci.32.101901.155803
10. M. Asta, C. Beckermann, A. Karma, W. Kurz, R. Napolitano, M. Plapp, G. Purdy, M. Rappaz, R. Trivedi, Solidification microstructures and solid-state parallels: recent developments, future directions. Acta Materialia **57** (2009)

11. A. Meyer, Atomic transport in dense multicomponent metallic liquids. Phys. Rev. B **66**(13), 134205 (2002). https://doi.org/10.1103/PhysRevB.66.134205
12. J. Horbach, A. Griesche, M.-P. Macht, G. Frohberg, A. Meyer, Self-diffusion and interdiffusion in Al_{80} Ni_{20} melts: simulation and experiment. Phys. Rev. **75**, 1–8 (2007)
13. S. Stüber, D. Holland-Moritz, T. Unruh, A. Meyer, Ni self-diffusion in refractory Al–Ni melts. Phys. Rev. B **81**(2) (2010). https://doi.org/10.1103/physrevb.81.024204
14. B. Zhang, A. Griesche, A. Meyer, Diffusion in Al–Cu melts studied by time-resolved X-ray radiography. Phys. Rev. Lett. **104**(3), 035902 (2010). https://doi.org/10.1103/PhysRevLett.104.035902
15. J. Brillo, A.I. Pommrich, A. Meyer, Relation between self-diffusion and viscosity in dense liquids: new experimental results from electrostatic levitation. Phys. Rev. Lett. **107**(16), 165902 (2011). https://doi.org/10.1103/PhysRevLett.107.165902
16. F. Kargl, M. Engelhardt, F. Yang, H. Weis, P. Schmakat, B. Schillinger, A. Griesche, A. Meyer, In situ studies of mass transport in liquid alloys by means of neutron radiography. J. Phys.: Condens. Matter **23**(25), 254201 (2011). https://doi.org/10.1088/0953-8984/23/25/254201
17. E. Sondermann, F. Kargl, A. Meyer, Influence of cross correlations on interdiffusion in Al-rich Al–Ni melts. Phys. Rev. B **93**(18), 184201 (2016)
18. F. Kargl, E. Sondermann, H. Weis, A. Meyer, Impact of convective flow on long-capillary chemical diffusion studies of liquid binary alloys. High Temp. High Press. **42**, 3–21 (2013)
19. P. Kuhn, J. Horbach, F. Kargl, A. Meyer, T. Voigtmann, Diffusion and interdiffusion in binary metallic melts. Phys. Rev. B **90**(2) (2014). https://doi.org/10.1103/physrevb.90.024309
20. R. Zwanzig, *Nonequilibrium Statistical Mechanics* (Oxford University Press Inc., NY, 2001)
21. R. Kubo, The fluctuation-dissipation theorem. J. Rep. Prog. Phys. **29**, 255–284 (1966)
22. R. Zwanzig, Statistical Mechanics of Irreversibility. Lect. Theor. Phys. **3**, 106–141 (1961)
23. R. Zwanzig, Memory effects in irreversible thermodynamics. Phys. Rev. **124**(4), 983–992 (1961). https://doi.org/10.1103/PhysRev.124.983
24. H. Mori, Transport, collective motion, and Brownian motion. Prog. Teor. Phys. **33**, 423–455 (1965)
25. H. Mori, Continued fraction representation of the time correlation functions. Prog. Teor. Phys. **34**, 399–416 (1965)
26. L.S. Darken, Diffusion, mobility and their interrelation through free energy in binary metallic systems. Trans. AIME **175**, 184–194 (1948)
27. J.R. Manning, Diffusion in a chemical concentration gradient. Phys. Rev. **124**(2), 470–482 (1961). https://doi.org/10.1103/PhysRev.124.470
28. H. Mehrer, *Diffusion in Solids*. Springer, Berlin (2007). https://doi.org/10.1007/978-3-540-71488-0
29. L. Onsager, Reciprocal relations in irreversible processes. I. Phys. Rev. **37**(4), 405–426 (1931). https://doi.org/10.1103/PhysRev.37.405
30. L. Onsager, Reciprocal relations in irreversible processes. II. Phys. Rev. **38**(12), 2265–2279 (1931). https://doi.org/10.1103/PhysRev.38.2265
31. E.V. Levchenko, A.V. Evteev, T. Ahmed, A. Kromik, R. Kozubski, I.V. Belova, Z.-K. Liu, G.E. Murch, Influence of the interatomic potential on thermotransport in binary liquid alloys: case study on NiAl. Phil. Mag. **96**(29), 3054–3074 (2016). https://doi.org/10.1080/14786435.2016.1223893
32. J.P. Hansen, I.R. McDonald, Preface to the third edition. In: *Theory of Simple Liquids*, 3rd edn. (Academic Press, Burlington, 2006), p. v
33. A.V. Evteev, E.V. Levchenko, L. Momenzadeh, Y. Sohn, I.V. Belova, G.E. Murch, Molecular dynamics study of phonon-mediated thermal transport in a $Ni_{50}Al_{50}$ melt: case analysis of the influence of the process on the kinetics of solidification. Phil. Mag. **95**(1), 90–111 (2015). https://doi.org/10.1080/14786435.2014.984006
34. A.V. Evteev, E.V. Levchenko, I.V. Belova, R. Kozubski, Z.-K. Liu, G.E. Murch, Thermotransport in binary system: case study on $Ni_{50}Al_{50}$ melt. Phil. Mag. **94**(31), 3574–3602 (2014). https://doi.org/10.1080/14786435.2014.965236

35. R. Zwanzig, Time-correlation functions and transport coefficients in statistical mechanics. Annu. Rev. Phys. Chem. **16**, 67–102 (1965)
36. A. Kromik, E.V. Levchenko, C. Massobrio, A.V. Evteev, Diffusion in Ni–Zr melts: insights from statistical mechanics and atomistic modeling. Adv. Theory Simul. (1800109), 1800109-1800101-1800109-1800114 (2018). https://doi.org/10.1002/adts.201800109
37. G.P. Purja Pun, Y. Mishin, Development of an interatomic potential for the Ni–Al system. Phil. Mag. **89**(34–36), 3245–3267 (2009). https://doi.org/10.1080/14786430903258184
38. M.S. Green, Markoff random processes and the statistical mechanics of time-dependent phenomena. II. Irreversible processes in fluids. J. Chem. Phys. **22**(3), 398–413 (1954). https://doi.org/10.1063/1.1740082
39. E.V. Levchenko, T. Ahmed, A.V. Evteev, Composition dependence of diffusion and thermotransport in Ni–Al melts: a step towards molecular dynamics assisted databases. Acta Mater. **136**, 74–89 (2017). https://doi.org/10.1016/j.actamat.2017.06.056
40. L. Verlet, Computer "experiments" on classical fluids. I. Thermodynamical properties of Lennard-Jones molecules. Phys. Rev. **159**(1), 98–103 (1967). https://doi.org/10.1103/physrev.159.98
41. P.L. Williams, Y. Mishin, J.C. Hamilton, An embedded-atom potential for the Cu–Ag system. Modell. Simul. Mater. Sci. Eng. **14**(5), 817–833 (2006). https://doi.org/10.1088/0965-0393/14/5/002
42. J.G. Kirkwood, F.P. Buff, The statistical mechanical theory of solutions. I. J. Chem. Phys. **19** (6), 774–777 (1951). https://doi.org/10.1063/1.1748352

Chapter 8
Advanced Monte Carlo Simulations for Ion-Channeling Studies of Complex Defects in Crystals

Przemyslaw Jozwik, Lech Nowicki, Renata Ratajczak, Cyprian Mieszczynski, Anna Stonert, Andrzej Turos, Katharina Lorenz and Eduardo Alves

Abstract This chapter describes the most important features of computational software called 'McChasy', which is a Monte Carlo (MC) simulation code developed for the evaluation of the Rutherford Backscattering Spectrometry data, in particular, recorded in the channeling mode (RBS/C). RBS/C is an experimental technique used in the analysis of defects in single crystals. Lattice distortions affect materials modified by ion beams or exposed to irradiation. Therefore, the analysis of damage in crystals is of high importance in materials science. Various types of defects can be created due to the interaction of ions with targets. However, RBS/C has different sensitivity to each of them so the analytical analysis of experimental data is hardly possible. MC simulations are a powerful tool used to overcome this limitation. The McChasy code simulates the movement of light ions in crystals. The software provides a fitting procedure of RBS/C spectra based on independent depth profiles of different defect types: interstitials, edge dislocations, substitutions, stacking faults or grain boundaries. The code works well not only with materials containing complex defects but also with heterostructures and superlattices. Recent

---

Przemyslaw Jozwik: On leave from the Institute of Electronic Materials Technology, Warsaw, Poland.

P. Jozwik (✉) · K. Lorenz · E. Alves
IPFN, Instituto Superior Técnico, Universidade de Lisboa, Estrada Nacional 10, 2695-066 Bobadela, Portugal
e-mail: p.jozwik@ctn.tecnico.ulisboa.pt

L. Nowicki · R. Ratajczak · C. Mieszczynski · A. Stonert · A. Turos
National Centre for Nuclear Research, A. Soltana 7, 05-400 Otwock-Swierk, Poland

A. Turos
Institute of Electronic Materials Technology, Wolczynska 133, 01-919 Warsaw, Poland

K. Lorenz
Instituto de Engenharia de Sistemas de Computadores-Microsistemas e Nanotecnologias (INESC-MN), Rua Alves Redol 9, 1000-029 Lisbon, Portugal

E. V. Levchenko et al. (eds.), *Theory and Simulation in Physics for Materials Applications*, Springer Series in Materials Science 296,
https://doi.org/10.1007/978-3-030-37790-8_8

improvements of the code include a unique approach of 3D-interaction between ions and target atoms. Application of the McChasy code in the analysis of crystal defects is described and possible ways of its further development are pointed out.

8.1 Introduction

8.1.1 Interaction of Ion Beams with Materials

Ion beams are a powerful tool widely used in materials science. Their practical application depends primarily on energy and atomic number Z of ions [1]:

(i) ion deposition (~10–100 eV): ions can rest on a substrate surface and grow into an epitaxial layer;
(ii) sputtering (~1 keV heavy ions): by absorbing energy from ions, some target atoms localized on the surface or near the surface can be ejected;
(iii) ion implantation (~100–350 keV): ions are injected into a substrate up to low concentrations in order to modify its properties;
(iv) ion beam analysis (light ions of the order of MeV): ions penetrate the material and interact with the target giving rise to a family of different techniques. In the Rutherford Backscattering Spectrometry (RBS) technique discussed in this chapter light projectiles are backscattered by target nuclei; energy spectra of backscattered ions give information about the scattering elements and (in some conditions) also about lattice distortions;
(v) swift heavy ions (mass >40 amu and energies >1 MeV/amu): ions naturally existing in the cosmic radiation or created in accelerators or nuclear fission—they interact with the target by electronic excitations, in some cases eventually leaving characteristic tracks behind.

This chapter is focused on ion-solid interactions mentioned above in (iii) and (iv).

When penetrating a material, ions lose their kinetic energy due to interactions with target atoms. The process of energy loss can be understood as a series of collisions of the ions with nuclei and electrons, leading also to deflections of the ions from their primary directions of motion. Two mechanisms of the interactions are usually distinguished and their separation is a good approximation [2]:

- electronic (dominant for high energy and low Z of ions): electrons absorb small amounts of energy from the ions and as a result, they are excited or ejected; ion trajectories are only slightly changed, lattice disorder created in this process is negligible;
- nuclear (dominant for low energy and high Z of the ions): large amounts of energy are transferred from the ions to the target atoms as a whole; deflections of the ions are considerable, lattice disorder created in this process is significant;

Nuclear interactions are dominant for the ion implantation process. This results in the formation of different types of defects in implanted structures. Nevertheless, this technique is of high importance in materials science being a powerful tool for materials modification (e.g. doping by ion implantation). Analysis of damage buildup upon ion bombardment is thus an important problem.

8.1.2 Ion Beams in Material Analysis: RBS and RBS/C

One of the fundamental methods used in ion beam analysis of damage in monocrystalline structures is RBS in channeling conditions (RBS/C). It has been used in studies of post-implantation lattice distortion in different materials for more than 50 years [3]. The method uses beams of light ions (usually $^4He^+$, $^3He^+$, $^1H^+$ or $^2H^+$) with energies of the order of MeV.

When penetrating a material, the ions can be scattered by target atoms if their impact parameters are small enough to allow a nuclear interaction (e.g. $\sim 10^{-14}$ m for MeV ^4He ions). Participating in such a collision, the ion transfers a part of its kinetic energy E_i to a target atom. Its kinetic energy after an event with the atom is $E_f = k(m, M, \theta)E_i$, where the parameter k is called kinematic factor and is given by [2]:

$$k(m, M, \theta) = \left(\frac{m \cos\theta + \sqrt{M^2 - m^2 \sin^2\theta}}{m + M} \right)^2, \tag{8.1}$$

where m, M are the masses of the ion and the target atom, respectively, and θ is the deflection angle of the ion. The kinetic energy T absorbed by the atom is given by [2]:

$$T = \left[\frac{4mM}{(m+M)^2} \sin^2\frac{\theta}{2} \right] E_i. \tag{8.2}$$

The backscattered ions can be detected at a certain angle. Subsequently, their energy spectrum (backscattering yield as a function of the energy) can be plotted. According to formula (8.1), the energies of backscattered ions depend on the mass of the target atom and the chosen angle of detection (which refers to the backscattering angle θ). The mass resolution increases with the backscattering angle up to 180°. Sufficient resolutions are usually obtained for detection angles between 140° and 170°.

The maximum energy that can be detected in a backscattering experiment using an ion beam of the energy E_0 interacting with target atoms of the mass M is $E_{\max} = k(m, M, \theta)E_0$, provided that the target atom is localized on the surface. The detected energy of the ion scattered at some depth from the surface is always smaller than E_s due to the energy loss that the ion experiences before the collision and during the way to the detector.

For amorphous or polycrystalline structures the intensity of the backscattering yield (called *random*) is high. In the case of monocrystalline samples, the random spectrum can be accomplished e.g. by rotating the crystal during the measurement. If one of the main crystallographic directions remains aligned with the beam, the ions can move along channels formed by surrounding atomic rows. In such channeling mode, the probability of backscattering significantly drops—the backscattering yield in the *aligned* spectrum is low. The aligned spectrum recorded for non-ion-bombarded monocrystals is called *virgin*. It usually drops as low as 2–5% of the random spectrum yield for some crystallographic directions of good crystalline quality samples.

8.1.3 Studying of Defects by Ion Channeling

The movement along channels can be kept as long as the angle between the ion momentum and the channeling direction is kept below the critical angle ψ_c, which is of the order of $\sim 1°$. The presence of defects significantly affects the channeling process causing deflections of the beam much exceeding the critical angle. As a result, the count rate in the aligned spectrum significantly increases. The shape of the spectrum can give qualitative information about the defect kind, e.g. a characteristic damage peak is a manifestation of point defects, uncorrelated clusters or amorphous regions.

In order to quantitatively evaluate RBS/C spectra recorded for ion-bombarded crystals, the Two-Beam Approximation (TBA) is often applied [4]. In this approach, the ion beam used in the RBS analysis is assumed to consist of two fractions: the random fraction and the aligned (channeled) one. In the presence of defects two mechanisms contribute to increasing the intensity of the measured RBS spectrum:

- dechanneling: ions from the aligned fraction are deflected into a random direction thus increasing the random fraction; the probability of dechanneling depends on the defect density at a given depth $n_D(z)$ and can be described by a *defect dechanneling factor* σ_D;
- direct scattering: ions from the aligned fraction are instantly backscattered by displaced atoms; the intensity of the process depends on $n_D(z)$ and on an associated *defect scattering factor* f.

Both factors σ_D and f are characteristic for every type of defects. For example, dislocations have a very low contribution to direct scattering (f approaching zero) and reveal the high impact on dechanneling of ions, while point defects contribute to both mechanisms (with the value of f reaching unity). In contrast to point defects, dislocations provide few scattering centers and cause rather a strained distortion of the surrounding structure.

Moreover, σ_D shows a high dependence on the energy of the ions: $\sim E^{1/2}$ for dislocations, $\sim E^{-1}$ for point defects and approximately constant in the case of stacking faults [1]. Therefore, because the ions lose their energy while penetrating the substrate, the probability of dechanneling resulting from dislocations drops, while it increases in the case of point defects.

8.1.4 Computer Simulations for Ion Beam Analysis

RBS/C spectra contain combined signals from all defects in the analyzed structures. Detected ions do not carry enough information to reproduce precisely the backscattering occurrences. Therefore, the analytical evaluation of the channeling spectra by means of the Two-Beam Approximation is highly limited. It is often performed on the assumption that only one type of defect is present or dominant. However, it is not true in most cases of post-implantation damage. Moreover, even if the depth-profile of defects is obtained from backscattering yield, it is hard to assign to it any physical unit. Instead, the term "relative disorder" is often in use. It was required to involve other computational techniques in order to provide a proper evaluation of RBS/C spectra.

Obtaining quantitative information about the presence of different types of defects from RBS/C data is a challenge, especially in the case of heterostructures or superlattices. This task becomes feasible by means of computer simulations.

Like games in the world-famous Casino de Monte-Carlo in Monaco, the Monte Carlo (MC) simulations are a class of computational algorithms based on randomness. The obtained result is the average over many individual events. The MC simulations have been successfully used for years in materials science, in particular in the ion beam analysis, e.g. as a powerful computational tool for evaluation of RBS and RBS/C data [3].

The use of the MC simulations in ion channeling studies was inspired by Barrett [5]. In his pioneering work, he proposed the sampling of scattering probability distribution to calculate the integral RBS spectrum. This probability distribution is known as *nuclear encounter probability* (*NEP*). It is the measure of a chance that the ion is scattered by an encountered atom and is calculated for every single ion-atom interaction. The cumulated distribution of the scattering probability can be eventually converted into an RBS spectrum. Moreover, NEP can also be used in simulations of such phenomena as nuclear reactions or atomic recoils.

Computer codes developed to simulate ion trajectories in crystalline structures consider the ion movement as a series of collisions with target atoms in frames of a screened Coulomb potential with a carefully chosen screening function. Binary Collision Approximation (BCA) is usually applied for this purpose. For each iteration, the ion interacts with the nearest atom selected among all atoms whose impact parameters are below the predefined limit [6]. After the collision, the ion loses some amount of its kinetic energy, changes its direction of motion and moves along the new straight path with the origin in the point of the closest approach. The influence of more distant atoms can be taken into account by applying additional models of the interaction.

MC simulations require prior creation of virtual cells of target structures, containing coordinates of atoms. The cells are periodically repeated during the simulation process. The size of the cells varies depending on the MC software used to work with them. In relatively small virtual structures (of the size of few/few tens of lattice constants) defects can be introduced during the simulation process as modifications of the structure. Models of different types of defects must be carefully developed and implemented into the source code of MC software.

On the other hand, Molecular Dynamics (MD) allows creating much larger virtual structures (of the order of hundreds of nm), based on known interatomic potentials for multi-body interactions. Such arbitrary structures can contain different types of defects of known nature and concentration. By simulating the movement of ions in such structures, the associated MC software can reproduce channeling spectra. The main limitation of this technique is the usage of time and huge computer memory.

Many MC and MD codes have been developed for use in RBS/C spectra evaluation [5, 7–16]. Reviews of some of them are provided in [3, 17, 18]. The main limitation of the majority of these codes is that they take into account only simple defects. However, simulations by Zhang et al. on MD-created arbitrary structures are the evidence that extended defects have a big impact on the intensity of RBS/C spectra and should not be neglected in the analysis [8].

One of the most powerful MC codes to evaluate ion channeling spectra is called ‘McChasy’. The acronym stands for **M**onte **C**arlo **CHA**nneling **SY**mulation (where the letter ‘Y’ comes from the Polish translation of the word “simulation”). The greatest virtues of the code are computing of separate contributions from point defects and edge dislocations to the simulated spectra as well as the possibility to run it on common PCs [16].

This chapter is devoted to describing the technical aspects of MC simulations available in the McChasy code, which has been recently significantly improved. Its last features include:

- 3D computing of thermal vibrations of target atoms,
- 3D determination of impact parameters of channeling ions,
- use of a rotation matrix for different orientations of dislocations,
- use of a script for Microsoft Excel to obtain dislocation parameters of dislocations from high-resolution Transmission Electron Microscopy (HRTEM) micrographs,
- consideration of Xe bubbles formed in UO_2.

The chapter is organized in sections describing features of the McChasy software: basic principles (Sect. 8.2), 3D interactions (Sect. 8.3), defect models

(Sect. 8.4) and selected examples (Sect. 8.5). Section 8.6 summarizes the latest improvements in the code. The previous states of the McChasy code have been reported in [10, 13, 16].

8.2 McChasy—The Principles

The McChasy code has been written to run on common PCs in the Windows environment. It uses MC simulations to reproduce trajectories of light MeV ions (^{4}He, ^{2}H, and ^{1}H) in monocrystalline structures. It is run by an input file prepared by the user (a SIP file: Simple Input Protocol). The file contains commands that regulate simulations (if not provided, the default values are loaded). The most important input parameters to be provided to the code are presented below:

- the energy of the beam (0.8–3.5 MeV) and energy resolution,
- calibration of the energy channels width and offset,
- backscattering angle,
- the sample structure,
- defect profiles,
- initial dispersion and straggling of the beam,
- scattering on electrons and local energy loss corrections.

8.2.1 Structure Preparation

The mechanism used in the McChasy code to calculate interactions of channeled ions with target atoms is based on modified BCA. The ion interacts in sequence with monolayers of atoms as shown in Fig. 8.1. The code takes into account not only the influence of the nearest atom but also the more distant ones within the

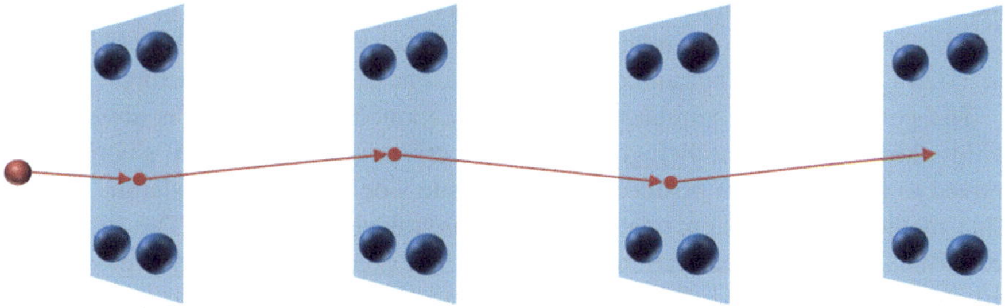

Fig. 8.1 The scheme of Planar Scattering Approximation. The ion (red ball) undergoes a series of interactions with monolayers of atoms (navy balls)

monolayer. This mechanism can be called Planar Scattering Approximation (PSA) and is described in detail in Sects. 8.2.2 and 8.2.3.

It is required to create a small virtual cell prior to beginning MC simulations with the McChasy code. The cell contains a certain number of monolayers and is periodically repeated while the movement of the channeled ion is simulated. The code modifies the cell 'in situ' by applying thermal vibrations of atoms (described in Sect. 8.2.4) and certain defects at the desired depth, if defined by the user. The types of defects available in the McChasy code are described in Sect. 8.4.

In order to create the virtual cells, the McChasy code is associated with a software called *Structure Preparation*. The code performs symmetry operations based on a defined unit cell and a given space group to calculate the positions of atoms. Subsequently, the atoms of the same z-coordinate are grouped in monolayers perpendicular to a chosen crystallographic direction and their x, y, z coordinates are saved in a separate file.

The number of monolayers necessary to describe a certain crystal structure is the minimum value required to provide the periodicity of the virtual cell. It depends on the complicacy of the structure and the chosen crystallographic direction. It varies from two monolayers for [001] in cubic crystals up to e.g. 76 monolayers for [011] in U_4O_9 or 67 monolayers for [−201] in Ga_2O_3.

The radius of the monolayers in the plane orthogonal to the chosen direction usually does not exceed 1 nm. Exemplary virtual cells for [0001] ZnO and [001] $SrTiO_3$ are shown in Fig. 8.2. In the case of ZnO four monolayers consisting of the series of Zn, O, Zn, O are required. The virtual cell of $SrTiO_3$ consists of two monolayers, the first one is built of Ti and O atoms, the second of Sr and O atoms.

8.2.1.1 Virtual Channel of Motion

There is a virtual central parallelogram defined in every structure file. Its x and y coordinates are calculated based on the translation vectors $\boldsymbol{u}$ and $\boldsymbol{w}$ of the structure. The vectors $\boldsymbol{u}$ and $\boldsymbol{w}$ are determined by the Structure Preparation code for an XY plane perpendicular to a given crystallographic direction. The origin of the XY coordinate system is located at the intersection of the parallelogram's diagonals and the coordinates of its corners are calculated in relation to the origin of the XY coordinate system from the relations: ($-1/2\boldsymbol{u}-1/2\boldsymbol{w}$; $-1/2\boldsymbol{u}+1/2\boldsymbol{w}$; $+1/2\boldsymbol{u}+1/2\boldsymbol{w}$; $+1/2\boldsymbol{u}-1/2\boldsymbol{w}$). The parallelogram contains at least one atom of the structure.

The position of the simulated ion at every monolayer of atoms is always kept within the area of the parallelogram. If the ion crosses a parallelogram's border, it is moved to the equivalent position at the opposite side using the translation vectors. This method allows using relatively small monolayers of atoms. The radius not exceeding 1000 pm is enough in all the cases, not exceeding 500 pm in most of the structures. However, the Structure Preparation code is prepared to create much larger monolayers, if needed.

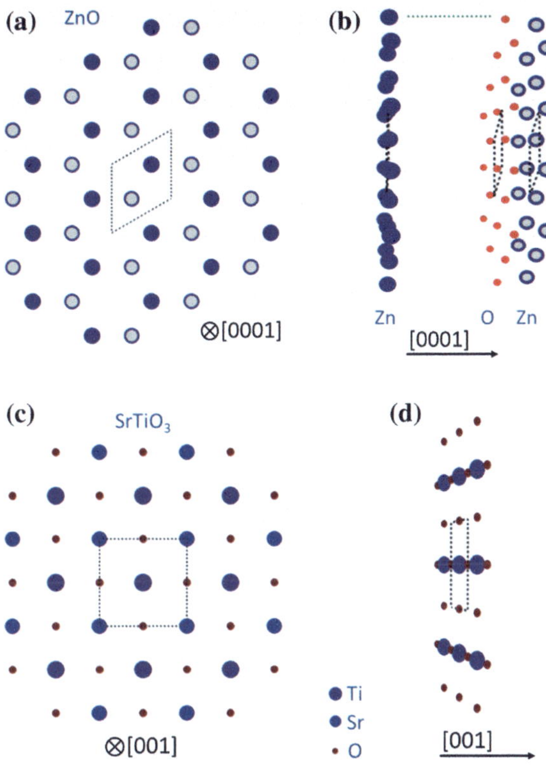

Fig. 8.2 Virtual cells created by the Structure Preparation code associated with the McChasy code: **a** projection along the [0001] channeling direction for ZnO, **b** side view of the cell created for [0001] ZnO, **c** projection along the [001] channeling direction for $SrTiO_3$, **d** side view of the cell created for [001] $SrTiO_3$

8.2.2 *Deflections*

The movement of an ion in the matter can be considered as a series of many events (collisions) involving two bodies (the ion and subsequent target atoms) with a central force interaction between them. In the center of mass frame, the atoms are motionless and the ion impinges on each of them with a specific impact parameter b. Depending on b, the potential of interaction and relative energy, the ion can be deflected from the initial direction at an angle θ.

It is possible to find the dependency $\theta(b)$ analytically only in particular situations, like potentials proportional to r^{-2}, where r is the distance from the scattering center, e.g. the Coulomb potential $V_C(r)$. In the case of interatomic interaction, the Coulomb potential of nuclei is screened by electrons; therefore the problem becomes more complicated.

The interatomic potentials are connected to the Coulomb potential by the screening function $\chi(r)$:

$$V(r) = V_c(r)\chi(r) \tag{8.3}$$

Different screening functions have been reviewed in [2, 10]. Scattering of ions in the McChasy code is based on the Ziegler-Biersack-Littmark (ZBL), also called the Universal potential with [19, 20]:

$$\begin{aligned}\chi_{\mathrm{ZBL}}(x) = {} & 0.1818\exp(-3.2x) + 0.5099\exp(-0.9423x) \\ & + 0.2802\exp(-0.4028x) + 0.02817\exp(-0.2016x),\end{aligned} \tag{8.4}$$

where $x = r/a_{\mathrm{U}}$ is the reduced distance defined by the universal screening length a_{U}:

$$a_{\mathrm{U}} = \frac{0.8854 a_0}{\left(Z_1^{0.23} + Z_2^{0.23}\right)}, \tag{8.5}$$

where a_0 = 0.05292 nm is the Bohr radius.

The solutions of the deflection function $\theta(b)$ were found for closely 12 thousand values of b for every element between He and U and tabularized for further use in the McChasy code. Two ways of calculating $\theta(b)$ werse used. For far interactions (small deflection angles, e.g. below $2.35 \times 10^{-4\circ}$ for Li and 0.037° for U) the momentum approximation based on the Gauss–Mehler quadratures was applied [21]. For close interactions (large deflection angles) the numerical method of integration of the equations of motion using the Euler–Cromer approach was used [22, 23].

During the simulation process, the code calculates impact parameters of ions and from the table of the deflection angles picks the corresponding value of θ. The final deflection angle is averaged over the interactions with atoms in the monolayer of atoms to determinate the new direction of the ion movement. Subsequently, the point of impact onto the next monolayer is found.

In the PSA approach, impact parameters are calculated in 2D, within a plane perpendicular to the channeling direction and associated with a given monolayer of atoms. Thermal vibrations of atoms are applied also in 2D, within the plane. More details about this procedure are provided in Sect. 8.3.1.

8.2.3 Energy Loss

The McChasy code uses tabularized values of the stopping power as calculated by the SRIM code (used by default) [9] or by the RUMP code (optionally) [11]. The energy of ions is updated at every monolayer, after the determination of deflection angles. In the case of simulations of ion channeling, the code takes into account the classical Lindhard approach: the stopping power depends on the local concentration of atoms as well as the local concentration of electrons. For the channeled ions in compounds, the Bragg's rule is assumed: cross-sections of the random stopping from core electrons, valence electrons, and plasmon excitations contribute additively for every element [24].

8.2.4 Thermal Vibrations

Thermal vibrations of the target atoms are applied by the McChasy code as additional random displacements added to the equilibrium positions of the atoms. They are calculated for all three dimensions x_i from the commonly used Gaussian distribution [25]:

$$T(x_i) = \frac{1}{\sqrt{2u_i^2}} \exp\left(-\frac{x_i^2}{-2u_i^2}\right), \tag{8.6}$$

where u_i is the standard deviation of atomic displacement in the given direction. Its value should be calculated before the simulations, e.g. based on the Debye theory.

Once u is known, the amplitude of the thermal vibrations can be applied as an input parameter provided in the SIP file. In this case, also the initial density of defects in unimplanted crystals can be evaluated by fitting the virgin spectrum with MC simulations made with an assumption of some low defect density.

However, the Debye theory can hardly be applied for compound structures so in most cases the real amplitude of thermal vibrations remains unknown. This problem can be solved in the McChasy code by assuming the amplitude of the thermal vibrations as a free parameter that can be adjusted before the simulations of damaged samples.

For this purpose, the channeling spectrum recorded for the high-quality pure crystal upon a possibly best alignment of the beam must be simulated with different values of the amplitude of the thermal vibrations until the best fit is obtained. The corresponding value is kept for all other simulations performed for this material, in particular, the simulations of the spectra recorded for ion-bombarded samples. The default value of the amplitude of the thermal vibrations is 10 pm and should be reasonably changed around this value.

8.3 McChasy—3D Interactions

The newest version of the McChasy code has been modified taking into account 3-dimensional interactions between channeling ions and target atoms. The model of 3D interactions is an attempt to combine advantages of the BCA and PSA computational methods described in Sects. 8.1.4 and 8.2.1 resulting in the more realistic determination of trajectories of channeling ions.

McChasy performs MC simulations on the structure file containing monolayers of atoms. The code reads the structure file and performs a series of procedures that are repeated when the channeling ion reaches each subsequent monolayer:

I. Based on the current position of the ion and its momentum, the impact point onto the closest monolayer is calculated. For the interaction with the surface, this point is picked randomly within the virtual central parallelogram.
II. For every atom in the monolayer, the distance between the nucleus and the ion impact point is calculated in order to reject those atoms that are too far to interact with the ion. The limit is determined separately for every element based on the maximum impact parameters included in the deflection tables.
III. If requested by the user the positions of the atoms are modified according to a desired type of defect—applying defects during the simulation process is described thoroughly in Sect. 8.4.
IV. For all atoms selected to interact with the ion, their thermal vibrations are applied according to the formula (8.6).
V. Taking into account the direction of the ion, its impact parameters for each selected atom are determined. In contrary to the PSA method, the impact parameters are now calculated in 3D.
VI. Dependent on the impact parameter, the corresponding deflection angle is read from the deflection table for a certain element. The final deflection is averaged over all interacting atoms in the monolayer.
VII. The new ion direction is now used to determine the impact point on the subsequent monolayer.
VIII. Independently, for every atom selected to calculate the interaction, the NEP is determined and cumulated in memory.

8.3.1 Impact Parameters in 2- and 3-Dimensions

The impact parameter calculated in 3D is the smallest distance between the current position of the atom (after the thermal vibrations applied) and the line determined by the direction of the ion motion. Selected 3D impact parameters are denoted as green lines in Fig. 8.3. The corresponding impact parameter evaluated in the PSA approach is the distance between the position of the atom (nucleus) and the impact point of the ion on the plane of the monolayer of atoms (red lines in Fig. 8.3). However, in the PSA approach, the thermal vibrations of atoms are calculated also only in 2D (over the XY plane orthogonal to the crystallographic direction z chosen as the channeling direction). Therefore, the 2D position of the atom refers to the plane projection of the 3D position of the atom with the 3D thermal vibrations applied.

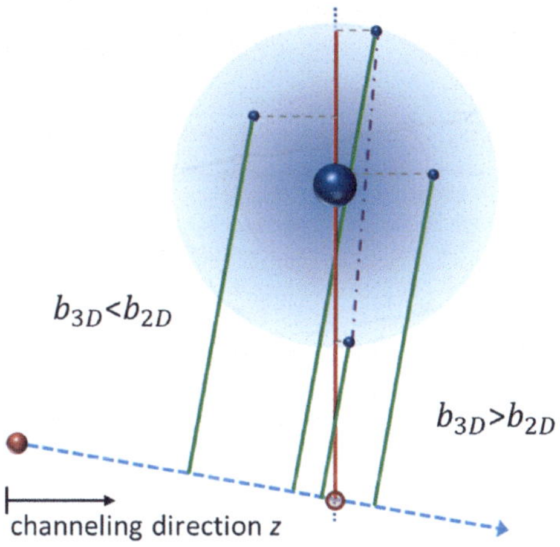

Fig. 8.3 Exemplary impact parameters of the ion (red ball) relative to different positions of the atom (small blue balls) as coming from the thermal vibrations around the equilibrium position (big blue ball). The blue gradient circle around the equilibrium position refers to the area of possible thermal vibrations (applied with Gaussian distribution). The impact point of the ion onto the monolayer is denoted by an empty red circle. Green solid lines refer to 3D impact parameters, while red solid lines refer to 2D impact parameters. The purple dash-dotted line refers to the positions of the atom, at which 2D and 3D impact parameters are equal

Because of this fact, in most occurrences of the ion-target interactions during the ion channeling process the 3D impact parameters are smaller than those obtained in the 2D approximation. This should result in stronger dechanneling of channeled ions and higher count rates in the RBS/C spectra (if all other parameters remain unchanged).

Indeed, Fig. 8.4 shows RBS/C spectra simulated by the McChasy code in frames of the 2D and the 3D approach for [0001] ZnO with an arbitrary depth-profile of randomly displaced atoms shown in Fig. 8.5 applied. The defect profile was also multiplied by factors 2, 3 and 4 to obtain different maximum values of defects: 25, 50, 75 and 100%. The principles of MC simulations in the presence of defects is described in Sect. 8.4.

The intensity of the RBS/C spectra shown in Fig. 8.4 simulated in frames of the 3D approach is higher, especially for low concentration of defects. In the energy range corresponding to a damage peak (around 1.25–1.5 MeV), differences between spectra simulated in the 2D and the 3D approach become smaller when defect density increases and vanish in the case of 100% defects. However, stronger dechanneling is responsible for the higher intensity of the 3D spectra behind the damage peak (below ~1.25 MeV). The differences between the 2D and the 3D approach become even more visible with the increase of thermal vibrations amplitude.

Despite some differences between simulations performed using the 2D and the 3D approach, the 3D method does not invalidate previous defect analysis done with

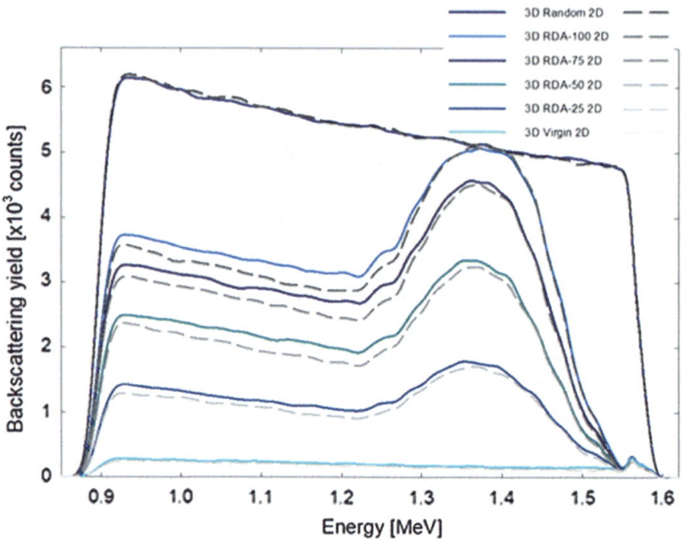

Fig. 8.4 Channeling spectra simulated for [0001] ZnO with the McChasy code using 2D and 3D approach. Simulations were performed for 2 MeV ^{4}He ions and scattersing angle 165°. Thermal vibrations were set to 7.0 pm for Zn and 17.5 pm for O. The thickness of the sample was set to 800 nm. The random spectrum was simulated in the mode of rotation around the [0001] axis with the tilt angle 4°. For aligned spectra four defect profiles were applied: one of them, with the maximum concentration of 25% of defects, is shown in Fig. 8.5 (RDA 25), the others were created by multiplying the RDA 25 by 2 (RDA 50 with maximum 50% of defects), 3 (RDA 75 with maximum 75% of defects) and 4 (RDA 100 with maximum 100% of defects). Solid lines refer to 3D simulations, dashed lines refer to 2D simulations. For the sake of clarity, only the signal from Zn atoms is shown

pure PSA. Some defect profiles obtained previously using only 2D approach may be considered slightly overestimated; however, the main conclusions relating to damage buildup in crystals remain valid.

8.4 McChasy—Defects

As mentioned in Sect. 8.3, defects are considered during the simulation process before the determination of the impact parameters of the channeled ion (point III on the list). Several types of defects can be taken into account in the McChasy code.

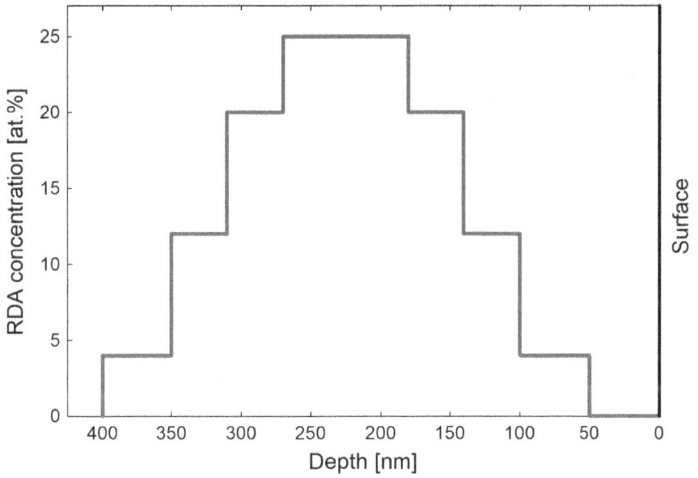

Fig. 8.5 Depth profile of randomly displaced atoms applied for MC simulations to obtain the RDA 25 spectrum shown in Fig. 8.4. The profile was also multiplied by 2, 3 and 4 to obtain other maximum concentrations of defects: 50% (the RDA 50 spectrum), 75% (the RDA 75 spectrum) and 100% (the RDA 100 spectrum)

8.4.1 Point Defects

8.4.1.1 Substitutional Atoms

Based on a given depth-distribution the McChasy code substitutes a certain number of target atoms by the desired element (elements). The impurity atoms always occupy substitutional lattice site locations. This feature works well for simulations in the random spectrum mode and allows getting information about the impurity concentration. In the aligned spectrum mode, some signals can be missing due to a lack of impurity atoms in interstitial positions of the lattice.

However, if other types of defects are applied then the impurity atoms are treated equally as the atoms they substituted—the same percentage of them is affected depending on the concerned defect.

The depth profile of substituting atoms can be obtained by complementary computing software (e.g. SRIM) or with the McChasy code using the trial-and-error procedure. In the latter case, the code calculates the random spectrum based on a given profile of impurities and the user compares it with experimental data until a satisfactory fit is obtained.

8.4.1.2 Interstitials

If this type of defect is to be taken into account, the McChasy code displaces some single atoms from the monolayer. At first, for every atom, the code determines the probability that the atom is shifted from its equilibrium position. The probability is based on the given depth-distribution defined in the SIP file (and expressed as the percentage of interstitials vs. depth). If the atom is selected as defected, the code calculates its additional displacement in one of the four ways:

- Randomly Displaced Atoms (RDA): the atomic displacements are random values calculated over the atomic monolayer; this is the default kind of point defects in the McChasy code,
- Gaussian distribution: the displacements are calculated in 2D around the equilibrium position of the atom with the probability based on the Gaussian distribution; the amplitude of the function is defined by the user within the range 0.01–50 pm,
- Distribution over a disk: atoms are tossed with equal probability over a disk of the radius defined by the user within the range 0.01–50 pm,
- Distribution over a ring: atoms are tossed with equal probability over a ring of the radius defined by the user within the range 0.01–100 pm.

Interstitials should be used in the McChasy code with caution and with support from other techniques since it is possible to fit practically any experimental RBS/C spectrum using only point defects. However, the results may not have any physical meaning.

8.4.1.3 Vacancies

Vacancies or voids can be taken into account as a combination of substitutions and interstitials. The McChasy code contains a table of elements and one of the elements is called Vo. It has zero mass and zero charge and was introduced to use as a vacancy substituting a target atom.

To use this type of defects during simulations, the SIP file must contain a depth-profile determining distribution of Vo. Based on the postulated percentage, random target atoms will disappear from the monolayer.

8.4.2 Edge Dislocations

8.4.2.1 The Model

The model of the edge dislocations (DIS) implemented in the McChasy code is based on the Peierls–Nabarro model [26–28]. An extra half-plane of atoms affects adjacent atomic planes causing a characteristic bending. The shape of the bent planes can be parametrized by the arctan function. Appropriate atomic displacements follow the equation:

$$u(z) = \frac{D}{\pi} \arctan(g(z-p)) + q, \tag{8.7}$$

where z refers to the axis of the channeling direction, the vector $[p, q]$ refers to the position of the inflection point of the function, D is the distance between the asymptotes of the arctan function and g is the parameter related to D by the relation [29]:

$$g = \pi \tan\eta / D, \tag{8.8}$$

where η is the bending angle defined as the angle between the tangent to the arctan function at the inflexion point and the arctan asymptotes. The geometrical interpretation of the parameters D and η is shown in Fig. 8.6.

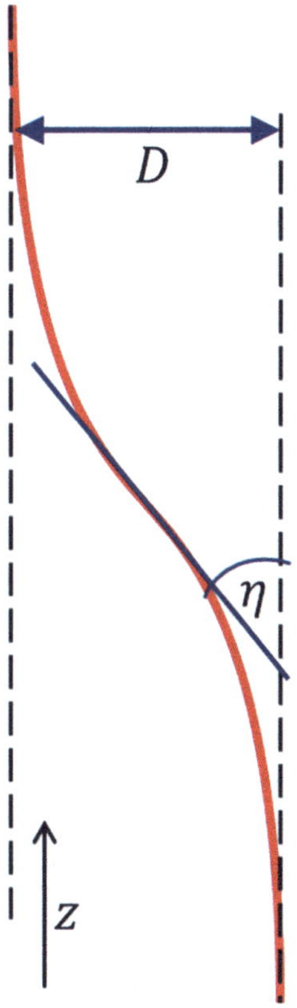

Fig. 8.6 A scheme of an arctan function showing the geometrical interpretation of the parameters D and η required to determine the arctan function using formulas (8.7) and (8.8)

The parameters D and η can be found based on high-resolution Transmission Electron Microscopy (HRTEM) images or structures created by MD. The bent atomic planes, deformed due to the presence of a dislocation, are visible on the HRTEM cross-sectional micrographs as bent atomic planes. Dislocation-containing structures created by MD can also be adjusted by any graphical software to show a desired crystallographic orientation. Such 2D images can be analyzed by a script dedicated to working in the Microsoft Excel environment which is distributed with the McChasy code [30, 31]. The script allows the fitting of the arctan function to atomic planes bent due to the presence of dislocation.

The bending radius of the atomic planes adjacent to the dislocation extra half-plane tends to become smaller when the plane distance in the normal direction from the dislocation line increases. This tendency results in a decrease of the parameters D and η and was measured for three structures: $Al_{0.4}Ga_{0.6}N$ (AGN), $SrTiO_3$ (STO) and ZnO. As reported in [30], the decays of the parameters D and η for these three materials follow the following relations:

$$D_{\mathrm{STO}}(r) = 194\exp\left(-\frac{r-1}{42}\right), \tag{8.9}$$

$$\eta_{\mathrm{STO}}(r) = 9.7^{\circ} r^{-0.96}, \tag{8.10}$$

$$D_{\mathrm{AGN}}(r) = 187\exp\left(-\frac{r-1}{7.7}\right), \tag{8.11}$$

$$\eta_{\mathrm{AGN}}(r) = 32^{\circ} r^{-0.91}, \tag{8.12}$$

$$D_{\mathrm{ZnO}}(r) = 166\exp\left(-\frac{r-1}{10.9}\right), \tag{8.13}$$

$$\eta_{\mathrm{ZnO}}(r) = 60^{\circ} r^{-1.04}, \tag{8.14}$$

where r refers to the distance from the dislocation line in the direction normal to the extra half-plane of atoms and is an integer multiple of the planar spacing for this direction. The values of D are expressed in pm. More details, in particular values of uncertainty, can be found in [30].

There are some differences between the three studied structures as it appears from formulas (8.9)–(8.14). The most important is related to the decays of the parameter D, which decreases to a value D_0e^{-1} at very different distances from the dislocation. In the case of hexagonal structures, AGN and ZnO the decay is very rapid, while in the case of STO the D decreases relatively slow, reaching the value D_0e^{-1} on the 43rd plane from the dislocation edge.

A concentration of dislocations is usually expressed in cm^{-2} (or m^{-2}). The unit comes from a typically used method of evaluation of their concentration. Chemical etching can reveal some dislocations. Due to different dissolution rates, there are pits formed around the dislocation edges crossing the exposed surface. By counting them, the dislocation density σ can then be expressed as the number of dislocations per unit area.

8.4.2.2 Implementation in the McChasy Code

The McChasy code converts the dislocation concentration postulated by the user in the SIP file from cm^{-2} into a certain number of dislocation lines per μm λ. This parameter can be understood by analyzing Fig. 8.7. There is a cylinder constructed around the virtual channel of the motion of the channeled ion. The length of the cylinder is L and its radius is $n \cdot a$, where a is the lattice constant for a chosen direction perpendicular to the channeling direction z and n is its integer multiple defined by the user.

For $L = 1$ μm the motion of every ion traveling inside the virtual channel is affected by the presence of λ dislocations, which distort the structure around the virtual channel. The dislocation lines are represented by blue solid lines in Fig. 8.7.

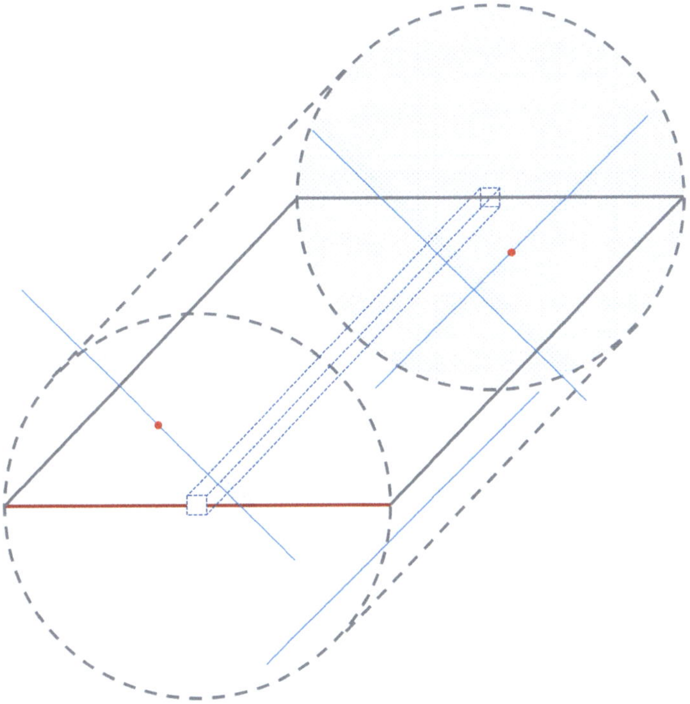

Fig. 8.7 The scheme of the dislocations as implemented in the McChasy code. The dislocation lines (blue solid lines) intersect a cylinder created around the virtual channel of the channeled ion motion. Some dislocation lines intersect the axial cross-section of the cylinder and can be detected in an etching experiment as pits (red dots). However, some dislocation lines do not intersect the axial cross-section but still distort the structure that influences the ion movement

At the current state of the McChasy code, only limited orientation of dislocations are taken into account: two possible orientations in cubic structures (dislocation lines are perpendicular to the channeling direction and relative to each other) and three orientations in hexagonal structures (dislocation lines are perpendicular to the channeling direction and at 120° relative to each other).

Let now the axial cross-section of the cylinder be an etched surface. The pits created by the chemical etching are marked as red dots in Fig. 8.7. However, some dislocations intersecting the cylinder do not intersect the etching surface. Therefore, they would not show up in an etching experiment but they do influence the ion channeling. Since the dislocations in the McChasy code are tossed uniformly, the relation between the dislocation density σ and λ is:

$$\sigma = \frac{\lambda\sqrt{2}}{4na}(\cos\alpha + \cos(90^\circ - \alpha)) \tag{8.15}$$

for cubic structures and

$$\sigma = \frac{\lambda}{3na}(\cos(60^\circ + \alpha) + \cos\alpha + \cos(60^\circ - \alpha)) \tag{8.16}$$

for hexagonal structures. In both cases, α is the smallest angle between the etching surface (the axial cross-section of the cylinder) and the chosen dislocation line. The maximum values of σ are obtained for $\alpha = 45^\circ$ for cubic and $\alpha = 0^\circ$ for hexagonal structures, respectively.

Based on a depth-profile of dislocations postulated by the user, the McChasy code calculates the number of dislocation lines per defined depth intervals. Then, within each depth interval, the code randomly draws for every dislocation: its depth, its distance from the virtual channel of motion and its orientation.

During simulations, for every target atom considered as interacting with the channeled ion, the code calculates the parameters of the arctan function referring to the atomic row containing the considered atom. Then, the atom displacement is determined from the formula (8.7).

The displacement vector is rotated using a rotation matrix according to the drawn orientation of the dislocation. Eventually, the atom is shifted from its equilibrium position and subsequent points from the list discussed in Sect. 8.3 are executed (IV–VIII).

8.4.3 Grain Boundaries

If desired, the grain structure in textured samples can be considered during the MC simulations. The user must define the size of the grains as well as the shift and the tilt angle between the grains at the grain boundary. This defect is available in the McChasy code in limited condition: only two different orientations of the grains are possible to be defined.

8.4.4 Stacking Faults

Stacking faults are implemented in the McChasy code as a lateral displacement between two subsequent monolayers of atoms. The user defines values x, y of the shift (in the plane orthogonal to the channeling direction) and the distance between the successive stacking faults.

The faulted monolayer is either simply displaced by the defined values x and y, or shifted and also rotated by 90°. The probability of both occurrences is 0.5. The depth of the first stacking fault is picked at random. The next stacking fault is applied always after the channeled ion overcomes a distance defined by the user.

8.4.5 Xe-Bubbles in UO_2

The presence of Xe-bubbles can be postulated in UO_2 as input data in the McChasy code. The option was developed to study nuclear materials under extreme conditions (bombarded by swift heavy ions). By using this option the code simulates the movement of He-ions channeling in the UO_2 structure, where a volume fraction is occupied by spherical precipitations containing Xe atoms (bubbles).

The main features of the option are:

- Xe bubbles are assumed to be amorphous;
- two parameters describe the bubbles:
  - the volume fraction of the bubble (expressed as % of the structure volume),
  - the bubble diameter (given in nm);
- bubbles are randomly and uniformly distributed within the UO_2 structure;
- defects in the UO_2 structure can be applied independently of the bubbles.

So far the model is only implemented for the UO_2 structure and it is limited to spherical Xe precipitations.

8.5 Selected Examples

8.5.1 Interstitials and Dislocations

Figure 8.8a shows the RBS/C spectra recorded for [0001] ZnO implanted with 300 keV Er ions to a fluence of 2.0×10^{15} cm^{-2}. The Er profile obtained from SRIM is shown for comparison in Fig. 8.8b. The spectra were fitted by the McChasy code. Simulations referring to the random and the virgin spectrum are denoted as navy dashed lines. The solid black line is the outcome of simulations run for depth-distributions of RDA and DIS shown in Fig. 8.8b. Simulations performed under the assumption that only RDA or only DIS are present are denoted as the dashed blue and the dotted-dashed red lines, respectively.

The spectrum obtained by the assumption that only RDA are present in the structure (the blue dashed line) reveals that this model fits well in the energy region from 1.35 to 1.23 MeV; however, for lower energies corresponding to deeper regions in the sample the backscattering yield is underestimated by the simulation. Further increase of the RDA distribution within the deeper region could allow fitting the experimental spectrum without the need to introduce dislocations. However, such a procedure would have no physical meaning since the implanted ions do not reach this depth and it is barely possible that single target atoms kicked-off from their lattice sites could migrate so deeply.

The fact that the backscattering yield in deeper regions is underestimated by the RDA model is rather the evidence that extended defects were formed during ion implantation, such as dislocations or dislocation loops, which have higher dechanneling yields than simple point defects. Therefore, it is necessary to apply another model including extended defects along with the RDA.

8.5.2 Xe-Bubbles

Some simulation tests for the Xe bubbles in UO_2 were performed. The results are presented in Fig. 8.9. The intensity of the aligned spectra increases with the volume of the bubbles and their diameter kept unchanged (Fig. 8.9a). When the volume is kept constant but the diameter of bubbles is changed, then only a small influence of the bubble diameter on the intensity is observed (Fig. 8.9b). Bigger bubbles provide a slightly higher scattering rate.

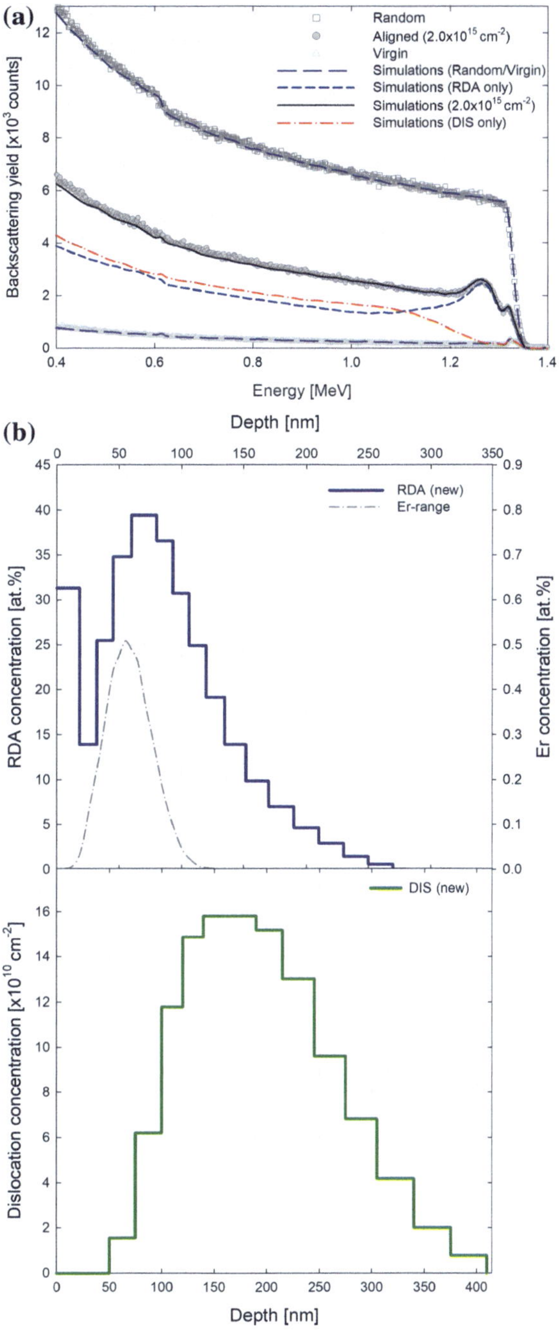

Fig. 8.8 **a** RBS/C spectra recorded for ZnO samples reported in [34]: virgin (referring to the unimplanted sample), random and aligned (referring to the sample bombarded with 300 keV Er ions to a fluence of 2.0×10^{15} cm^{-2}). Fitting lines are the results of MC simulations performed in the McChasy code: navy dashed lines refer to the virgin and random spectra; solid black line refers to the aligned spectrum fitted under an assumption of coexistence of RDA and DIS with depth-profiles shown in (**b**); the blue dashed line refers to the simulations performed under an assumption of the presence of only RDA; the red dash-dotted line refers to the simulations performed under an assumption of the presence of only DIS; **b** depth profiles of RDA and DIS corresponding to the simulations of the Er-implanted sample

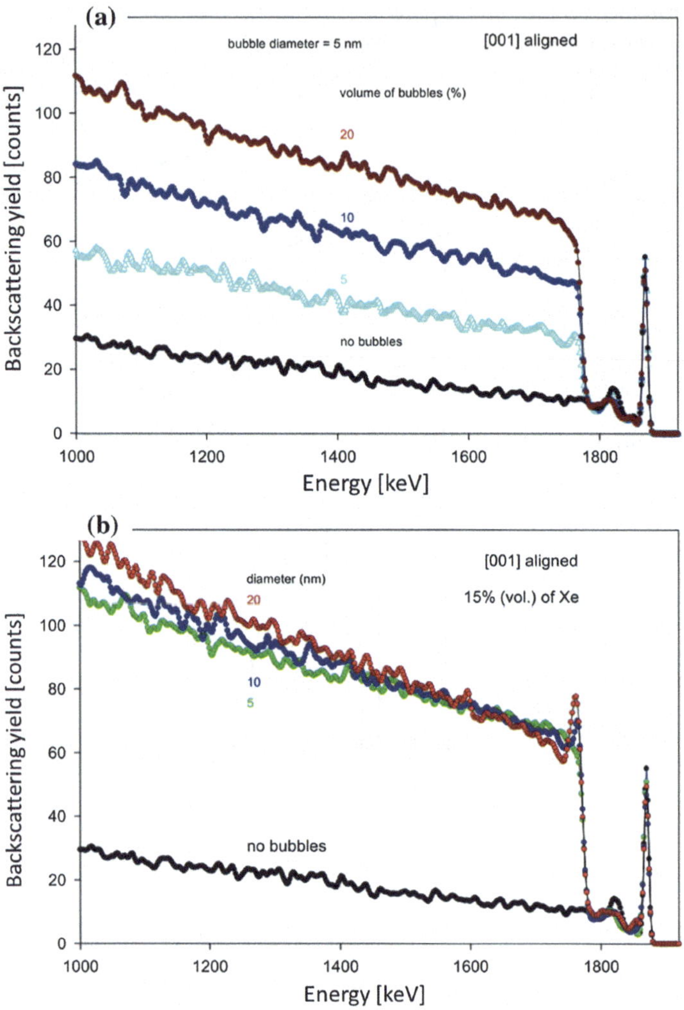

Fig. 8.9 Simulated aligned spectra of [001] UO_2 containing Xe bubbles: **a** constant diameter of bubbles and different volume as specified on the plot; **b** constant volume and different diameter of bubbles as specified on the plot. Simulations made with the McChasy code with 2 MeV ^{4}He channeled ions and backscattering angle of 170°

8.6 Summary

Ion channeling is a powerful technique used in the analysis of post-implanted damage in crystals. However, different types of defects contribute differently to the backscattering yield. It is hardly possible to distinguish the defect types by analytical evaluation of RBS/C spectra. The task becomes even more complicated in the case of heterostructures and superlattices.

Monte Carlo simulations are a useful tool used in the evaluation of RBS/C data. One of the most powerful MC computer codes used for that purpose is the McChasy software. It reproduces the movement of light ions in a target and considers different types of defects during simulations.

This chapter describes the most important features of the code and explains the principles of its operation. The mechanism of 3-dimensional computing of interactions between the channeled ion and target atoms grouped in monolayers is described in Sect. 8.3. Types of defects possible to be applied in the code are pointed out in Sect. 8.4. Selected results of simulations performed for structures containing complex defects are shown in Sect. 8.5.

The McChasy code allows fitting RBS/C spectra by the trial and error simulations based on postulated depth-profiles of defects. Results obtained by the code can be useful in determining the quantitative and qualitative information about the radiation damage in the macroscopic scale. This can help to understand the complicated processes occurring upon ion bombardment.

Performing MC simulations under the assumption that different types of defects are present might be a tricky task. A single RBS/C spectrum can be fitted based on several different combinations of depth-profiles of defects. It can be understood as one equation with more than one unknown. To provide explicit results several RBS/C spectra should be measured at different energies of the analyzing beam. MC simulations performed with the same depth-profiles of defects applied to all the spectra recorded at different energies of the beam will lead to the explicit depth-distributions of defects.

The current version of the McChasy code is not its final state. The code is still improved, depending on the demand. The most important possible ways of its development are the following:

- *interstitial substitutions*: so far impurity atoms can only substitute target atoms; no atoms are displaced to an interstitial location, the substituted target atoms disappear from the monolayer,
- *dislocation loops*: the model of edge dislocations implemented in the code assumes the presence of half-planes of atoms; however, dislocation loops are rather observed among defects created in post-implanted damage [32, 33]; the model of dislocations implemented in the McChasy code will be updated in frames of dislocation loops,
- *impurity bubbles*: the model of spherical precipitations of impurity atoms will be extended to other structures, also including other elements than Xe,
- *swift heavy ion tracks*: the model of tracks created by swift heavy ions is to be developed and implemented into the McChasy code.

Even at its current state, the McChasy code is one of the most powerful MC codes using in the evaluation of RBS/C data. The code enables fast analysis of RBS/C spectra e.g. those recorded for structures implanted to different ion fluences. Such studies allow e.g. knowing the kinetics of damage buildup upon ion bombardment [34, 35]. Further analysis with complementary techniques (e.g. X-Ray

Diffraction) is highly recommended. Defect profiles obtained from the McChasy code can also be useful as input data for MD simulations to create large structures containing complex defects to provide a more detailed analysis.

Acknowledgements Presented work was supported by Instituto Superior Tecnico (Lisbon, Portugal) from post-doctoral grant 1018P.03026 and by the Institute of Electronic Materials Technology (Warsaw, Poland) from the prize for young scientists awarded to the first author. Parts of this research were presented at the European Materials Research Society 2018 Fall Meeting, Symposium J: Modelling. Theory and simulation in physics for materials applications, talk J.5.2.

References

1. L.C. Feldman, J.W. Mayer, S.T. Picraux, *Materials Analysis by Ion Channeling: Submicron Crystallography* (Academic Press, New York, 1982)
2. M. Nastasi, J.W. Mayer, J.K. Hirvonen, *Ion-Solid Interactions: Fundamentals and Applications* (Cambridge University Press, Cambridge, 1996)
3. A. Vantomme, 50 years of ion channeling in materials science. Nucl. Instrum. Methods Phys. Res. Sect. B Beam Interact. Mater. Atoms **371**, 12–26 (2016). https://doi.org/10.1016/j.nimb.2015.11.035
4. E. Bøgh, Defect studies in crystals by means of channeling. Can. J. Phys. **46**, 653–662 (1968). https://doi.org/10.1139/p68-081
5. J.H. Barrett, Monte carlo channeling calculations. Phys. Rev. B. **3**, 1527–1547 (1971). https://doi.org/10.1103/PhysRevB.3.1527
6. M.T. Robinson, I.M. Torrens, Computer simulation of atomic-displacement cascades in solids in the binary-collision approximation. Phys. Rev. B. **9**, 5008–5024 (1974). https://doi.org/10.1103/PhysRevB.9.5008
7. O.S. Oen, M.T. Robinson, Computer studies of the reflection of light ions from solids. Nucl. Instrum. Methods **132**, 647–653 (1976). https://doi.org/10.1016/0029-554X(76)90806-5
8. S. Zhang, K. Nordlund, F. Djurabekova, Y. Zhang, G. Velisa, T.S. Wang, Simulation of Rutherford backscattering spectrometry from arbitrary atom structures. Phys. Rev. E **94**, 043319 (2016). https://doi.org/10.1103/PhysRevE.94.043319
9. P.J.M. Smulders, D.O. Boerma, Computer simulation of channeling in single crystals. Nucl. Inst. Methods Phys. Res. B **29**, 471–489 (1987). https://doi.org/10.1016/0168-583X(87)90058-9
10. A. Dygo, A. Turos, Surface studies of AIIIBV compound semiconductors by ion channeling. Phys. Rev. B **40**, 7704–7713 (1989). https://doi.org/10.1103/PhysRevB.40.7704
11. E. Albertazzi, M. Bianconi, G. Lulli, R. Nipoti, M. Cantiano, Different methods for the determination of damage profiles in Si from RBS-channeling spectra: a comparison. Nucl. Instrum. Methods Phys. Res. Sect. B Beam Interact. Mater. Atoms **118**, 128–132 (1996). https://doi.org/10.1016/0168-583x(95)01489-6
12. G. Lulli, E. Albertazzi, M. Bianconi, G.G. Bentini, R. Nipoti, R. Lotti, Determination of He electronic energy loss in crystalline Si by Monte-Carlo simulation of Rutherford backscattering-channeling spectra. Nucl. Instrum. Methods Phys. Res. Sect. B Beam Interact. Mater. Atoms. **170**, 1–9 (2000). https://doi.org/10.1016/s0168-583x(00)00089-6
13. L. Nowicki, A. Turos, R. Ratajczak, A. Stonert, F. Garrido, Modern analysis of ion channeling data by Monte Carlo simulations. Nucl. Instrum. Methods Phys. Res. Sect. B Beam Interact. Mater. Atoms **240**, 277–282 (2005). https://doi.org/10.1016/j.nimb.2005.06.129

14. K. Gärtner, Modified master equation approach of axial dechanneling in perfect compound crystals. Nucl. Instruments Methods Phys. Res. Sect. B Beam Interact. Mater. Atoms **227**, 522–530 (2005). https://doi.org/10.1016/j.nimb.2004.10.087
15. F. Schiettekatte, Fast Monte Carlo for ion beam analysis simulations. Nucl. Instrum. Methods Phys. Res. Sect. B Beam Interact. Mater. Atoms **266**, 1880–1885 (2008). https://doi.org/10.1016/j.nimb.2007.11.075
16. A. Turos, P. Jozwik, L. Nowicki, N. Sathish, Ion channeling study of defects in compound crystals using Monte Carlo simulations. Nucl. Instruments Methods Phys. Res. Sect. B Beam Interact. Mater. Atoms **332**, 50–55 (2014). https://doi.org/10.1016/j.nimb.2014.02.028
17. M. Mayer, W. Eckstein, H. Langhuth, F. Schiettekatte, U. Von Toussaint, Computer simulation of ion beam analysis: spossibilities and limitations. Nucl. Instrum. Methods Phys. Res. Sect. B Beam Interact. Mater. Atoms **269**, 3006–3013 (2011). https://doi.org/10.1016/j.nimb.2011.04.066
18. J.H. Barrett, Methods of channeling simulation. Nucl. Inst. Methods Phys. Res. B **44**, 367–372 (1990). https://doi.org/10.1016/0168-583X(90)90652-B
19. J.F. Ziegler, J.P. Biersack, The stopping and range of ions in matter, in *Treatise on Heavy-Ion Science*, ed. by D.A. Bromley (Springer, Boston, MA, 1985), pp. 93–129
20. J.P. Biersack, J.F. Ziegler, Refined universal potentials in atomic collisions. Nucl. Instrum. Methods Phys. Res. **194**, 93–100 (1982). https://doi.org/10.1016/0029-554X(82)90496-7
21. A. Dygo, A. Turos, Small angle scattering by screened coulomb fields. Radiat. Eff. **85**, 237–242 (1984). https://doi.org/10.1080/01422448508209697
22. A. Cromer, Stable solutions using the Euler approximation. Am. J. Phys. **49**, 455–459 (1981). https://doi.org/10.1119/1.12478
23. P. Jozwik, N. Sathish, L. Nowicki, J. Jagielski, A. Turos, L. Kovarik, B. Arey, Monte Carlo simulations of backscattering process in dislocation-containing $SrTiO_3$ single crystal. Nucl. Instrum. Methods Phys. Res. Sect. B Beam Interact. Mater. Atoms **326**, 234–237 (2014). https://doi.org/10.1016/j.nimb.2013.11.022
24. A. Dygo, A. Turos, Monte Carlo calculation of energy loss of channeled ions. Phys. Lett. A **127**, 281–284 (1988). https://doi.org/10.1016/0375-9601(88)90697-4
25. B.T.M. Willis, A.W. Pryor, *Thermal Vibrations in Crystallography* (University Press, Cambridge, 1975)
26. R. Peierls, The size of a dislocation. Proc. Phys. Soc. **52**, 34–37 (1940). https://doi.org/10.1088/0959-5309/52/1/305
27. F.R.N. Nabarro, Dislocations in a simple cubic lattice. Proc. Phys. Soc. **59**, 256–272 (1947). https://doi.org/10.1088/0959-5309/59/2/309
28. Y. Xiang, Modeling dislocations at different scales. Commun. Comput. Phys. **1**, 383–424 (2006)
29. P. Jóźwik, N. Sathish, L. Nowicki, J. Jagielski, A. Turos, L. Kovarik, B. Arey, S. Shutthanandan, W. Jiang, J. Dyczewski, A. Barcz, Analysis of crystal lattice deformation by ion channeling. Acta Phys. Pol., A **123**, 828–830 (2013). https://doi.org/10.12693/APhysPolA.123.828
30. P. Jozwik, L. Nowicki, R. Ratajczak, A. Stonert, C. Mieszczynski, A. Turos, K. Morawiec, K. Lorenz, E. Alves, Monte Carlo simulations of ion channeling in crystals containing dislocations and randomly displaced atoms. J. Appl. Phys. **126**, 195107 (2019). https://doi.org/10.1063/1.5111619. Supplementary material: a movie illustrating the Microsoft Excel script for measuring the dislocation parameters, which is distributed with the McChasy code, available at https://aip.scitation.org/doi/suppl/10.1063/1.5111619/suppl_file/mcchasydislocationexcelscript.mp4 (2019)
31. P. Jóźwik, L. Nowicki, C. Mieszczyński, R. Ratajczak, A. Stonert, A. Turos, J. Jagielski, K. Lorenz, E. Alves, Advanced Monte Carlo simulations for ion-channeling studies of defects in crystals. In: E-MRS 2018 Fall Meeting, Symposium J: Theory and Simulation in Physics for Materials Applications, p. (talk) J.5.2., Warsaw (2018)

32. G. Perillat-Merceroz, P. Gergaud, P. Marotel, S. Brochen, P.H. Jouneau, G. Feuillet, Formation and annealing of dislocation loops induced by nitrogen implantation of ZnO. J. Appl. Phys. **109**, 023513 (2011). https://doi.org/10.1063/1.3537917
33. K. Lorenz, E. Wendler, A. Redondo-Cubero, N. Catarino, M.P. Chauvat, S. Schwaiger, F. Scholz, E. Alves, P. Ruterana, Implantation damage formation in a-, c- and m-plane GaN. Acta Mater. **123**, 177–187 (2017). https://doi.org/10.1016/j.actamat.2016.10.020
34. P. Jozwik, S. Magalhães, R. Ratajczak, C. Mieszczynski, M. Sequeira, A. Turos, R. Böttger, R. Heller, K. Lorenz, E. Alves, RBS/C, XRR, and XRD studies of damage buildup in Er-implanted ZnO. Phys. Status Solidi Basic Res. **256**, 1800364 (2018). https://doi.org/10.1002/pssb.201800364
35. A. Turos, P. Jóźwik, M. Wójcik, J. Gaca, R. Ratajczak, A. Stonert, Mechanism of damage buildup in ion bombarded ZnO. Acta Mater. **134**, 249–256 (2017). https://doi.org/10.1016/j.actamat.2017.06.005

Part III
Recent Progress in Electronic Transport and Device Simulation, Optical Properties

Chapter 9
Electronic and Optical Properties of Polypyrrole as a Toxic Carbonyl Gas Sensor

Francisco C. Franco Jr.

Abstract Chronic formaldehyde and acetaldehyde exposure is known to cause various health problems and the detection of these toxic carbonyl gases is very important and a subject of interest both experimentally and theoretically. In this study, the interaction of various oligomers (n = 1, 3, 5, 7, and 9) of polypyrrole towards toxic carbonyl species: acetaldehyde and formaldehyde, and less toxic carbonyl species: acetone and butanone were studied using density functional theory (DFT). The interactions of the carbonyl species with oligopyrrole lead to differences in interaction energies and changes in structural features: H-bond distances, bond angles, and dihedral angles. The changes resulted in variations in the electronic properties of the pyrrole-gas complexes: HOMO/LUMO energies, ionization potentials (IP), electron affinity (EA), and energy gap (E_{Gap}). The pyrrole-carbonyl complexes resulted in higher HOMO energies due to electron charge donation from the carbonyl gases and lower LUMO energies resulting in smaller E_{Gap} values compared to pyrrole. It was observed that the smallest carbonyl molecule, formaldehyde (For), had the lowest LUMO energy and lowest E_{Gap} value, while the largest carbonyl molecule, butanone (MEK), had the highest LUMO energy and highest E_{Gap} value. Furthermore, simulated UV–Vis absorption studies showed red-shifted first singlet excited state, $\lambda_{\mathbf{1st}}$, for the pyrrole-gas complexes. The results do not only demonstrate the potential of polypyrrole as a toxic carbonyl gas sensor but also its selectivity towards different carbonyl species.

9.1 Introduction

Chronic exposure to toxic carbonyl gas species such as formaldehyde and acetaldehyde are known to cause various health problems including irritation, asthma, cancer, genetic deficiency, among others [1, 2]. Formaldehyde is a colorless and flammable gas with a distinct strong unpleasant odor. It is known that chronic

F. C. Franco Jr. (✉)
Chemistry Department, De La Salle University, 2401 Taft Avenue, 0922 Manila, Philippines
e-mail: francisco.franco@dlsu.edu.ph

E. V. Levchenko et al. (eds.), *Theory and Simulation in Physics for Materials Applications*, Springer Series in Materials Science 296,
https://doi.org/10.1007/978-3-030-37790-8_9

exposure to formaldehyde may cause several adverse health effects including eye, skin, throat and nasal irritation, cancer, asthma, and allergy [3]. Formaldehyde is commonly used and found in resins, building materials, household products, preservatives, cosmetics, and agricultural products [2, 4]. Acetaldehyde is a colorless and flammable liquid with a suffocating smell. Acetaldehyde is an irritant, carcinogen and may result in damage to the lungs, kidney, liver, and skin for prolonged exposure [5]. Acetaldehyde is naturally occurring and may be found in plants, vegetables, perfumes, drugs, gasoline, and cigarette smoke and a product of ethanol oxidation in the liver [2, 6]. Less toxic carbonyl compounds such as acetone and methyl ethyl ketone may cause health problems at high doses and mainly causes irritation [7, 8]. Therefore, it is imperative that these toxic gaseous compounds be easily detected to reduce their damaging health effects. Common gas sensors for toxic gases are metal oxide semiconductors, field-effect transistors, surface acoustic wave, and organic conducting polymers [9–11]. Conducting polymers (CPs) offer various advantages over inorganic semiconductors, such as ease of fabrication, flexibility, light-weight, low-cost, stability, sensitivity and adjustable optoelectronic properties [12–14].

Since its discovery, CPs have attracted a great deal of interest due to its wide scope of applications, such as optoelectronic materials, solar cells, batteries, fuel cells, actuators, and sensors [15–18]. Some of these interesting properties of CPs mainly stem from its tunable electrical conductivity by chemical modification, doping, and de-doping processes [19]. Various CPs have been introduced and used for gas sensing including polythiophene, polyfuran, polyacetylene, polyaniline, and polypyrrole [12–14]. Polypyrrole is one of the most important and well-studied biosensors and gas sensors due to its stability, sensitivity to pH, and high conductivity [20]. Polypyrrole has been studied both experimentally and theoretically as a sensor for various gaseous molecules, such as H_2, CH_4, NO_3^-, NO_2, H_2O_2, CO_2, NH_3 [21–26]. Polypyrrole has been previously studied to detect acetone as a gas sensor by chemical oxidation-casting (COC), chemical vapor deposition (CVD), and impregnated oxidation (IO) synthesized on a gold/aluminum oxide surface [27]. However, theoretical and experimental studies on polypyrrole as a gas sensor for toxic carbonyl gases such as formaldehyde and acetaldehyde, and less toxic ones, acetone, and methyl ethyl ketone has been very limited so far. Thus, more theoretical studies on the interaction of polypyrrole to these toxic carbonyl gases must be carried out to understand its sensitivity, selectivity, and potential as a carbonyl gas sensor.

Predicting material properties and trends via computational methods is a convenient, fast, and affordable way of determining the potential of a material of interest for the intended application. From these methods, density functional theory (DFT) is arguably the most popular method to understand the electronic and optical properties of materials at the atomic level due to its balance between computational cost and accuracy [28, 29]. DFT studies on π-conjugated systems have been carried out extensively in the literature for various material applications and even though its limitations are well-known, the trends for similar systems resulting from these calculations are typically the same with the experimental values and are thus meaningful [30–33]. Furthermore, structure-property relationships between the structural changes in the system and its resulting electronic properties are well-addressed by DFT [34–36]. Understanding and predicting properties using DFT can aid in the development of polypyrrole as potential gas sensors for toxic carbonyl gases.

In this study, the interaction of several carbonyl gases: formaldehyde (For), acetaldehyde (Actl), acetone (Ace), and methyl ethyl ketone (MEK) are studied by DFT. First, the structural changes in the oligopyrrole (*n*Py) due to the presence of the four gases were determined including the bond lengths, bond angles, and dihedral angles. The interaction energies were then determined and related to the changes in the structural parameters and subsequent analysis. Second, the charge transfer occurring between the adsorbed gas and *n*Py were then analyzed and related to the strength of the interaction between the gas and oligopyrrole. Third, the electronic properties, the density of states, and excitation energies were determined to understand the resulting electrical and optical properties of the oligopyrrole gas sensor towards the carbonyl gases.

9.2 Materials and Methods

Figure 9.1 shows the chemical structures of the pyrrole (*n*Py) complexes with several carbonyl gases: formaldehyde (For), acetaldehyde (Actl), acetone (Ace), and methyl ethyl ketone (MEK) used in this study. All quantum-chemical calculations were carried out using GAMESS(US) quantum chemical package [37]. The equilibrium geometries in gas phase of the oligomers were determined using DFT/B3LYP-D3(BJ)/6-31G(d) with gradient minimization at Becke three-parameter exchange, Lee, Yang, and Parr (B3LYP) [38], a hybrid functional with 20% HF exchange and 6-31G(d) as the basis set. Dispersion forces, although weak, may have a significant impact at the molecular level and corrections have been carried out with the third implementation of Grimme's electron dispersion correction (D3) with modification for damping (BJ) [39–41]. The equilibrium geometries were then confirmed by frequency analysis to make sure that these structures correspond to true minima. Total energies were calculated also at the same level with geometrical counterpoise correction, DFT/B3LYP-gCP-D3(BJ)/6-31G(d), to remove the error in energy calculations due to the overlap of basis sets generally referred to as the basis set superposition error (BSSE). The gCP corrections were calculated from Grimme's Webservice [42]. The uncorrected binding energies, E_{bind}, and gCP-D3 corrected binding energies, ${E_{\mathrm{bind}}}^{\mathrm{gCP-D3}}$, were calculated as:

$$E_{\mathrm{bind}} = E_{\mathrm{complex}} - E_{\mathrm{Py}} - E_{\mathrm{X}} \quad (9.1)$$

$${E_{\mathrm{bind}}}^{\mathrm{gCP-D3}} = {E_{\mathrm{complex}}}^{\mathrm{gCP-D3}} - {E_{\mathrm{Py}}}^{\mathrm{gCP-D3}} - {E_{\mathrm{X}}}^{\mathrm{gCP-D3}} \quad (9.2)$$

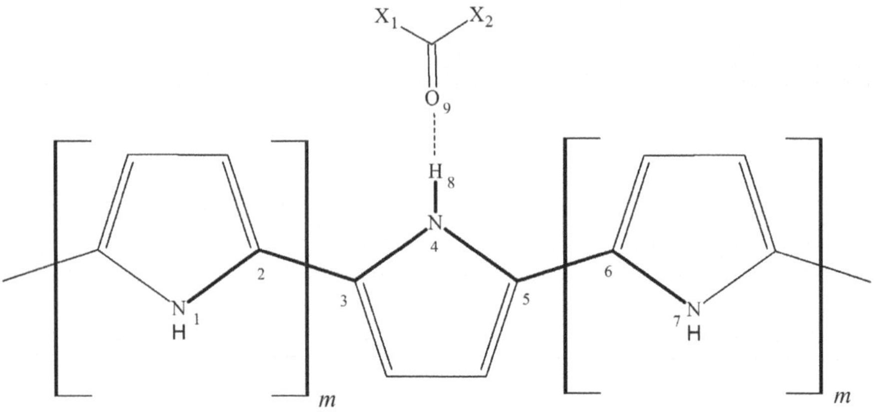

where $m = 0, 1, 2, 3, 4$; and $n = 2m + 1$

For: $X_1 = H, X_2 = H$ **Ace**: $X_1 = CH_3, X_2 = CH_3$

Actl: $X_1 = H, X_2 = CH_3$ **MEK**: $X_1 = CH_3, X_2 = CH_2CH_3$

Fig. 9.1 Chemical structures of the (*n*Py–X) pyrrole-gas complexes and the numbering scheme used in this study

where $E_{complex}$, E_{Py}, and E_X are the total energies of the pyrrole-carbonyl gas complex (Py–X), isolated pyrrole (Py), and isolated carbonyl gas (X), respectively. The UV–Vis spectrum was simulated by TDDFT calculations at the B3LYP/6-31G (d) level. NBO population analysis was carried out using the NBO program [43], and the molecular orbitals (MOs) were viewed analyzed using MacMolPlt [44] and Chemcraft [45]. The UV–Vis spectrum and density of states (DOS) were plotted using the GaussSum package [46].

9.3 Results and Discussion

9.3.1 *Structural Parameters for the* n*Py–X Complexes*

The interaction of carbonyl gases with pyrrole introduces structural changes in its bond distances (d), bond angles (θ), and dihedral angles (Φ) as shown in Table 9.1. The values and trends for the structural properties are virtually the same from 3Py–X to 9Py–X complexes. The distance between N_4–H_8 elongates slightly to 1.02 Å from 1.01 Å when Py interacts with X. For the H_8–O_9 hydrogen bond (H-bond), the shortest distance occurs for *n*Py–Ace followed by *n*Py–Actl/*n*Py–MEK, and lastly *n*Py–For. The trend indicates that Ace has the strongest interaction with Py, and For has the weakest interaction. The angle between C_3–N_4–C_5 slightly decreases by a few tenths upon the interaction of Py with X. On the other hand, the N_4–H_8–O_9 angle is also indicative of the strength of the H-bond between Py and X, where the

Table 9.1 Structural parameters for the optimized *n*Py–X complexes (X = For, Actl, Ace, and MEK) at the B3LYP-D3(BJ)/6-31G(d) level

	d (Å)		θ (°)		Φ (°)		
*n*Py–X	$H_8 \cdot O_9$	N_4–H_8	$N_4H_8O_9$	$C_3N_4C_5$	$N_1C_2C_3N_4$	$C_2C_3N_4C_5$	$N_4C_5C_6N_7$
1Py	–	1.01	–	109.9	–	–	–
1Py–For	2.03	1.02	148.5	109.6	–	–	–
1Py–Actl	1.97	1.02	154.2	109.5	–	–	–
1Py–Ace	1.94	1.02	162.4	109.5	–	–	–
1Py–MEK	1.97	1.02	151.3	109.6	–	–	–
3Py	–	1.01	–	110.7	156.9	−179.5	−156.3
3Py–For	1.97	1.02	158.3	110.5	153.8	−178.5	173.0
3Py–Actl	1.92	1.02	164.6	110.5	178.2	−178.9	−159.0
3Py–Ace	1.91	1.02	169.7	110.5	169.1	−178.0	−167.4
3Py–MEK	1.93	1.02	159.7	110.6	163.4	−177.3	177.2
5Py	–	1.01	–	110.6	159.1	−179.3	−159.1
5Py–For	1.97	1.02	158.7	110.5	154.9	−179.0	177.3
5Py–Actl	1.93	1.02	165.0	110.5	176.5	−179.1	−160.9
5Py–Ace	1.91	1.02	169.2	110.4	169.9	−178.0	−168.2
5Py–MEK	1.93	1.02	160.0	110.5	164.6	−177.7	−179.8
7Py	–	1.01	–	110.6	159.3	−179.3	−159.0
7Py–For	1.97	1.02	159.2	110.4	154.8	−178.9	177.1
7Py–Actl	1.92	1.02	166.0	110.4	175.6	−178.9	−161.0
7Py–Ace	1.91	1.02	169.4	110.5	170.2	−178.1	−168.1
7Py–MEK	1.92	1.02	160.1	110.5	164.7	−177.6	−179.6
9Py	–	1.01	–	110.6	159.0	−179.2	−158.7
9Py–For	1.97	1.02	159.1	110.4	154.5	−178.8	177.3
9Py–Actl	1.93	1.02	166.3	110.4	175.9	−178.8	−160.9
9Py–Ace	1.90	1.02	169.8	110.4	170.8	−178.2	−166.8
9Py–MEK	1.93	1.02	159.9	110.6	164.6	−177.5	−178.6

more linear H-bond indicates stronger H-bond. 9Py–Ace (169.8°) has the most linear H-bond followed by 9Py–Actl (166.3°), 9Py–MEK (159.9°), and 9Py–For (159.1°). For the dihedral angles, C_2–C_3–N_4–C_5 decreases by about ~2° or less when Py interacts with X. However, large changes in the dihedral angles were observed for N_1–C_2–C_3–N_4 and N_4–C_5–C_6–N_7, two important dihedral angles that reveal changes in the planarity of the Py backbone. The changes in the dihedral angles resulted in increased planarity for Py after its interaction with X as shown in Fig. 9.2 for 9Py–X structures. The N_1–C_2–C_3–N_4 dihedral angle for 9Py (159.0°), slightly becomes out-of-plane to 154.5° for 9Py–For but increased in planarity for 9Py–Actl (175.9°), 9Py–Ace (170.8°), and 9Py–MEK (164.6°). On the other hand, for the N_4–C_5–C_6–N_7 dihedral angle, 9Py (−158.7°) becomes more planar for 9Py–For (177.3°), 9Py–Actl (−160.9°), 9Py–Ace (−166.8°), and 9Py–MEK (−178.6°). Even though bulky molecules interacted with the polymer backbone, which could

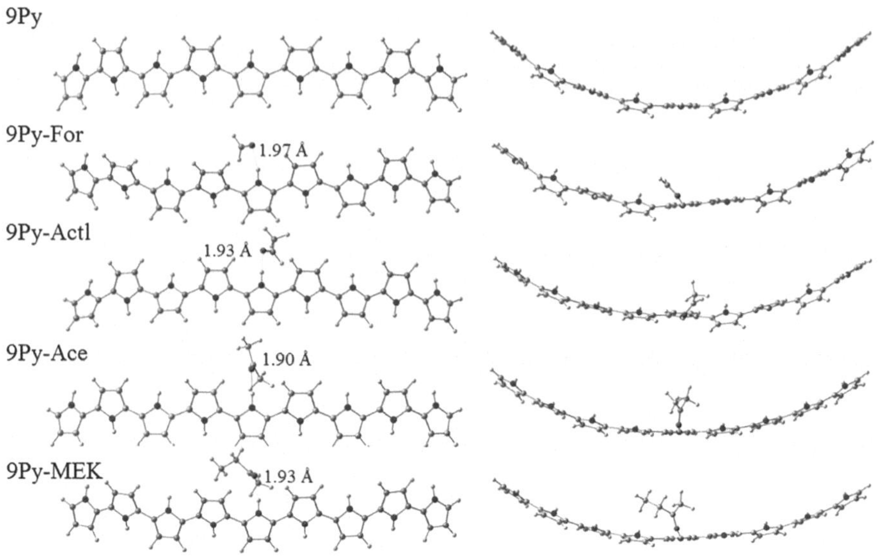

Fig. 9.2 Optimized structures for 9Py–X complexes (X = For, Actl, Ace, and MEK) at the B3LYP-D3(BJ)/6-31G(d) level

result to decrease in planarity due to steric effect, the greater electronic effect was observed for Py resulting in increased planarity upon the interaction of carbonyl gases with Py.

9.3.2 *Vibrational Analysis of the* n*Py–X Complexes*

The vibrations for the 9Py–X complexes were then analyzed and it was observed that the peak at 3687 cm^{-1} (N_4–H_8 stretching) for the isolated pyrrole (9Py) decreased significantly as it interacts with the carbonyl molecules. The most significant shift was observed for 9Py–Ace (3439 cm^{-1}) followed by 9Py–MEK (3478 cm^{-1}), 9Py–Actl (3475 cm^{-1}), and 9Py–For (3501 cm^{-1}). The shift to lower frequencies is due to the weakening of the N_4–H_8 bond upon interaction with the carbonyl molecules. The stronger is the H-bonding between Py and X, the more significant is the shift to lower frequencies.

9.3.3 n*Py–X Binding Energies*

Table 9.2 shows the binding energies for the interaction of *n*Py and the carbonyl molecules for both uncorrected and gCP-D3 corrected energies. As with the

Table 9.2 Binding energies for *n*Py–X complexes (X = For, Actl, Ace, and MEK) at the B3LYP-D3(BJ)/6-31G(d) level

	E_{bind}, uncorrected (kcal/mol)					E_{bind}, gCP-D3 corrected (kcal/mol)				
*n*Py–X	1Py	3Py	5Py	7Py	9Py	1Py	3Py	5Py	7Py	9Py
*n*Py–For	−5.48	−5.93	−5.79	−5.81	−5.84	−5.78	−7.42	−7.32	−7.34	−7.37
*n*Py–Actl	−6.22	−6.97	−6.98	−7.06	−7.06	−6.52	−8.33	−8.35	−8.40	−8.36
*n*Py–Ace	−6.33	−7.45	−7.51	−7.60	−7.61	−7.13	−9.85	−9.99	−10.08	−10.06
*n*Py–MEK	−4.33	−5.47	−5.47	−5.55	−5.43	−5.87	−9.79	−9.95	−10.05	−9.94

structural parameters in Table 9.1, similar values and trends were observed from 3Py–X to 9Py–X complexes where the binding energies do not significantly change as the *n*Py chain length increases. These results show that the interactions for *n*Py–X complexes can be modeled at shorter chain lengths starting from 3Py. The gCP-D3 correction in the binding energies was included to minimize the effect of BSSE and include weak dispersion interactions between the *n*Py chains and the carbonyl molecules. The binding energies show that the BSSE and dispersion forces affect the complexes at varying degrees, from a correction of −1.5 kcal/mol for 9Py–For, and up to −4.5 kcal/mol for 9Py–MEK due to the size of the MEK molecule, which allows it to have significant weak interactions with the neighboring hydrogens. The trend for $E_{bind}{}^{gCP-D3}$ the is consistent with the structural and vibrational analyses in the previous sections where *n*Py–Ace (−10.06 kcal/mol) has the strongest interactions followed by *n*Py–MEK (−9.94 kcal/mol), *n*Py–Actl (−8.36 kcal/mol), and lastly, *n*Py–For (−7.37 kcal/mol). The structural, vibrational, and binding energy results show that the carbonyl gases have significant interaction with the Py chain resulting in electron charge transfer and changes in the electronic properties as shown in the next sections.

9.3.4 Charge Transfer Analysis

The electron charge transfer of an analyte with a CP results in changes to the electronic structure of the CP, thus greatly affecting its sensing capability. The charge transfer between Py and the carbonyl molecules were studied using Mulliken ($Q_{Mulliken}$) and NBO (Q_{NBO}) and are shown in Table 9.3. The charge transfer results show that all values are positive (*n*-type doping) and that electron charges were transferred from the carbonyl molecules to the *n*Py chain. It can also be observed that the magnitude of charge transfers are virtually constants as the chain length increases from 3Py to 9Py. Also, even though $Q_{Mulliken}$ and Q_{NBO} have different magnitudes, the trends are the same for both methods where Ace has the

Table 9.3 Population analysis for the carbonyl molecules (X) in the *n*Py–X complexes (X = For, Actl, Ace, and MEK) at the B3LYP/6-31G(d) level

	$Q_{Mulliken}$ (for X)					Q_{NBO} (for X)				
*n*Py–X	1Py	3Py	5Py	7Py	9Py	1Py	3Py	5Py	7Py	9Py
*n*Py–For	0.036	0.032	0.031	0.033	0.033	0.023	0.018	0.017	0.018	0.018
*n*Py–Actl	0.044	0.055	0.055	0.057	0.056	0.028	0.034	0.034	0.034	0.034
*n*Py–Ace	0.053	0.059	0.059	0.060	0.060	0.032	0.039	0.038	0.039	0.039
*n*Py–MEK	0.045	0.038	0.037	0.038	0.038	0.028	0.031	0.031	0.031	0.031

high electron donation, followed by Actl, MEK, and with For having the lowest charge transfer. In organic chemistry, alkyl groups are considered as electron donors, and their electron-donating ability increases as the chain length increases. The carbonyl Ace has two methyl groups donating electron charge to the O atom, which is then transferred to the Py chain resulting in strong H-bonding. On the other hand, MEK has a longer alkyl group on one side of its carbonyl group but has less charge donation to the Py chain due to the steric hindrance of its alkyl groups, resulting in weaker H-bonding compared to Ace. For has the weakest H-bonding between the carbonyl molecules and resulting in the least charge donation to the Py chain. In the next section, the charge transfer between Py and its interaction with carbonyl gases is then shown to affect its electronic energies, particularly the HOMO energy.

9.3.5 *Effect of Carbonyl Gases on the Electronic Properties of Pyrrole*

The sensing potential of CPs is based on their chemiresistive property, which highly depends on their electronic properties. Generally, a decrease (or increase) in the energy gap (E_{Gap}) of the CP after interacting with analyte results in an increase (or decrease) in the conductivity of the material. Thus, the better the interaction of a substance with a CP, the more pronounced are the changes in its electronic properties and the better its sensing capability. For various analytes interacting with the CP, the more varied are the changes in the electronic properties, the better is the selectivity of the CP towards different substances. Figure 9.3 shows the changes in the frontier orbital energies (E_{HOMO} and E_{LUMO}) of the *n*Py–X complexes. Figure 9.3a shows the changes in the E_{HOMO} and E_{LUMO} of the pyrrole CP for $n = 1, 3, 5, 7$, and 9, where it can be observed that the E_{Gap} decreases from 6.86 eV for 1Py and converges to 3.30 eV for 9Py as the chain length increases, which is consistent with the experimental E_{Gap} of 3.1 eV for the infinite chain length polypyrrole [47]. The trend in the E_{Gap} is attributed to the increase in the delocalization of electrons for longer polymer chain resulting to increase in the E_{HOMO}

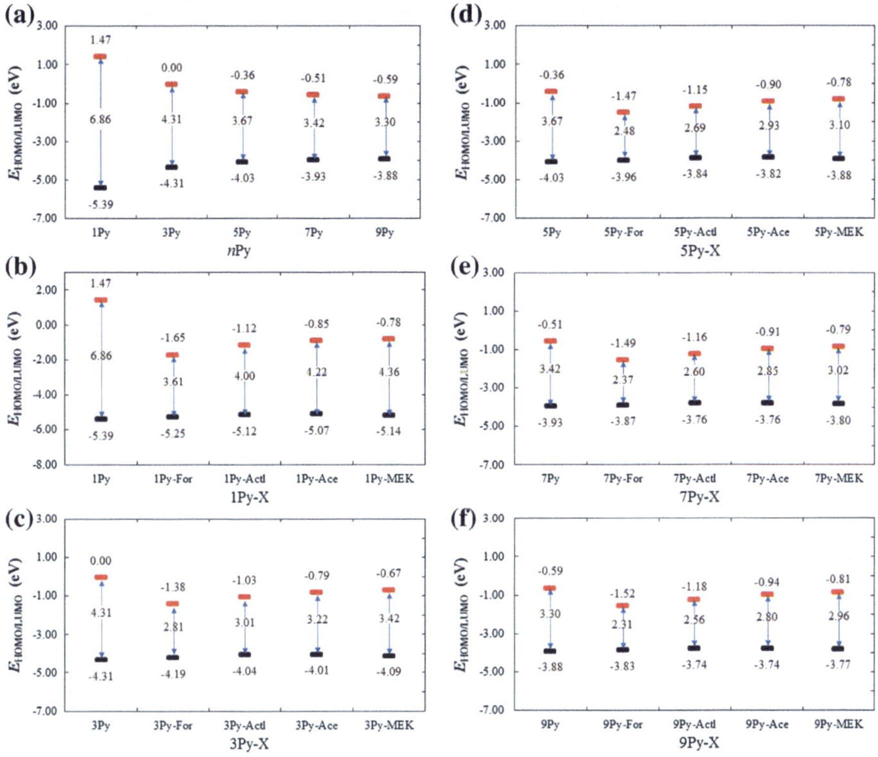

Fig. 9.3 Frontier orbital energies: E_{HOMO} (black marker), E_{LUMO} (red marker), and E_{Gap} (arrow) for the *n*Py–X complexes (X = For, Actl, Ace, and MEK) at the B3LYP-gCP-D3(BJ)/6-31G(d) level

(from −5.39 eV for 1Py to −3.88 eV for 9Py) and decrease in E_{LUMO} (from 1.47 eV for 1Py to −0.59 eV for 9Py).

For the *n*Py–X complexes, shown in Fig. 9.3b–f, reduction in the E_{Gap} values compared to the isolated *n*Py was observed due to the increase (and decrease) in E_{HOMO} (and E_{LUMO}) of pyrrole upon its interaction with the carbonyl gases. It can be observed from Fig. 9.3 that all *n*Py–X chain lengths (n = 1, 3, 5, 7, and 9) have the same trends in the frontier orbital energies. A summary of the frontier orbital energies (E_{HOMO}, E_{LUMO}, E_{Gap}), ionization potential (IP), and electron affinity (EA) for 9Py–X are shown in Table 9.4. The values for IP and EA were estimated using Koopman's theorem: IP = $-E_{\mathrm{HOMO}}$ and EA = $-E_{\mathrm{LUMO}}$. The values show that 9Py–Ace followed 9Py–Actl have the highest increase in E_{HOMO} resulting to decrease in IP due to the highest electron charge transfer of Ace and Actl to the Py chain, while 9Py–For has the smallest increase in E_{HOMO} due to the least electron charge transfer to Py. As will be shown below, the HOMO orbitals for the *n*Py–X complexes mainly have pyrrole character, and that the increase in E_{HOMO}, therefore, is attributed to the donation of electron charge to pyrrole by the carbonyl gases as

Table 9.4 Summary of the frontier orbital energies (E_{HOMO}, E_{LUMO}, E_{Gap}), IP, and EA for 9Py–X complexes

9Py–X	E_{HOMO} (eV)	E_{LUMO} (eV)	E_{Gap} (eV)	IP (eV)	EA (eV)
9Py	−3.88	−0.59	3.30	3.88	0.59
9Py–For	−3.83	−1.52	2.31	3.83	1.52
9Py–Actl	−3.74	−1.18	2.56	3.74	1.18
9Py–Ace	−3.74	−0.94	2.80	3.74	0.94
9Py–MEK	−3.77	−0.81	2.96	3.77	0.81

discussed in the previous section. On the other hand, the LUMO orbitals of the *n*Py–X mainly have carbonyl gas character and the E_{LUMO} and EA trend reflects the trend of the E_{LUMO} of the isolated carbonyl gases: $E_{LUMO} = -1.05$ eV (For), −0.50 eV (Actl), −0.20 eV (Ace), and −0.16 eV (MEK). The changes in the frontier orbital energies for the *n*Py–X complexes resulted to a large decrease in the energy gaps of pyrrole where For had the greatest effect on *n*Py and reduced the E_{Gap} to 2.31 eV from 3.30 eV for 9Py. The trend in the E_{Gap} values is the same for all chain lengths (1Py to 9Py): For < Actl < Ace < MEK. The values from Fig. 9.3 and Table 9.4 show that the energy gaps are significantly different among the carbonyl gases modeled in this study. These observations demonstrate the sensitivity (large reduction in energy gaps compared to pyrrole) and selectivity (significant differences in energy gaps for various complexes) of pyrrole as a potential sensor toxic carbonyl gas.

9.3.6 Density of States

The density of states (DOS) provides a further understanding of the changes in the electronic structure of the *n*Py–X complexes. The TDOS for 9Py–X complexes and PDOS for X are shown in Fig. 9.4a–d, while the frontier molecular orbital (FMO) surfaces were plotted and shown in Fig. 9.5. Compared to the isolated Py chain, additional peaks were observed for all the 9Py–X complexes due to the frontier orbital energies contributed by the carbonyl gases. These peaks correspond to the LUMO orbitals of the carbonyl gases as shown in Fig. 9.5. The LUMO orbitals of the 9Py–X complexes ($LUMO_{Py-X}$) have mostly the LUMO orbital character of the carbonyl gases ($LUMO_X$). The smallest E_{Gap} was observed for the Py–For complexes due to the additional LUMO orbital energy from For, reducing the difference between the HOMO and LUMO energies. On the other hand, the LUMO orbital energies of the rest of the complexes are closer in energy to the LUMO of isolated Py resulting in larger E_{Gap}. It can also be observed that the HOMO orbital energies of the Py–X complexes move to higher energy values due to the electron donation from the carbonyl gases. The DOS and FMO in this section further explained the observed changes in the electronic energies of Py upon interaction with the carbonyl gases.

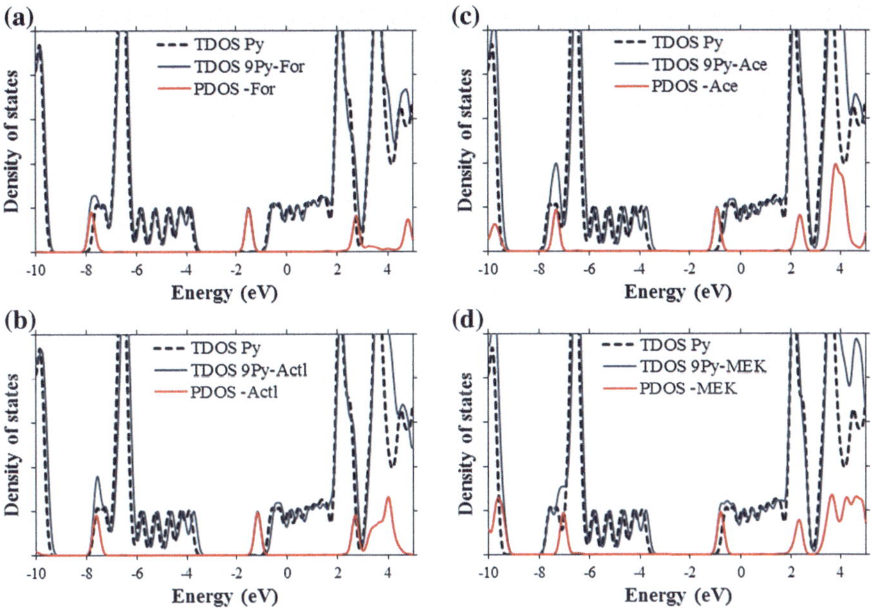

Fig. 9.4 The total density of states (TDOS) for 9Py–X complexes (solid blue line) and projected density of states (PDOS) for X (X = For, Actl, Ace, and MEK) shown as a red line. TDOS for 9Py is also shown as a dashed black line

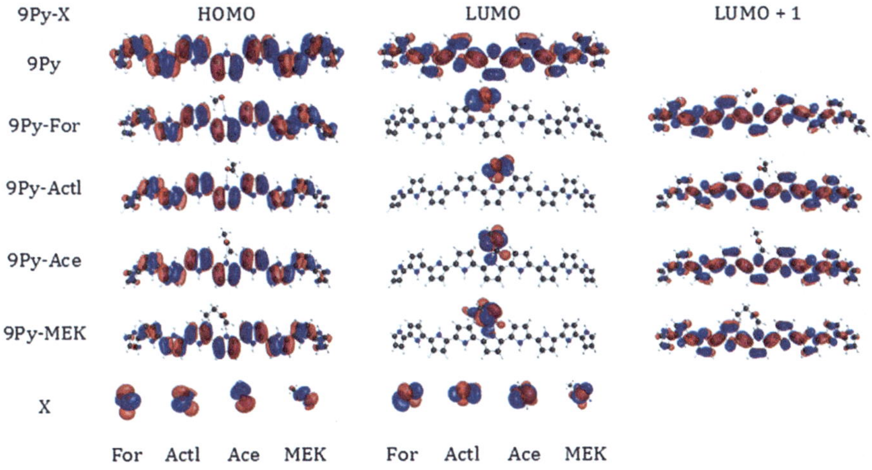

Fig. 9.5 Frontier molecular orbitals for 9Py–X complexes (X = For, Actl, Ace, and MEK) (isovalue = 0.02 a.u.)

9.3.7 *Simulated UV–Vis Absorption Spectra of the* n*Py–X Complexes*

In the above section, it was shown that the formation of pyrrole-carbonyl gas complexes results in a large decrease in the E_{Gap} compared to the isolated pyrrole chain. In addition to detecting the changes in the conductivity/resistivity of CPs in sensing analytes, CPs may also be used as optical sensors by detecting the changes in the absorption spectrum (absorbance and absorption wavelength). Figure 9.6 summarizes the simulated UV–Vis absorption spectra of the *n*Py–X complexes. For the isolated pyrrole chain from 3Py to 9Py (Fig. 9.6a–d), three major peaks were observed that shift to lower energies (red-shift) as the chain length increases, similar with previous studies on polypyrrole UV–Vis absorption [48–51]. For 9Py, peaks were observed at 428 nm, 364 nm (shoulder), and 336 nm, with their corresponding oscillator strengths 2, 53, 0.25, and 0.52, respectively. The first singlet excited state ($E_{excited,1st}$) was observed at 428 nm ($\pi \rightarrow \pi*$), consistent with the experimental value of 442 nm for polypyrrole [52]. As observed in Fig. 9.6 and discussed in the previous section, the energy gap and λ_{max} decrease as the chain length increases, thus it should be expected that further red-shift will occur at infinite chain length.

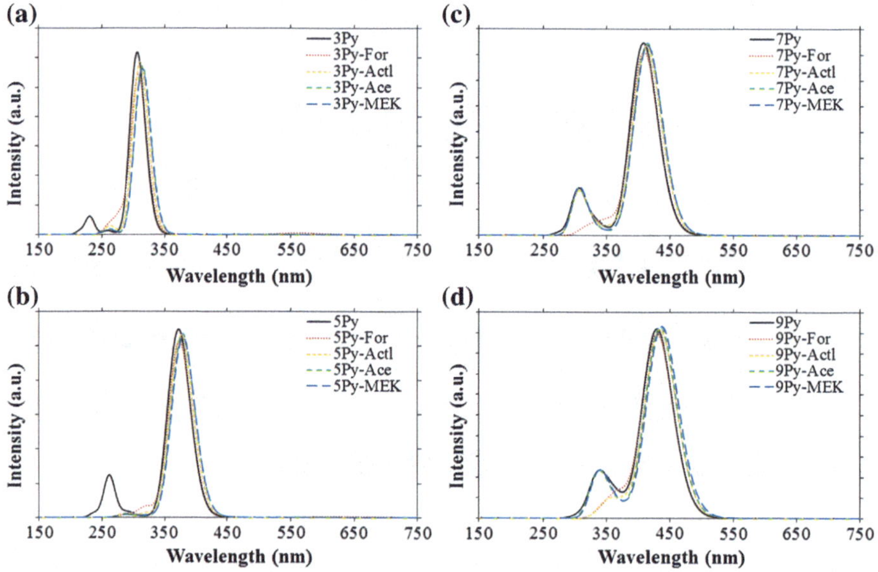

Fig. 9.6 Simulated UV–Vis absorption spectra for the *n*Py–X complexes. *n*Py–For (dotted red line), *n*Py–Actl (dotted yellow line), *n*Py–Ace (dashed green line), and *n*Py–MEK (dashed blue line) at the TD-DFT/B3LYP/6-31G(d) level (FWHM = 3000 cm^{-1}). The UV–Vis spectra for *n*Py is also shown as a black solid line

Table 9.5 Excitation energies for the maximum absorbance ($E_{excited,max}$) and first excited state ($E_{excited,1st}$), oscillator strengths (f_{Osc}) of 9Py–X complexes calculated at the TD-DFT/B3LYP/6-31G(d) level. The corresponding main electronic configurations (>5%) are also shown

9Py–X	$E_{excited,max}$, eV (λ_{max}, nm)	f_{Osc}	$E_{excited,1st}$, eV (λ_{1st}, nm)	f_{Osc}
9Py	2.89 (429) H → L	2.5396	2.89 (429) H → L	2.5396
9Py–For	2.88 (430) H → L + 1	2.4715	1.82 (678) H → L	0.0067
9Py–Actl	2.86 (434) H → L + 1	2.5326	2.06 (602) H → L	0.0052
9Py–Ace	2.86 (434) H → L + 1	2.5529	2.30 (540) H → L	0.0016
9Py–MEK	2.84 (436) H → L + 1	2.5751	2.45 (506) H → L	0.0061

Figure 9.6 also shows the changes in the UV–Vis absorption spectra for the *n*Py–X complexes. The λ_{max} for all the *n*Py–X complexes red-shifted, consistent with the reduction in E_{Gap} for the complexes as discussed above. Table 9.5 summarizes the excitation energies for the 9Py–X complexes and their oscillator strengths. Unlike the isolated 9Py chain, the first excited singlet transitions for the 9Py–X complexes occur at much longer wavelengths: 678 nm (9Py–For), 602 nm (9Py–Actl), 602 nm (9Py–Ace), and 506 nm (9Py–MEK), with very weak oscillator strengths (low absorption coefficients): 0.0067, 0.0052, 0.0016, and 0.0061, respectively, and not observable in Fig. 9.6. These excitations are assigned to $\pi(Py) \rightarrow \pi^*(X)$ transitions, where $\pi(Py)$ is the bonding π-orbital ($HOMO_{Py}$) for the isolated Py chain and $\pi^*(X)$ is the antibonding π-orbital ($LUMO_X$) for the carbonyl gas and discussed in the previous section. The trend in the $E_{excited,1st}$ is the same with the trends in the E_{Gap} where For < Actl < Ace < MEK. However, due to the weak absorbance of the λ_{1st}, it would be more practical to measure their λ_{max}. The maximum absorbance ($E_{excited,max}$) is assigned to the $\pi(Py) \rightarrow \pi^*(Py)$ transitions. Although the peak values are lower, they are all very close to *n*Py's λ_{max} which would make it difficult to differentiate between the isolated Py and the Py–X complex. On the other hand, only a single molecule of carbonyl gas was allowed to interact with the pyrrole chain and an increase in the concentration of the interacting gas may be able to increase the absorbance of the first excited singlet since there will be more available $\pi^*(X)$ states at higher concentration.

9.4 Conclusions

In this study, the sensing potential of polypyrrole (Py), a conducting polymer (CP) was studied using oligomers of various chain lengths (n = 1, 3, 5, 7, and 9) against several toxic carbonyl gases (X): formaldehyde (For), acetaldehyde (Actl), acetone (Ace), and butanone (MEK). The pyrrole chain increased in planarity upon interaction with the carbonyl gases and various changes in the structural parameters of pyrrole were observed due to the different interactions between the carbonyl gases and pyrrole. Acetone was found to have the strongest H-bonding and electron charge donation (*n*-doping) to the pyrrole chain resulting in the largest increase in the HOMO energy. The LUMO orbital energies of the pyrrole-carbonyl gas complexes have been observed to be contributed mainly by the LUMO orbitals of the carbonyl molecules. Formaldehyde has the lowest LUMO energy among the carbonyl gases resulting in the lowest LUMO energies for the *n*Py–For complexes. The changes in the HOMO and LUMO energies resulted in a significant variation in the energy gap values for the *n*Py–X complexes, showing the potential selectivity of polypyrrole due to the varying chemiresistive response for toxic carbonyl gases. UV–Vis simulations of the Py–X complexes showed that the first singlet excited states, $\lambda_{\mathbf{1st}}$ are greatly red-shifted compared to the isolated pyrrole chain and caused by the $\pi(\mathrm{Py}) \rightarrow \pi^*(\mathrm{X})$ transitions. This study has shown the potential of polypyrrole as toxic carbonyl gases sensor and its selectivity, and the results in this study may aid in future experiments and designs of conducting polymers for toxic carbonyl gas sensing applications.

Acknowledgements F. Franco, Jr. would like to thank the University Research Coordination Office of De La Salle University-Manila, Philippines (DLSU-URCO) and the Philippine Council for Industry, Energy, and Emerging Technology Research and Development of the Department of Science and Technology, Philippines (DOST-PCIEERD) for the support and computational resources.

References

1. K. Fujioka, T. Shibamoto, Determination of toxic carbonyl compounds in cigarette smoke. Environ. Toxicol. **21**, 47–54 (2006). https://doi.org/10.1002/tox.20153
2. D.J. Luecken, W.T. Hutzell, M.L. Strum, G.A. Pouliot, Regional sources of atmospheric formaldehyde and acetaldehyde, and implications for atmospheric modeling. Atmos. Environ. **47**, 477–490 (2012). https://doi.org/10.1016/j.atmosenv.2011.10.005
3. A. Heck d'Henry, M. Casanova, T.B. Starr, Formaldehyde toxicity—new understanding. Crit. Rev. Toxicol. **20**, 397–426 (1990). https://doi.org/10.3109/10408449009029329
4. M. Soffritti, C. Maltoni, F. Maffei, R. Biagi, Formaldehyde: an experimental multipotential carcinogen. Toxicol. Ind. Health **5**, 699–730 (1989). https://doi.org/10.1177/074823378900500510
5. J.B. Morris, Dosimetry, toxicity and carcinogenicity of inspired acetaldehyde in the rat. Mutat. Res. Mol. Mech. Mutagen. **380**, 113–124 (1997). https://doi.org/10.1016/S0027-5107(97)00130-9
6. M.A. Korsten, S. Matsuzaki, L. Feinman, C.S. Lieber, High blood acetaldehyde levels after ethanol administration. N. Engl. J. Med. **292**, 386–389 (1975). https://doi.org/10.1056/NEJM197502202920802
7. D. Grosjean, E. Grosjean, A.W. Gertler, On-road emissions of carbonyls from light-duty and heavy-duty vehicles. Environ. Sci. Technol. **35**, 45–53 (2001). https://doi.org/10.1021/es001326a
8. C. Gao, R. Govind, H.H. Tabak, Application of the group contribution method for predicting the toxicity of organic chemicals. Environ. Toxicol. Chem. **11**, 631–636 (1992). https://doi.org/10.1002/etc.5620110506
9. J.W. Grate, Acoustic wave microsensor arrays for vapor sensing. Chem. Rev. **100**, 2627–2648 (2000). https://doi.org/10.1021/cr980094j
10. K. Natori, Ballistic metal-oxide-semiconductor field effect transistor. J. Appl. Phys. **76**, 4879–4890 (1994). https://doi.org/10.1063/1.357263

11. S.K. Shukla, C.S. Kushwaha, N.B. Singh, Recent developments in conducting polymer based composites for sensing devices. Mater. Today Proc. **4**, 5672–5681 (2017). https://doi.org/10.1016/j.matpr.2017.06.029
12. H. Yoon, Current trends in sensors based on conducting polymer nanomaterials. Nanomater. **3**, 524–549 (2013). https://doi.org/10.3390/nano3030524
13. S. Park, C. Park, H. Yoon, Chemo-electrical gas sensors based on conducting polymer hybrids. Polym. (Basel) **9**, 155 (2017). https://doi.org/10.3390/polym9050155
14. A. Gusain, N.J. Joshi, P.V. Varde, D.K. Aswal, Flexible NO gas sensor based on conducting polymer poly[N-9′-heptadecanyl-2,7-carbazole-alt-5,5-(4′,7′-di-2-thienyl-2′,1′,3′-benzothiadiazole)] (PCDTBT). Sensors Actuators B Chem. **239**, 734–745 (2017). https://doi.org/10.1016/j.snb.2016.07.176
15. K.D. Kreuer, On the development of proton conducting polymer membranes for hydrogen and methanol fuel cells. J. Memb. Sci. **185**, 29–39 (2001). https://doi.org/10.1016/S0376-7388(00)00632-3
16. K.-S. Park, S.B. Schougaard, J.B. Goodenough, Conducting-polymer/iron-redox- couple composite cathodes for lithium secondary batteries. Adv. Mater. **19**, 848–851 (2007). https://doi.org/10.1002/adma.200600369
17. F.C. Franco Jr., A.A.B. Padama, On the structural and optoelectronic properties of chemically modified oligothiophenes with electron-withdrawing substituents for organic solar cell applications: a DFT/TDDFT study. J. Phys. Soc. Japan **86**, 064802 (2017). https://doi.org/10.7566/JPSJ.86.064802
18. F.C. Franco Jr., A.A.B. Padama, DFT and TD-DFT study on the structural and optoelectronic characteristics of chemically modified donor-acceptor conjugated oligomers for organic polymer solar cells. Polym. (Guildf) **97**, 55–62 (2016). https://doi.org/10.1016/j.polymer.2016.05.025
19. H. Ullah, A.-H.A. Shah, S. Bilal, K. Ayub, Doping and dedoping processes of polypyrrole: DFT study with hybrid functionals. J. Phys. Chem. C **118**, 17819–17830 (2014). https://doi.org/10.1021/jp505626d
20. P. Galar, B. Dzurnak, P. Maly, J. Cermak, A. Kromka, M. Omastova, B. Rezek, Chemical changes and photoluminescence properties of UV modified polypyrrole. Int. J. Electrochem. Sci. **8**, 57–70 (2013)
21. L. Al-Mashat, C. Debiemme-Chouvy, S. Borensztajn, W. Wlodarski, Electropolymerized polypyrrole nanowires for hydrogen gas sensing. J. Phys. Chem. C **116**, 13388–13394 (2012). https://doi.org/10.1021/jp3015854
22. M. Campos, F.R. Simões, E.C. Pereira, Influence of methane in the electrical properties of polypyrrole films doped with dodecylbenzene sulfonic acid. Sens. Actuators B Chem. **125**, 158–166 (2007). https://doi.org/10.1016/j.snb.2007.02.001
23. M.R. Mahmoudian, Y. Alias, W.J. Basirun, P. MengWoi, F. Jamali-Sheini, M. Sookhakian, M. Silakhori, A sensitive electrochemical nitrate sensor based on polypyrrole coated palladium nanoclusters. J. Electroanal. Chem. **751**, 30–36 (2015). https://doi.org/10.1016/j.jelechem.2015.05.026
24. G. Li, Y. Wang, H. Xu, A hydrogen peroxide sensor PREPARED by electropolymerization of pyrrole based on screen-printed carbon paste electrodes. Sensors **7**, 239–250 (2007). https://doi.org/10.3390/s7030239
25. S. Paul, N.N. Chavan, S. Radhakrishnan, Polypyrrole functionalized with ferrocenyl derivative as a rapid carbon monoxide sensor. Synth. Met. **159**, 415–418 (2009). https://doi.org/10.1016/j.synthmet.2008.10.013
26. A. Joshi, S.A. Gangal, S.K. Gupta, Ammonia sensing properties of polypyrrole thin films at room temperature. Sens. Actuators B Chem. **156**, 938–942 (2011). https://doi.org/10.1016/j.snb.2011.03.009
27. J.-S. Do, S.-H. Wang, On the sensitivity of conductimetric acetone gas sensor based on polypyrrole and polyaniline conducting polymers. Sens. Actuators B Chem. **185**, 39–46 (2013). https://doi.org/10.1016/j.snb.2013.04.080

28. G.R. Schleder, A.C.M. Padilha, C.M. Acosta, M. Costa, A. Fazzio, From DFT to machine learning: recent approaches to materials science–a review. J. Phys. Mater. **2**, 032001 (2019). https://doi.org/10.1088/2515-7639/ab084b
29. P. Rogl, R. Podloucky, W. Wolf, DFT calculations: a powerful tool for materials design. J. Phase Equilib. Diffus. **35**, 221–222 (2014). https://doi.org/10.1007/s11669-014-0309-7
30. J. Roncali, Molecular engineering of the band gap of π-conjugated systems: facing technological applications. Macromol. Rapid Commun. **28**, 1761–1775 (2007). https://doi.org/10.1002/marc.200700345
31. X. Jia, Z. Chen, C. Duan, Z. Wang, Q. Yin, F. Huang, Y. Cao, Polythiophene derivatives compatible with both fullerene and non-fullerene acceptors for polymer solar cells. J. Mater. Chem. C. **7**, 314–323 (2019). https://doi.org/10.1039/C8TC04746D
32. F. Franco Jr., Computational study on the structural and optoelectronic properties of a carbazole-benzothiadiazole based conjugated oligomer with various alkyl side-chain lengths. Mol. Simul. **43**, 222–227 (2017). https://doi.org/10.1080/08927022.2016.1250267
33. F.C. Franco Jr., Computational study on optoelectronic property Tuning of oligothiophenes via chemical modifications for solar cell applications. Asian J. Chem. **30**, 329–332 (2018). https://doi.org/10.14233/ajchem.2018.20948
34. L. Pandey, C. Risko, J.E. Norton, J.-L. Brédas, Donor-acceptor COPOLYMERS of relevance for organic photovoltaics: a theoretical investigation of the impact of chemical structure modifications on the electronic and optical properties. Macromol. **45**, 6405–6414 (2012). https://doi.org/10.1021/ma301164e
35. J.M. Granadino-Roldán, A. Garzón, G. García, M. Moral, A. Navarro, M. P. Fernández-Liencres, T. Peña-Ruiz, M. Fernández-Gómez, Theoretical study of the effect of alkyl and Alkoxy lateral chains on the structural and electronic properties of π-conjugated polymers consisting of phenylethynyl-1,3,4-thiadiazole. J. Phys. Chem. C **115**, 2865–2873 (2011). https://doi.org/10.1021/jp108910j
36. C. Risko, M.D. McGehee, J.-L. Brédas, A quantum-chemical perspective into low optical-gap polymers for highly-efficient organic solar cells. Chem. Sci. **2**, 1200–1218 (2011). https://doi.org/10.1039/C0SC00642D
37. M.W. Schmidt, K.K. Baldridge, J.A. Boatz, S.T. Elbert, M.S. Gordon, J.H. Jensen, S. Koseki, N. Matsunaga, K.A. Nguyen, S. Su, T.L. Windus, M. Dupuis, J.A. Montgomery, General atomic and molecular electronic structure system. J. Comput. Chem. **14**, 1347–1363 (1993). https://doi.org/10.1002/jcc.540141112
38. J. Tirado-Rives, W.L. Jorgensen, Performance of B3LYP density functional methods for a large set of organic molecules. J. Chem. Theory Comput. **4**, 297–306 (2008). https://doi.org/10.1021/ct700248k
39. S. Grimme, S. Ehrlich, L. Goerigk, Effect of the damping function in dispersion corrected density functional theory. J. Comput. Chem. **32**, 1456–1465 (2011). https://doi.org/10.1002/jcc.21759
40. S. Grimme, J. Antony, S. Ehrlich, H. Krieg, A consistent and accurate ab initio parametrization of density functional dispersion correction (DFT-D) for the 94 elements H-Pu. J. Chem. Phys. **132**, 154104 (2010). https://doi.org/10.1063/1.3382344
41. S. Grimme, Density functional theory with London dispersion corrections. Wiley Interdiscip. Rev. Comput. Mol. Sci. **1**, 211–228 (2011). https://doi.org/10.1002/wcms.30
42. S. Grimme, H. Kruse, gCP-D3 Webservice. http://wwwtc.thch.uni-bonn.de/
43. E.D. Glendening, C.R. Landis, F. Weinhold, NBO 6.0 : natural bond orbital analysis program. J. Comput. Chem. **34**, 1429–1437 (2013). https://doi.org/10.1002/jcc.23266
44. B.M. Bode, M.S. Gordon, Macmolplt: a graphical user interface for GAMESS. J. Mol. Graph. Model. **16**, 133–138 (1998). https://doi.org/10.1016/S1093-3263(99)00002-9
45. Chemcraft—Graphical Software for Visualization of Quantum Chemistry Computations. https://www.chemcraftprog.com
46. N.M. O'boyle, A.L. Tenderholt, K.M. Langner, cclib: a library for package-independent computational chemistry algorithms. J. Comput. Chem. **29**, 839–845 (2008). https://doi.org/10.1002/jcc.20823

47. L. Dai, Conjugated and fullerene-containing polymers for electronic and photonic applications: advanced syntheses and microlithographic fabrications. J. Macromol. Sci. Part C Polym. Rev. **39**, 273–387 (1999). https://doi.org/10.1081/mc-100101421
48. F. Wasim, T. Mahmood, K. Ayub, An accurate cost effective DFT approach to study the sensing behaviour of polypyrrole towards nitrate ions in gas and aqueous phases. Phys. Chem. Chem. Phys. **18**, 19236–19247 (2016). https://doi.org/10.1039/C6CP02271E
49. H. Ullah, K. Ayub, Z. Ullah, M. Hanif, R. Nawaz, A.-H.A. Shah, S. Bilal, Theoretical insight of polypyrrole ammonia gas sensor. Synth. Met. **172**, 14–20 (2013). https://doi.org/10.1016/j.synthmet.2013.03.021
50. H. Ullah, Inter-molecular interaction in polypyrrole/TiO_2: a DFT study. J. Alloys Compd. **692**, 140–148 (2017). https://doi.org/10.1016/j.jallcom.2016.08.169
51. H. Ullah, A.A. Tahir, T.K. Mallick, Polypyrrole/TiO_2 composites for the application of photocatalysis. Sens. Actuators B Chem. **241**, 1161–1169 (2017). https://doi.org/10.1016/j.snb.2016.10.019
52. S.R. Nalage, S.T. Navale, V.B. Patil, Polypyrrole-NiO hybrid nanocomposite: structural, morphological, optical and electrical transport studies. Measurement **46**, 3268–3275 (2013). https://doi.org/10.1016/j.measurement.2013.06.049

Chapter 10
Thermoelectric Power Factor Under Strain-Induced Band-Alignment in the Half-Heuslers NbCoSn and TiCoSb

Chathurangi Kumarasinghe and Neophytos Neophytou

Abstract Band convergence is an effective strategy to improve the thermoelectric performance of complex bandstructure thermoelectric materials. Half-Heuslers are good candidates for band convergence studies because they have multiple bands near the valence bad edge that can be converged through various band engineering approaches providing power factor improvement opportunities. Theoretical calculations to identify the outcome of band convergence employ various approximations for the carrier scattering relaxation times (the most common being the constant relaxation time approximation) due to the high computational complexity involved in extracting them accurately. Here, we compare the outcome of strain-induced band convergence under two such scattering scenarios: (i) the most commonly used constant relaxation time approximation and (ii) energy dependent inter- and intra-valley scattering considerations for the half-Heuslers NbCoSn and TiCoSb. We show that the outcome of band convergence on the power factor depends on the carrier scattering assumptions, as well as the temperature. For both materials examined, band convergence improves the power factor. For NbCoSn, however, band convergence becomes more beneficial as temperature increases, under both scattering relaxation time assumptions. In the case of TiCoSb, on the other hand, constant relaxation time considerations also indicate that the relative power factor improvement increases with temperature, but under the energy dependent scattering time considerations, the relative improvement weakens with temperature. This indicates that the scattering details need to be accurately considered in band convergence studies to predict more accurate trends.

C. Kumarasinghe (✉) · N. Neophytou
School of Engineering, University of Warwick, Coventry CV4 7AL, UK
e-mail: Chathu.Kumarasinghe@warwick.ac.uk

E. V. Levchenko et al. (eds.), *Theory and Simulation in Physics for Materials Applications*, Springer Series in Materials Science 296,
https://doi.org/10.1007/978-3-030-37790-8_10

10.1 Introduction

Thermoelectric materials are capable of direct conversion of thermal energy into electricity and vice versa. They have drawn interest as an environment-friendly and highly reliable energy conversion technology [1–4]. The efficiency of a thermoelectric material is quantified by the thermoelectric figure of merit (ZT) given by:

$$\mathrm{ZT} = \left(\sigma S^2 T\right)/\kappa, \tag{10.1}$$

where σ is the electrical conductivity, S is the Seebeck coefficient, T is the temperature and κ is the thermal conductivity of the material. To obtain a high ZT, a high electrical conductivity, a high Seebeck coefficient, i.e. a high power factor (σS^2), and a low thermal conductivity κ is needed, however, simultaneous optimization of these parameters is difficult due to their complex inter-dependencies.

Bandstructure engineering approaches such as 'band-convergence' also known as 'band-alignment', can overcome these inter-dependencies and improve the power factor (PF) [5, 6]. The aim of band-convergence is to increase the number of carriers contributing to electronic transport by increasing the valley or orbital degeneracy near conduction or valence bands edges. In bulk materials, band structures can be manipulated to achieve this by applying strain, doping, alloying, and second phasing with other suitable structures [7–9]. For example, half-Heusler (HH) alloys [10–14] have complex bandstructures with the potential for beneficial band convergence. They are also known to have good thermal and mechanical stability, large scale reproducibility, and made out of relatively inexpensive, non-toxic elements.

One of the most commonly used theoretical approaches for obtaining thermoelectric properties of materials is applying the Boltzmann transport theory in the relaxation time approximation using detailed material bandstructures obtained from ab initio density functional theory (DFT) [15, 16]. The carrier relaxation time is an important input in this calculation, but due to the high computational complexity in calculating accurate carrier relaxation times, a constant relaxation time is generally assumed (usually $\tau \approx 10^{-14}$s is used). However, such simplifications are known to give less accurate predictions particularly during band-convergence optimization studies [17]. This is because, while band-alignment can increase the number of carriers available for conduction, it can also increase the number of states that carriers scatter into. Therefore, the energy dependence of the scattering mechanisms, as well as intra- or inter-valley scattering considerations are important in identifying if a given bandstructure engineering approach leads to an improved power factor. In addition, the scattering rate depends on the temperature as well with more scattering of electrons/holes by phonons expected at higher temperatures.

In this work we study the effect of considering different scattering time scenarios on the strain-induced band-convergence for two Co-based Half-Heuslers, NbCoSn and TiCoSb. We consider constant (τ_C), and energy dependent inter- and intra-valley scattering (τ_{IIV}) relaxation time approximations, and quantify the relative power factor improvements that each predict at different temperatures and strain levels. We finally stress the importance of accurate treatment of scattering in arriving to useful predictions.

10.2 Methods

10.2.1 Boltzmann Transport Theory

Thermoelectric coefficients are calculated using the Boltzmann transport formalism within the relaxation time approximation (RTA). We describe the material properties using bandstructures obtained from DFT calculations. The thermoelectric coefficients, electrical conductivity $\sigma_{\alpha\beta}(T, E_F)$ and Seebeck coefficient $S_{\alpha\beta}(T, E_F)$ tensors, can be written as [18, 19]:

$$\sigma_{\alpha\beta}(T, E_F) = e^2 \int \Xi_{\alpha\beta}(E)\left(-\frac{\partial f(E, E_F, T)}{\partial E}\right)\mathrm{d}E \tag{10.2}$$

$$S_{\alpha\beta}(T, E_F) = \frac{e\int \Xi_{\alpha\beta}(E)(E - E_F)\left(-\frac{\partial f(E,E_F,T)}{\partial E}\right)\mathrm{d}E}{T\sigma_{\alpha\beta}(T, E_F)} \tag{10.3}$$

where $f(E, E_F, T)$ is the fermi distribution function at a given temperature T and a chemical potential level E_F and e is the charge of an electron. $\Xi_{\alpha\beta}(E)$ is the transport distribution (TD) function, which is given by:

$$\Xi_{\alpha\beta}(E) = \sum_{i,\mathbf{k}} \tau_{i,\mathbf{k}}(E) v_\alpha(i, \mathbf{k}) v_\beta(i, \mathbf{k}) \delta(E - E_{i,\mathbf{k}}), \tag{10.4}$$

where i and $\mathbf{k}$ represent the band index and the k-point, respectively. $\tau_{i,\mathbf{k}}(E)$ is the electron relaxation time and $v_\alpha(i, \mathbf{k})(\alpha = \{x, y, z\})$ represents the αth component of the group velocity $\mathbf{v}(i, \mathbf{k})$, which can be derived from the gradient of the bands in the bandstructure as:

$$\mathbf{v}(i, k) = \frac{1}{\hbar}\nabla_{\mathbf{k}} E_{i,\mathbf{k}}. \tag{10.5}$$

10.2.2 Ab Initio Electronic Structure Calculations

Ab initio DFT calculations were performed for the Co-based HHs, NbCoSn and TiCoSb with the QUANTUM ESPRESSO package [20]. Projector augmented wave technique was used with the PBE-GGA functional. Throughout our calculations, a kinetic energy cutoff greater than 60 Ry was used for wavefunctions and an energy convergence criterion of 10^{-8} Ry was adopted for self-consistency. For transport property calculations, a $15 \times 15 \times 15$ Monkhorst–Pack k-point sampling was used for the primitive unit cell with three atoms. Calculations using denser k-points were carried out to confirm the convergence of the results.

10.2.3 Relaxation Time Approximation

By applying the Boltzmann transport theory using the DFT extracted bandstructures, we examine the influence of the nature of scattering times on the power factor for various temperatures and upon strain-induced band-convergence. We consider: (i) the commonly employed constant relaxation time ($\tau_i = \tau_C$) and (ii) scattering proportional to total density of states, allowing both inter- and intra-band scattering $\left(\tau_i(E) = \tau_{IIV} \propto 1/\sum_i \mathrm{DOS}_i(E)\right)$.

To obtain a general understanding about these two scattering scenarios, we first analyse them using a simple parabolic band model. Since under the parabolic band approximation, the velocity and density of states of each band is $v_i(E) = (2E/m_i)^{1/2}$, and $\mathrm{DOS}_i(E) = 2^{1/2} m_i^{3/2} N_i E^{1/2}/(\pi^2 \hbar^3)$, respectively, the TD function given in (10.4) for valence bands is reduced to:

$$\Xi(E) \propto \sum_i N_i \tau_i(E) m_i^{\frac{1}{2}} E^{\frac{3}{2}} (\mathrm{H}[-\Delta E_i]). \tag{10.6}$$

where the subscript i indicates the specific band, ΔE_i indicates the distance to the band edge from the valence band edge, N_i indicates the band degeneracy and $\mathrm{H}[\Delta E_i]$ indicates the Heaviside step function. (For conduction bands, $(\mathrm{H}[-\Delta E_i])$ should be replace by $\mathrm{H}[\Delta E_i]$.)

Under the constant relaxation time approximation, $\tau_i(E) = \tau_C$, where τ_C is a constant. Therefore, the TD function given by (10.6) can be further simplified to:

$$\Xi(E) \propto \sum_i m_i^{\frac{1}{2}} E^{\frac{3}{2}} \mathrm{H}[-\Delta E_i]. \tag{10.7}$$

This indicates that under a constant relaxation time, aligning a band of any mass will increase the TD function, resulting finally in an increased conductivity, with the larger masses resulting in larger improvements, a common scenario seen in most band-alignment theory literature. The magnitude and sign of the Seebeck coefficient are related to an asymmetry of the electron transport around the Fermi level [21, 22], which is indicated by the energy gradient of the TD function. If the band convergence does not additionally introduce significant asymmetry (significant change in gradient of the TD function), there will not be a significant change in the Seebeck coefficient. Therefore, in the case of a constant relaxation time, a power factor improvement is always achieved upon band alignment, driven by improvements in the conductivity, with a heavier aligning band being preferred, as it offers a higher density of conducting states (but not more scattering states under τ_C). As we discuss further below, this is not the case when $\tau = \tau_{IIV}(E)$.

In the case of inter- and intra-band scattering ($\tau_{IIV}(E)$) that we consider next, carriers are allowed to scatter elastically to the total density of states available at the energy under consideration, without any selection rules, i.e. both intra and inter-band (with inter and intra-valley in multi-valley materials) scattering is allowed $\left(\tau_i(E) = \tau_{IIV}(E) \propto 1/\sum_i \mathrm{DOS}_i(E)\right)$. The TD functions given by (10.6) in this case can be simplified to:

$$\Xi(E) \propto \frac{\sum_i m_i^{\frac{1}{2}} E^{\frac{3}{2}} \mathrm{H}[-\Delta E_i]}{\sum_i m_i^{\frac{3}{2}} E^{\frac{1}{2}} \mathrm{H}[-\Delta E_i]} \tag{10.8}$$

From (10.8), since the denominator (its appearance being a result of the scattering times being inversely proportional to the DOS), has a higher mass exponent, it can be deduced that upon full band alignment, the TD function will only increase when a light band is brought close to the band edge and is aligned with a heavier band. When additional bands are gradually brought close to the band edge to be aligned, three competing effects take place: (i) the presence of the additional conducting states tends to increase the TD function, (ii) the same states increase the scattering of already aligned bands, which tends to reduce the TD function, and (iii) scattering from newly aligned bands reduces since there are less states to scatter into at energies closer to valence band edge (due to the energy dependence $E^{1/2}$ of the DOS), increasing the TD function. These interdependences do not allow for significant improvements in the TD function, the conductivity, and the PF as in the previous scattering scenario. As a result of these competing effects, aligning bands is not always advantageous for the power factor under $\tau_{IIV}(E)$ [17].

10.2.4 Temperature Dependent Carrier Relaxation Time

When performing temperature dependent studies, and evaluating the trends under increasing temperature, the Fermi distribution broadens, which allows the occupation of more states, however, to satisfy charge neutrality the Fermi level will shift towards the bandgap, increasing the Seebeck coefficient. If one compares at the same Fermi level position, rather than carrier density, however, the conductivity increases, while not many changes are observed in the Seebeck coefficient. On the other hand, temperature increases the population of phonons, which increase carrier-phonon scattering. In the most common acoustic deformation potential scattering scenario, under the equipartition approximation, the scattering rate increases linearly with temperature. Thus, we account for the increase in carrier scattering with increase in temperature by scaling the relaxation times by $1/T$ [27].

We refer to this as 'T-normalized relaxation time'. In the case of the constant relaxation time, we take $\tau_C = 10^{-14}$s at 300 K, and linearly scale for other temperatures as $\tau_C = 10^{-14}(300/T)$s, to account for deviations from room temperature In the case of energy dependent relaxation time, temperature scaling is implemented as $\tau_{\mathrm{IIV}} \propto 1/\sum_i \mathrm{DOS}_i(E)(300/T)$.

10.3 Results and Discussion

Figure 10.1 shows the power factors calculated using both temperature normalized and non-normalized relaxation times. Under non-normalized conditions (solid lines), at the same Fermi level position, the power factor rises with the temperature because the derivative of the Fermi distribution function $((\partial f(E, E_F, T))/\partial E)$ in the (10.2) broadens (for a given Fermi level), increasing the conductivity. However, when we take into account the increase in carrier scattering by normalizing the carrier scattering time by $300/T$, the power factor difference almost disappears, and only modest differences exist between temperatures for both scattering scenarios. This illustrates that it is important to take increased carrier scattering into consideration when comparing thermoelectric parameters at different temperatures. (Note that it is often the case in experiments to observe that the power factor changes with temperature. That is because those are performed at a constant density, rather than a constant Fermi level position, which would have shifted the two lines compared to each other).

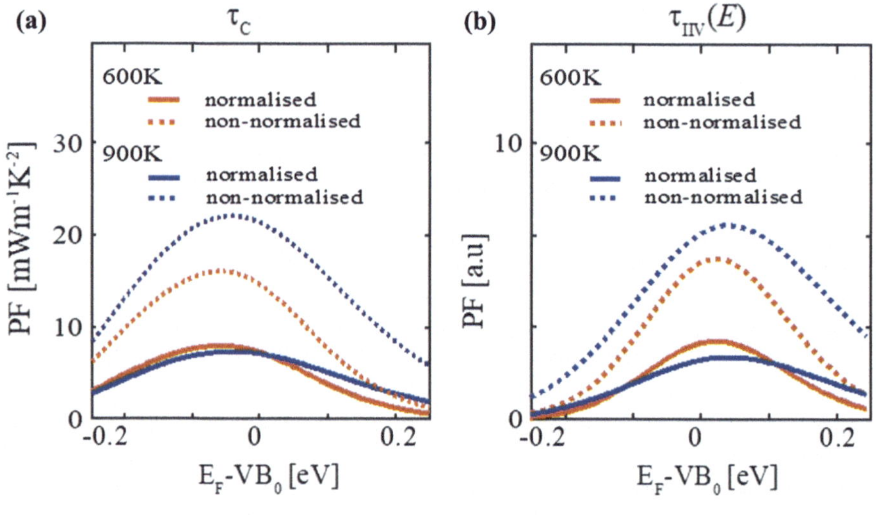

Fig. 10.1 Power factors calculated using normalized (solid-lines) and non-normalized (dashed-lines) carrier relaxation times at 600 K (red) and 900 K(blue) for (**a**) constant relaxation time (τ_C) and (**b**) energy dependent relaxation time (τ_{IIV}) for NbCoSn

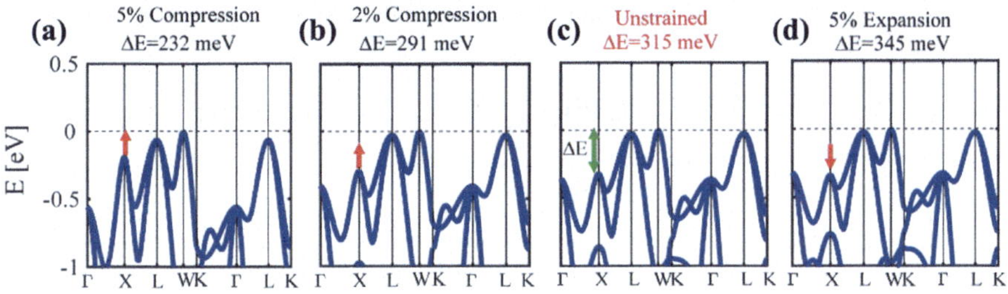

Fig. 10.2 a–d Application of compressive and tensile strain in NbCoSb to align the X valley with the valence band edge. Strain values and the energy separation ΔE between the X valley and VB_0 are noted above the sub-figures. Subfigure (**d**) shows the unstrained bandstructure

Band convergence can be achieved using a variety of methods including applying strain [17, 23, 24] and alloying [6, 25, 26]. Here, for the purposes of our investigation into the influence of band alignment on the PF under different scattering scenarios and temperatures, we use the easiest method within DFT, which is the use of hydrostatic strain, either compressive or tensile.

In Fig. 10.2c, in the unstrained NbCoSn bandstructure, it is apparent that multiple bands from several valleys are available close to the valence band edge VB_0. Bands at the L and W points are already aligned at VB_0. There also exist heavy and light bands at the X and Γ points within 0.3 eV of the VB_0. Aligning these bands that are in the vicinity of the VB_0, particularly the bands at the X point that have a large equivalent valley degeneracy of 3, can lead to an improved conductivity and power factor. As seen in Fig. 10.2a–d, the bands of NbCoSn can be manipulated with compression and expansion. When the material is compressed, the bands at the X point are brought closer to VB_0, reducing their energy separation ΔE, and when expanded, ΔE increases. It is important to note, however, that the curvatures of the bands, i.e. the carrier effective masses also change with strain. Compression increases the band curvatures (lower carrier effective mass) in general and expansion has the opposite effect. The fact that the masses are reduced with band convergence is unfavorable under a constant scattering rate as seen above in 10.7. It is, however, favorable under $\tau_{\text{IIV}}(E)$ under most situations [17] as it brings bands with higher velocities that scatter less within the transport window. We consider large strain values of up to 5% and achieve a degree of alignment, but a compressive strain larger than 5% is required to completely align the bands.

We now examine the thermoelectric coefficients of NbCoSn under strain, for three different temperatures 300, 600, and 900 K (Fig. 10.3). Figure 10.3 shows the TE coefficients conductivity, Seebeck coefficient and the power factor calculated under the assumption of a constant rate of scattering (τ_{C}) with temperature scaling. Later on, we will show how these observations are altered in the case of energy dependent τ_{IIV} scattering. At 300 K (Fig. 10.3a–c) we only see a 3% improvement in the PF with band convergence. With increasing temperature, the conductivity for

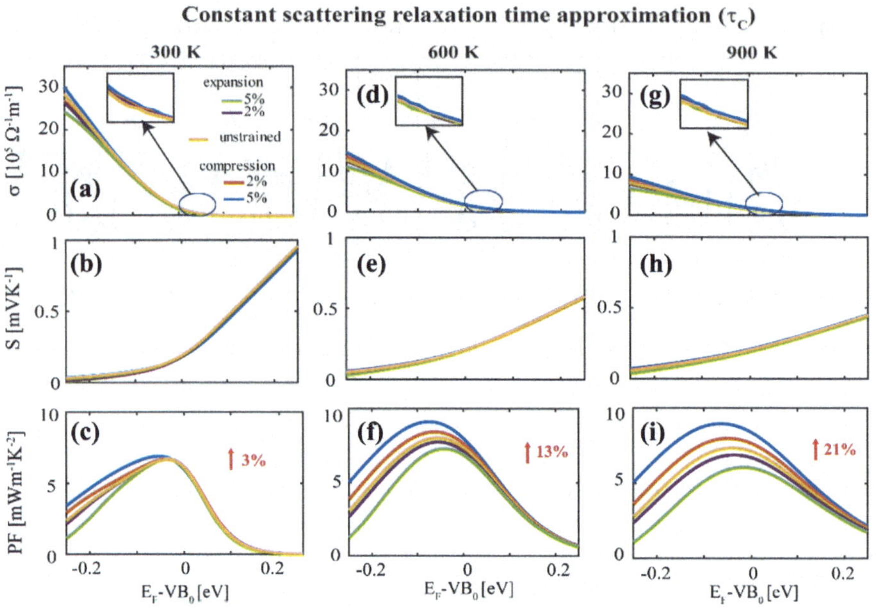

Fig. 10.3 Thermoelectric coefficients (σ, S, and PF) calculated for NbCoSn using the temperature-normalized constant relaxation time approximation: **a–c** for 300 K, **d–f** for 600 K and **g–i** for 900 K. The percentage improvement given is the peak to peak improvement between the unstrained ($\Delta E = 315$ meV) and 5% compressive strain ($\Delta E = 232$ meV), given by the yellow-solid lines and blue-solid lines, respectively

non-degenerate Fermi levels increases, while it decreases for higher Fermi levels. The opposite effect is seen in the Seebeck coefficient where its values decrease for non-degenerate conditions and increase for highly degenerate Fermi values (right versus left regions of the sub-figures). Also, the gradients of conductivity and Seebeck coefficient curves are reducing with increasing temperature. The combined effect is that with increasing temperature the power factor improves for all compressive strain value, which tend to align the bands, with the improvements at 5% compressive strain to increase from 3% to 13% to 21% at temperatures of 300 K, 600 K, and 900 K, respectively. Note that the absolute PF values do not necessarily increase with temperature, it is the relative increase between the unstrained and strained cases that increases.

Next, we examine the behavior of NbCoSn under the energy dependent τ_{IIV} scattering for the three temperatures of interest, 300, 600, and 900 K (Fig. 10.4, column-wise). In this case, the differences in the conductivity are more noticeable between the different strain conditions, compared to the τ_{C} case in Fig. 10.3. The Seebeck coefficients as in the τ_{C} case are almost unchanged with strain at the same

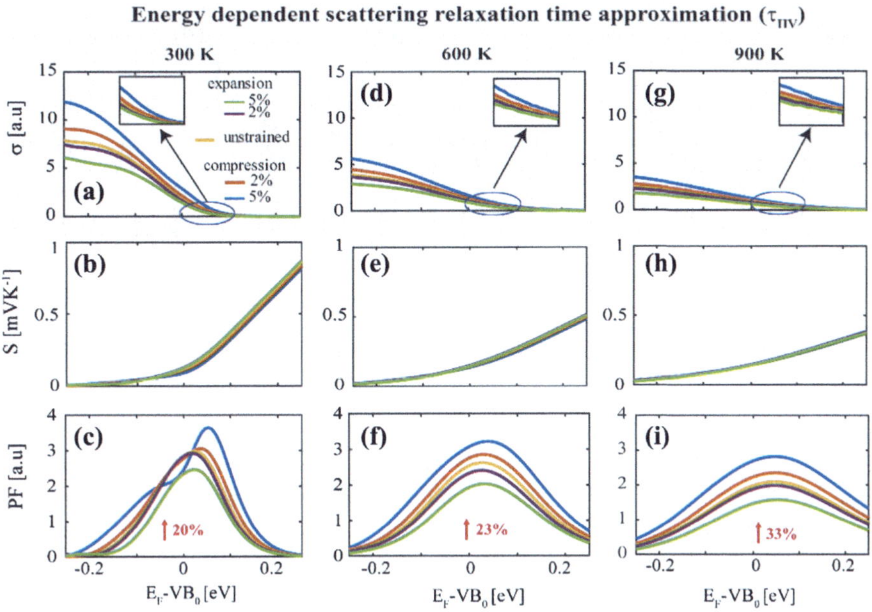

Fig. 10.4 Thermoelectric coefficients (σ, S, and PF) calculated for NbCoSn using the temperature-normalized energy dependent τ_{IIV} relaxation time approximation: **a–c** for 300 K, **d–f** for 600 K and **g–i** for 900 K. The percentage improvement given is the peak to peak improvement between the unstrained ($\Delta E = 315$ meV) and 5% compressive strain ($\Delta E = 232$ meV), given by the yellow-solid lines and blue-solid lines, respectively

Fermi level position, especially for the higher temperatures. The improvements to the PF, on the other hand, are larger with compressive strain, and they begin to appear and being significant even at 300 K. In this τ_{IIV} case, the PF improvements for the 5% compressive strain, at 300 K, 600 K, and 900 K are 20%, 23%, and 33%, respectively.

The larger improvements compared to the τ_C case, originate from the changes in the curvature of the bands in addition to the reduced band separation. The bands that are brought closer to the band edge to be aligned as well as the bands already at the band edge become lighter in the process in NbCoSn [17]. Specifically, the masses of the bands at X, changes from 0.48 m_0 to 0.29 m_0 while its separation is reduced from 315 to 232 meV upon 5% compression. Alignment tends to increase the PF in both τ_C and τ_{IIV}, but in the τ_C case the fact that the bands become lighter, mitigates this benefit. On the other hand, in the case of τ_{IIV}, light bands are beneficial, which provides an additional improvement to the PF. Finally, the reason the benefits increase with temperature in both cases, is due to the simple fact that the broadening of the Fermi distribution with temperature, allows for an exponentially

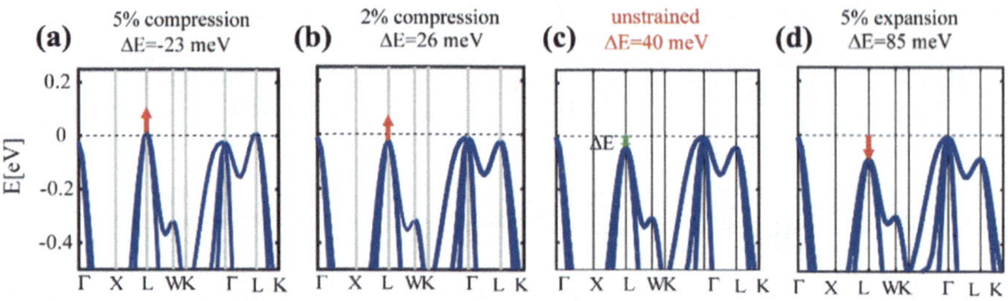

Fig. 10.5 **a–d** Application of compressive and tensile strain in TiCoSb to align the X valley with the valence band edge. Strain values and the energy separation ΔE between the X valley and VB_0 are noted above the sub-figures. Subfigure (**d**) shows the unstrained bandstructure

larger percentage contribution of the upper valleys to the total contribution. The separation ΔE is still large, even at 5% compression, accounting for several $k_B T$ even at 900 K. Thus, at the same Fermi level position, the overall relative PF improvement with temperature increases with T. For NbCoSn, either τ_C or τ_{IIV} indicate a higher benefits of band alignment at higher temperatures.

Next, we apply strain to TiCoSb in the same way. The smaller energy separation ΔE between bands at L and Γ points (see Fig. 10.5c) can be reduced by applying compressive strain, leading to band convergence, as shown in Fig. 10.5a–d. The unstrained ΔE value in this situation is only 40 meV ($1.55 k_B T$ at $T = 300$ K), i.e. the bands are almost aligned even without strain. The bands can be fully aligned by applying only ≈2% compressive strain as shown in Fig. 10.5b (a much more realistic value compared to the one needed for NbCoSn).

Figure 10.6 shows the TE coefficients conductivity, Seebeck coefficient and PF for the constant time assumption τ_C for 300, 600, and 900 K column-wise, in the same manner as earlier for the NbCoSn. We only see 6% improvement when bands are fully aligned using compressive strain (Fig. 10.6c), but the improvement increases for higher temperatures, reaching 13% and 19% at 600 K and 900 K, respectively (Fig. 10.6f, i). Again, the improvement in the PF with compression at a given temperature originates from the conductivity. Improvements increase in the higher temperatures because the positive contribution from the broadening of the $\partial f(E, E_F, T)/\partial E$ function overweight the negative effects on the TD function from increased scattering.

We now examine the case where the relaxation time in TiCoSb is energy dependent (τ_{IIV}). Figure 10.7 shows the TE coefficients conductivity, Seebeck coefficient and PF for this energy dependent scattering assumption for the three temperatures of interest, under 5% expansion, 2% compression, and 5% compression as before in Fig. 10.6. We observe a 37% improvement when bands are fully aligned using compressive strain (Fig. 10.7a) at 300 K. The fact that band curvatures also increase (carrier effective masses reduce) with alignment, is more

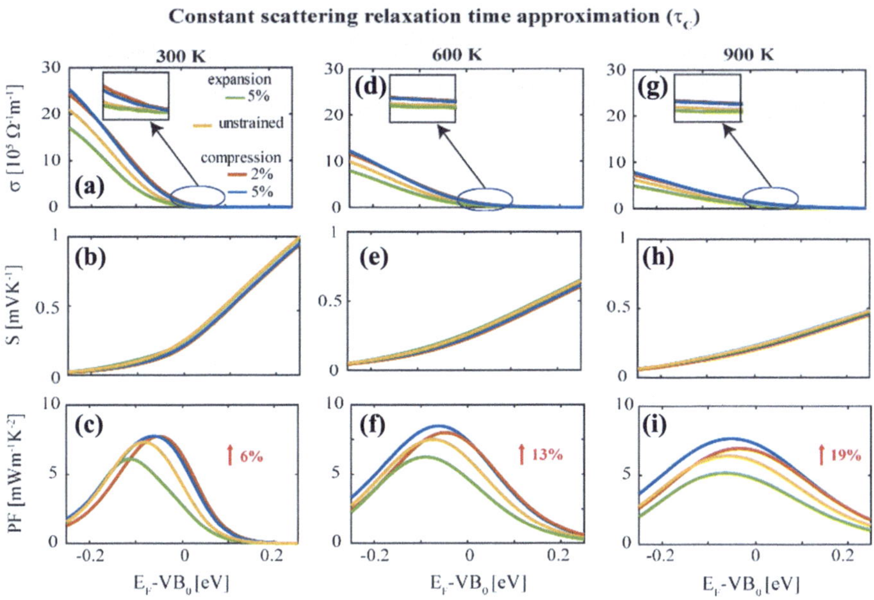

Fig. 10.6 Thermoelectric coefficients (σ, S, and PF) calculated for TiCoSb using the temperature-normalized τ_C approximation: **a–c** for 300 K, **d–f** for 600 K and **g–i** for 900 K. The percentage improvement given is the peak to peak improvement between the unstrained (ΔE = 40 meV) and 5% compressive strain (ΔE = −23 meV), given by the yellow-solid lines and blue-solid lines, respectively

favorable under the τ_{IIV} scattering scenario, which has contributed to this large improvement observed. Interestingly, as opposed to all previous scenarios the improvement reduces with temperature increase, down to 27% and 25% at 600 K and 900 K, respectively (Fig. 10.6f, i). The reason behind the smaller improvement variation with temperature upon band alignment for TiCoSb compared to the NbCoSn case, is of course the fact that the bands were closer together to begin with, and thus in all temperature cases the initial and the aligning bands contribute similarly. In other words, the influence of the broadening of $\partial f(E, E_F, T)/\partial E$ affects all cases similarly. The increase in scattering with temperature, however, overcomes any positive contribution from this broadening, reducing the PF improvement with temperature.

In summary, as seen in Fig. 10.8, band convergence is advantageous for both NbCoSn and TiCoSb (there is an improvement in the PF). Under both scattering scenarios we have considered, band convergence provides larger relative benefits for NbCoSn as the temperature increases (blue lines). Band convergence on the other hand, provides relative improvements in the PF for TiCoSb under τ_C, but reductions under τ_{IIV} as the temperature increases.

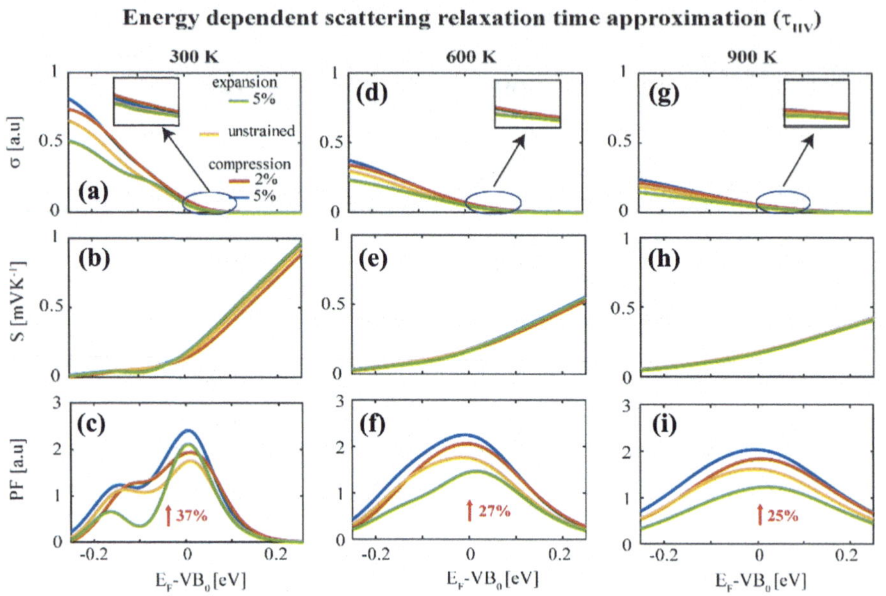

Fig. 10.7 Thermoelectric coefficients (σ, S, and PF) calculated for TiCoSb using the temperature-normalized τ_{IIV} relaxation time approximation: **a–c** for 300 K, **d–f** for 600 K and **g–i** for 900 K. The percentage improvement given is the peak to peak improvement between the unstrained ($\Delta E = 40$ meV) and 5% compressive strain ($\Delta E = -23$ meV), given by the yellow-solid lines and blue-solid lines, respectively

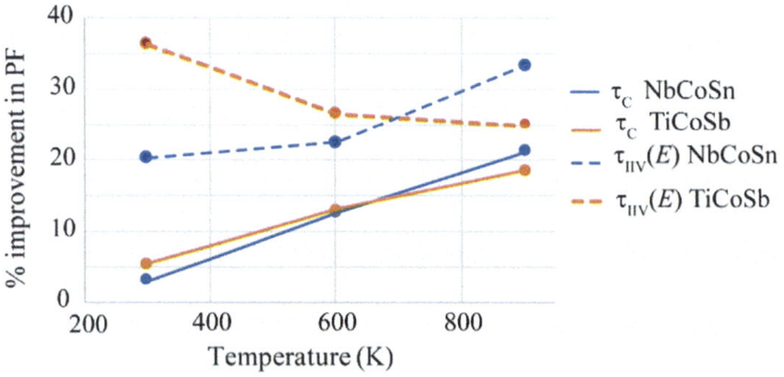

Fig. 10.8 Percentage improvement of the power factor (from peak to peak between unstrained and 5% compression) under a constant rate of scattering (solid-lines), and energy dependent τ_{IIV} scattering (dashed-lines) versus temperature for NbCoSn, and TiCoSb

10.4 Conclusion

In this work we have used strain to align the bands near the valence band edge of the p-type Co-based half-Heuslers NbCoSn and TiCoSb for thermoelectric power factor improvements. Using the Boltzmann transport equation under the relaxation time approximation, we explored the band alignment effect on the power factor under two different scattering conditions (as the detail scattering physics of half-Heuslers are still not known): (i) the constant relaxation time approximation—as is common in the literature, (ii) scattering rates depending on energy, proportional to the density of final states, with both inter- and intra-band\inter- and intra-valley scattering considerations. We showed that the outcome of band alignment can be different in each of the different scattering cases and at different temperatures for these materials. It depends on the nature of the bandstructure, particularly on how the curvatures change with band convergence and the relative separations between bands to begin with. When benefits of broadening of $\partial f(E, E_F, T)/\partial E$ overcome negative contributions from increased carrier scattering with temperature, we see increasing improvements in the PF with temperature. For NbCoSn, band convergence using strain is more beneficial in higher temperatures in both scattering scenarios, while for TiCoSb higher temperatures are only beneficial under a constant rate of scattering. Our work stresses the importance of more accurate theoretical treatment of carrier relaxation times and their temperature dependences.

Acknowledgements This work has received funding from the European Research Council (ERC) under the European Union's Horizon 2020 Research and Innovation Programme (Grant Agreement No. 678763).

References

1. D. Beretta et al., Thermoelectrics: from history, a window to the future. Mater. Sci. Eng. R Rep. (2018). https://doi.org/10.1016/J.MSER.2018.09.001
2. L.E. Bell, Cooling, heating, generating power, and recovering waste heat with thermoelectric systems. Science **321**, 1457–1461 (2008)
3. A. Shakouri, Recent developments in semiconductor thermoelectric physics and materials. Annu. Rev. Mater. Res. **41**, 399–431 (2011)
4. K. Koumoto et al., Thermoelectric ceramics for energy harvesting. J. Am. Ceram. Soc. **96**, 1–23 (2013)
5. P. Norouzzadeh, D. Vashaee, Classification of valleytronics in thermoelectricity. Sci. Rep. **6**, 22724 (2016)
6. Y. Pei et al., Convergence of electronic bands for high performance bulk thermoelectrics. Nature **473**, 66–69 (2011)
7. S. Bhattacharya, G.K.H. Madsen, High-throughput exploration of alloying as design strategy for thermoelectrics. Phys. Rev. B **92**, 85205 (2015)
8. G. Tan, L.-D. Zhao, M.G. Kanatzidis, Rationally designing high-performance bulk thermoelectric materials. Chem. Rev. **116**, 12123–12149 (2016)

9. J.-H. Lee, Significant enhancement in the thermoelectric performance of strained nanoporous Si. Phys. Chem. Chem. Phys. **16**, 2425–2429 (2014)
10. L. Huang et al., Recent progress in half-Heusler thermoelectric materials. Mater. Res. Bull. **76**, 107–112 (2016)
11. A. Page, P.F.P. Poudeu, C. Uher, A first-principles approach to half-Heusler thermoelectrics: accelerated prediction and understanding of material properties. J. Mater. **2**, 104–113 (2016)
12. W. Xie et al., Recent advances in nanostructured thermoelectric half-Heusler compounds. Nanomaterials **2**, 379–412 (2012)
13. J.-W.G. Bos, R.A. Downie, Half-Heusler thermoelectrics: a complex class of materials. J. Phys. Condens. Matter **26**, 433201 (2014)
14. F. Casper, T. Graf, S. Chadov, B. Balke, C. Felser, Half-Heusler compounds: novel materials for energy and spintronic applications. Semicond. Sci. Technol. **27**, 063001 (2012)
15. T.J. Scheidemantel, C. Ambrosch-Draxl, T. Thonhauser, J.V. Badding, J.O. Sofo, Transport coefficients from first-principles calculations. Phys. Rev. B **68**, 125210 (2003)
16. J. Yang et al., Evaluation of half-Heusler compounds as thermoelectric materials based on the calculated electrical transport properties. Adv. Funct. Mater. **18**, 2880–2888 (2008)
17. C. Kumarasinghe, N. Neophytou, Band alignment and scattering considerations for enhancing the thermoelectric power factor of complex materials: the case of Co-based half-Heusler alloys. Phys. Rev. B **99**, 195202 (2019)
18. G.D. Mahan, T. Balseiro, A.C. Bariloche, The best thermoelectric. Proc. Natl. Acad. Sci. **93**, 7436–7439 (1996)
19. N. Neophytou, M. Wagner, H. Kosina, S. Selberherr, Analysis of thermoelectric properties of scaled silicon nanowires using an atomistic tight-binding model. J. Electron. Mater. **39**, 1902–1908 (2010)
20. P. Giannozzi et al., QUANTUM ESPRESSO: a modular and open-source software project for quantum simulations of materials. J. Phys.: Condens. Matter **21**, 395502 (2009)
21. J. Corps et al., Interplay of metal-atom ordering, fermi level tuning, and thermoelectric properties in cobalt shandites $Co_3M_2S_2$ (M = Sn, In). Chem. Mater. **27**, 3946–3956 (2015)
22. S. Roychowdhury, U.S. Shenoy, U.V. Waghmare, K. Biswas, An enhanced Seebeck coefficient and high thermoelectric performance in p-type In and Mg Co-doped $Sn_{1-x}Pb_xTe$ via the co-adjuvant effect of the resonance level and heavy hole valence band. J. Mater. Chem. C **5**, 5737–5748 (2017)
23. G. Capellini et al., Tensile Ge microstructures for lasing fabricated by means of a silicon complementary metaloxide-semiconductor process. Opt. Express **22**, 399–410 (2014)
24. I. Jeong, J. Kwon, C. Kim, Y.J. Park, Design and numerical analysis of surface plasmon-enhanced fin Ge–Si light-emitting diode. Opt. Express **22**, 5927 (2014)
25. K.L. Low, Y. Yang, G. Han, W. Fan, Y.-C. Yeo, Electronic band structure and effective mass parameters of $Ge_{1-x}Sn_x$ alloys. Cit. J. Appl. Phys. **112**, 73707 (2012)
26. J. Zhou, S. Cheng, W.-L. You, H. Jiang, Effects of intervalley scattering on the transport properties in one-dimensional valleytronic devices. Sci. Rep. **6**, 23211 (2016)
27. Patrizio Graziosi, Chathurangi Kumarasinghe, Neophytos Neophytou, Impact of the scattering physics on the power factor of complex thermoelectric materials. J. Appl. Phys. **126**(15), 155701 (2019).

Part IV
Surfaces, Interfaces in Low–Dimensional Systems

Chapter 11
Prediction of Energy Gaps in Graphene—Hexagonal Boron Nitride Nanoflakes Using Artificial Neural Networks

Tudor Luca Mitran and George Alexandru Nemnes

Abstract Machine learning methods are currently applied in conjunction with ab initio density functional theory (DFT) simulations in order to establish computationally efficient alternatives for high-throughput processing in atomistic computations. The proposed method, based on artificial neural networks (ANNs), was used to predict the HOMO-LUMO energy gap in quasi-0D graphene nanoflake systems with randomly generated boron nitride embedded regions. Several artificial neural network (ANN) algorithms were tested in order to optimize the network parameters for the problem at hand. The trained ANNs prove to be computationally efficient at determining the energy gap with good accuracy and show a significant speedup over the classical DFT approach.

11.1 Introduction

The recent resurgence of machine learning (ML) methods in computer science in the last few years has lead to the production of efficient and user friendly software libraries that have become commonplace in many fields of research and even in production environments. ML has become an umbrella term for a vast array of numerical methods that vary in scope (e.g. classification, clustering, regression) and learning method (e.g. unsupervised, semi-supervised, supervised, reinforcement). Data driven

T. L. Mitran · G. A. Nemnes (✉)
Horia Hulubei National Institute for R&D in Physics and Nuclear Engineering, Str. Reactorului no. 30, P.O. BOX MG-6, Bucharest, Magurele, Romania
e-mail: alexandru.nemnes@nipne.ro

T. L. Mitran
e-mail: tudor.mitran@nipne.ro

G. A. Nemnes
Faculty of Physics, University of Bucharest, Str. Atomistilor, nr. 405, Magurele, Romania

Research Institute of the University of Bucharest (ICUB), 36-46 Mihail Kogalniceanu Blvd, 050107 Bucharest, Romania

E. V. Levchenko et al. (eds.), *Theory and Simulation in Physics for Materials Applications*, Springer Series in Materials Science 296,
https://doi.org/10.1007/978-3-030-37790-8_11

ML algorithms have been found useful in complex problems where specialized, man designed, software usually fails but where there is usually an abundance of data that ML methods can use as a scaffold.

ML methods have been applied in many fields of research, and have quite recently even found applicability in condensed matter (CM) physics. Some examples of ML use in CM are: classification of phases of matter [1], detection of phase transitions [2–4], classification of crystalline structures [5], materials discovery and design [6–8], determining charge density functionals [9], screening for new materials [10], modeling electronic quantum transport [11], atomic scale property prediction [12, 13], bond energy determination [14].

There are also studies that make use of ML methods as a speed-up solution, or even as a substitute to classical DFT methods, being able to introduce approximations for DFT Hamiltonians [15], predict the bond energies [16], determine density functionals [17–20], obtain ground state properties [21], emulate complex potential energy surfaces [22], completely bypass the Kohn-Sham equations [23], learn atomic potentials [24], optimize the basis sets [25], or determine the electron densities [26] and density functionals for the Hubbard model [27]. These recent results show that ML methods are a viable and useful enhancement to DFT and other molecular and atomistic simulation methods and could even substitute them in certain situations.

Graphene nanoflakes (GNF) are quasi-0D forms of graphene that have attracted interest during the last few years because of their mechanical [28, 29], optical [30–32], electronic [33, 34] and magnetic properties [35] that make them useful when considering biological [36], electrochemical detection [37] and tribological applications [38]. Taking advantage of their quasi-0D structure, the properties of GNFs can be tuned by manipulating their size or shape and by edge functionalization and chemical doping. Furthermore, GNFs can be experimentally produced through several methods: bottom-up, by mechanical extrusion, using magnetic field alignment and thermal annealing [39, 40], or top-down through plasma etching [41], cationic surfactant mediated exfoliation of graphite [42] or e-beam lithography [43].

In [44] ANN based methods were introduced to predict the energy gap in hybrid graphene-hexagonal boron nitride (G-hBN) nanoflake systems, combining DFT and ML techniques. GNFs with rectangular shaped hBN domains were considered. The ANNs were first trained on the results obtained by DFT simulations and, subsequently, their ability to predict the energy gap values was tested on a new set of systems and compared to reference DFT results. Two variants of ANNs were introduced, with respect to the structural information used as input. In *Method 1* the input layer of neurons contains one neuron for each atom in the structure and binary 0/1 values are assigned to graphene and B/N species, respectively, and in addition, one neuron encoding the proportion of each species. This method has a high accuracy, although it is bound to a given system size. In *Method 2*, the inputs are assigned based on chemical neighborhood and species proportions. This second method has the advantage that it is scalable, while still maintaining high accuracies.

Here we explore a larger class of G-hBN nanoflakes, by considering randomly distributed shapes and sizes of the hBN domains. As in [44], the property of interest is the LUMO-HOMO energy difference, or energy gap, which was investigated for a large set of systems. Several ANN architectures and specific ANN parameters were systematically investigated by using a grid search in order to optimize the final ANN configuration.

11.2 Description of Systems and Computational Methods

The atomic systems investigated were rectangular GNFs, passivated with hydrogen along the edges, with random substitutions by B and N atoms. In order to preserve the intrinsic character of the GNF, an equal number of B and N atoms were considered. The GNFs are composed of a total of 200 atoms, out of which 34 are H and the remaining 166 can range from all C to all BN. A total of 4333 unique atomic configurations were generated and ab initio DFT calculations were performed in order to obtain the values of their energy gaps. Out of the 4333 samples, 3000 are used for the initial training, 500 for the validation. For the final testing phase, the ANNs are trained on the total 3500 samples (training + validation) in order to increase the accuracy and are tested on a novel set of 833 samples (Fig. 11.1).

The SIESTA software package [45] was employed to perform the DFT simulations because it has the advantage of linear scaling of the computational time with the system size. This computational efficiency is achieved by using a localized basis set instead of the commonly used plane waves basis. Another advantage of a localized basis is the almost negligible computing overhead when empty space is present in the system's super-cell. The particular basis set was standard double-ζ polarized and the electron density was estimated in the local density approximation (LDA) of Ceperley and Alder [46]. The real space in the unit cell was set by 100 Ry grid cutoff and norm-conserving Troullier-Martins pseudopotentials [47] were used in the calculations. All GNFs were simulated in a cubic super-cell with 50 Å side length and since the systems are quasi-0D, the reciprocal space was only sampled in the Gamma point. In order to constrain the variability of the ANN input and to better correlate the prediction ability

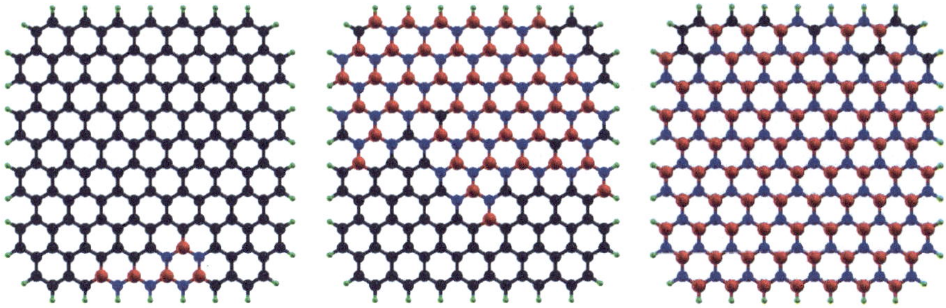

Fig. 11.1 Typical graphene nanoflake structures with embedded hBN domains, with different ratios of C to BN. Color codes: C—black, B—red, N—blue, H—green

to the chemical complexity of the samples, no structural relaxation was performed. After computing the ground state energy spectrum, the gap energies are obtained by subtracting the lowest unoccupied from the highest occupied energies: $E_{\text{gap}} = E_{\text{LUMO}} - E_{\text{HOMO}}$.

The ANNs were implemented with the TensorFlow library [48] which allows the implementation of complex network architectures as easily programmed tensor graphs, with interchangeable activation functions, optimization algorithms or variable number of neural layer with different widths.

Based on the the results obtained from the DFT simulations, training, validation and test sets were built for the ANNs. The training and validation sets are used for the optimization of the ANN architecture by determining the optimal configuration parameters in order to minimize the prediction error. The best performing ANNs are finally validated on the testing set.

The DFT results are preprocessed similarly to *Method 2* from [44] by classifying each atom and its first order neighbors (three atoms) in one of 22 possible classes. This 22 class distribution is normalized, and along with four other values that represent the normalized chemical composition (one input for each type of atomic species), is fed to the 26 input neurons. The output is a floating point value given by a single neuron that has no activation function and is compared to the energy gap value which is itself normalized to the maximum value of 4 eV. The parameters that were tuned in the ANNs concern the geometry (one, two or three hidden layers, with 100, 200 or 300 neurons on the first layer and 50 and 25 neurons in the second and third layers), the activation function (sigmoid or softmax) and learning rate (0.0001, 0.001 or 0.01). The three layer networks were also tested with a dropout approach that is used for increasing the generalization ability of the network by randomly turning off neurons in all thee hidden layers. Dropout ensures that the ANN does not just simply memorize all input data but forces it to learn the more fundamental governing rules of the system. The Adam optimization algorithm [49] was used to efficiently tune the weights and biases of the network in order to guide it to a state of minimal error. The mean squared error (MSE) is tracked for the training and validation sets in order to determine the optimum training point and over-fitting. Over-training or over-fitting occurs when the validation MSE and training MSE start to diverge, with the first increasing and the second continuing to decrease. Regularization by early-stopping can be implemented by interrupting the training process when the minimum validation error is reached. The number of training steps needed to reach a certain error value also impacts the performance of an algorithm if retraining is necessary for different system classes, but it is considered less relevant in practice since training is only performed once. The smoothness of the error graph is an indication of an algorithm's stability and a large error fluctuation from epoch to epoch makes it less able to generalize the results on new samples.

11.3 Results and Discussion

The difference between the LUMO and HOMO energies is computed by DFT simulations for the structures with B and N substitutions in the GNFs. The actual distribution of samples is represented as a function of the the energy gap and as a function of BN concentration in Fig. 11.2. The BN concentration shows a uniform distribution, as expected. The energy gap distribution (on a logarithmic scale) shows that almost 80% of samples have a gap below 0.1 eV, with the rest showing a sharp decrease in number up to 3.6 eV.

In Fig. 11.2 the distribution of the energy gap is represented as a function of the BN concentration. As expected, at low BN concentration the energy gap is close to zero but at high BN concentration there are only few cases with a high energy gap. Most samples have energy gaps below 0.1 eV and are distributed almost uniformly in BN concentration, also confirmed by the sample and energy gap distribution. By analyzing the atomic configuration of the GNFs with high BN concentration (greater than 90%), two distinct cases were detected: at low energy gaps (below 0.3 eV) the GNFs present isolated C atoms, while at high energy gaps (above 2 eV) the carbon atoms are distributed in pairs or clusters. Small energy gaps appear even at low concentrations of C with respect to the number of BN atoms (the lowest being 1.2% or 2 C atoms per 164 BN atoms) because of the new energy states located in the BN gap.

In order to evaluate the performance of the different ML algorithms and, at the same time, control whether over-fitting appears after a certain number of optimization epochs, the mean squared error (MSE) is sampled during the training process. The MSE is represented for ANNs with a single hidden layer in Fig. 11.3, two hidden layers in Fig. 11.4, three hidden layers in Fig. 11.5 and ANNs with three hidden layers and dropout in Fig. 11.6. Training is stopped after 125,000 steps.

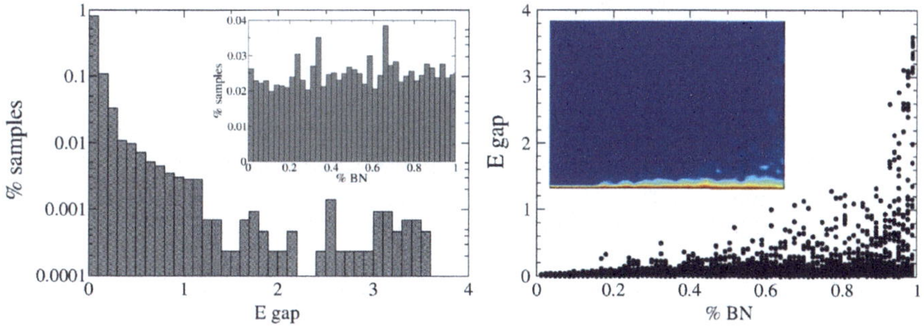

Fig. 11.2 Left: the distribution of samples according to their energy gap on a logarithmic scale and the uniform distribution of BN concentrations (inset). Right: the energy gap as a function of BN concentration, also shown as a density map on a logarithmic scale (inset)

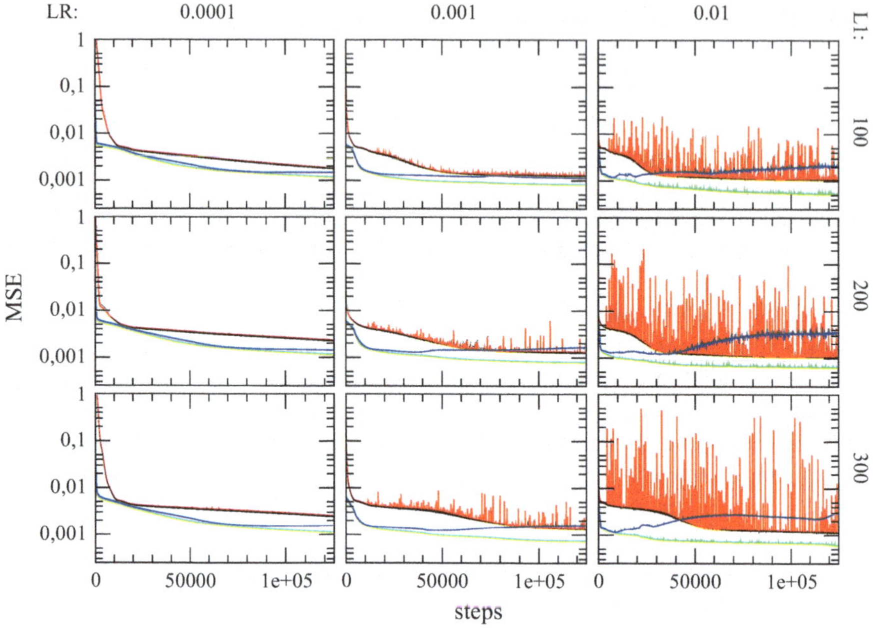

Fig. 11.3 The evolution of the MSE (mean squared error) with training for ANNs with a single hidden layer (L1) of 100, 200 or 300 neurons for thee learning rates (LR) of 0.0001, 0.001 and 0.01. In the case of the sigmoid activation function the training MSE is represented with black and red represents the MSE for the validation set. For the softmax activation function, the training MSE is represented in green and the validation MSE in blue

For all the cases of ANNs that do not implement dropout, the sigmoid activation function makes the ANN less stable in training and gives a much noisier MSE result from epoch to epoch than softmax. This makes it a less suitable choice since, even if it is able to attain low MSE values at certain training steps, it is not robust enough to ensure the same performance on new samples, even if using early stopping. A possible explanation for this behavior is given by the key difference between sigmoid and softmax: the softmax activated neurons have an output bound to one by design as opposed to the values produced by sigmoid which are independent and unconstrained.

The upper bound of softmax has an effective regularization effect. Another general observation is that higher learning rates also show more training noise, since the optimization algorithm performs larger steps in the networks configuration space and is less likely to be bound to local minima. MSE noise is also significant for sigmoid ANNs with only one hidden layer and increases with the layer's size in Fig. 11.3.

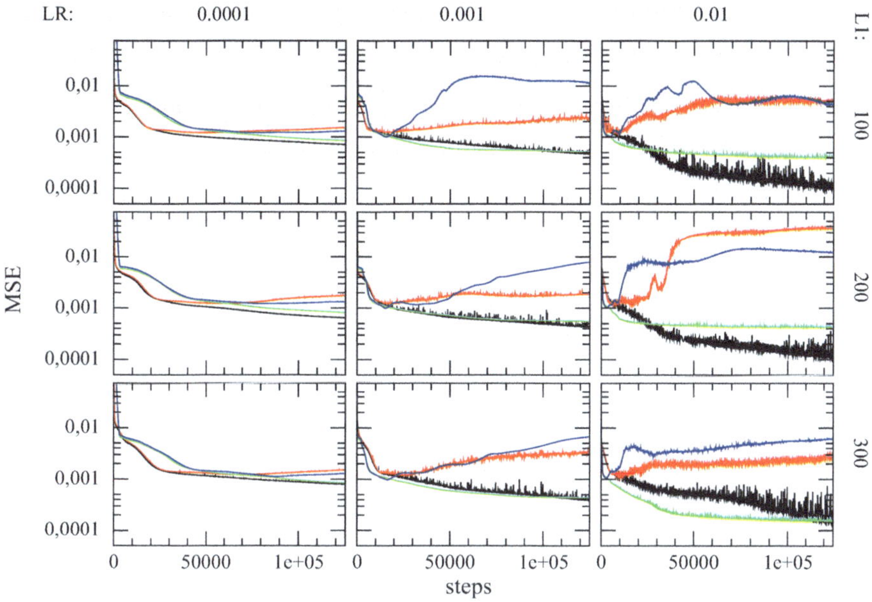

Fig. 11.4 The evolution of the MSE (mean squared error) with training for ANNs with two hidden layers (the first—L1—having 100, 200 or 300 neurons and the second 50 neurons) for thee learning rates (LR) of 0.0001, 0.001 and 0.01. In the case of the sigmoid activation function the training MSE is represented with black and red represents the MSE for the validation set. For the softmax activation function, the training MSE is represented in green and the validation MSE in blue

Figure 11.7 represents the lowest MSEs and the corresponding number of training steps that were attained by all the algorithms tested. As expected, the MSE minimum is reached after less training by increasing the learning rate. A high learning rate can prevent the algorithm from finding the optimum network configuration and it also has a narrower optimal interval in number of training epochs which increases the probability to under/over fit for a new sample. ANNs with two and three hidden layers converge to the optimum point faster than ones with a single hidden layer. For ANNs with sigmoid activation functions dropout manages to lower the error, but at the cost of increasing the training time when compared to three layer networks that do not use it. For ANNs with softmax activation functions, dropout has a negative effect, increasing both the MSE and the number of training steps required. Because of the high noise level in MSE of sigmoid based ANNs, only softmax ANNs are further considered. The best performing softmax ANNs are the ones with two and three hidden layers, having validation MSEs close to 0.001. For the final performance metric, these 12 networks are retrained on the initial training set plus the validation (3500 samples) set and are tested on a separate test set (833 samples).

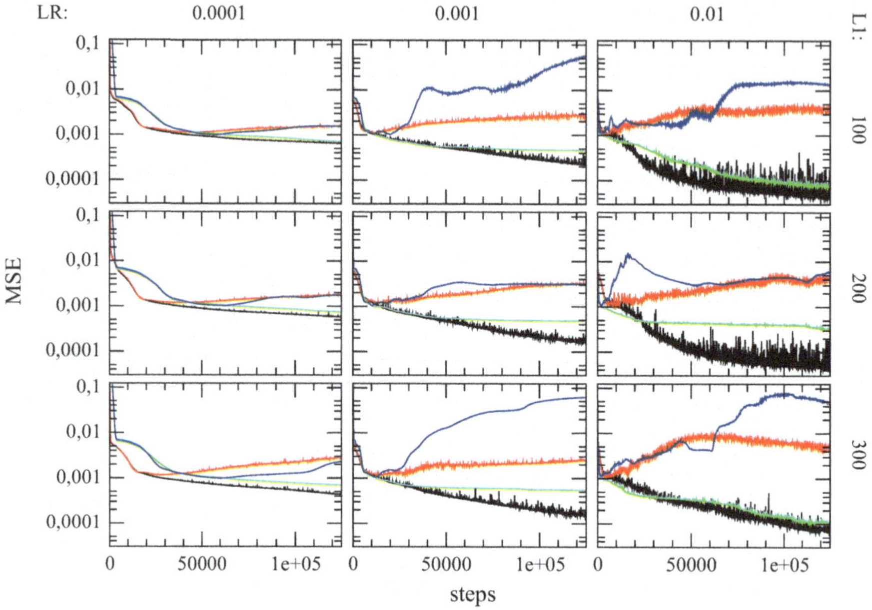

Fig. 11.5 The evolution of the MSE (mean squared error) with training for ANNs with three hidden layers (the first—L1—having 100, 200 or 300 neurons, the second 50 neurons and the third 25) for three learning rates (LR) of 0.0001, 0.001 and 0.01. In the case of the sigmoid activation function the training MSE is represented with black and red represents the MSE for the validation set. For the softmax activation function, the training MSE is represented in green and the validation MSE in blue

In order to compare the quality of prediction for the testing set, Fig. 11.8 showed the R^2 coefficient computed for the best performing ANNs. As expected from the MSE plot in Fig. 11.7, all 12 ANNs show comparable results, with R^2 coefficients ranging from 0.83 to 0.87 for the training set and from 0.71 to 0.82 for the test set. The best performing ANN has two hidden layers (with 300 and 50 neurons) uses a 0.01 learning rate and obtained a R^2 coefficient of 0.83 for training and 0.82 for testing. Figure 11.8 also shows the actual comparison between the individual DFT energy gap values versus predicted energy gap values for this network. The deviations of the predicted values from the computed DFT values are greatest in the region of low energy gaps (below 0.25 eV) mostly because of the large diversity of atomic configurations that can be realized. However, the large energy gaps are accurately identified.

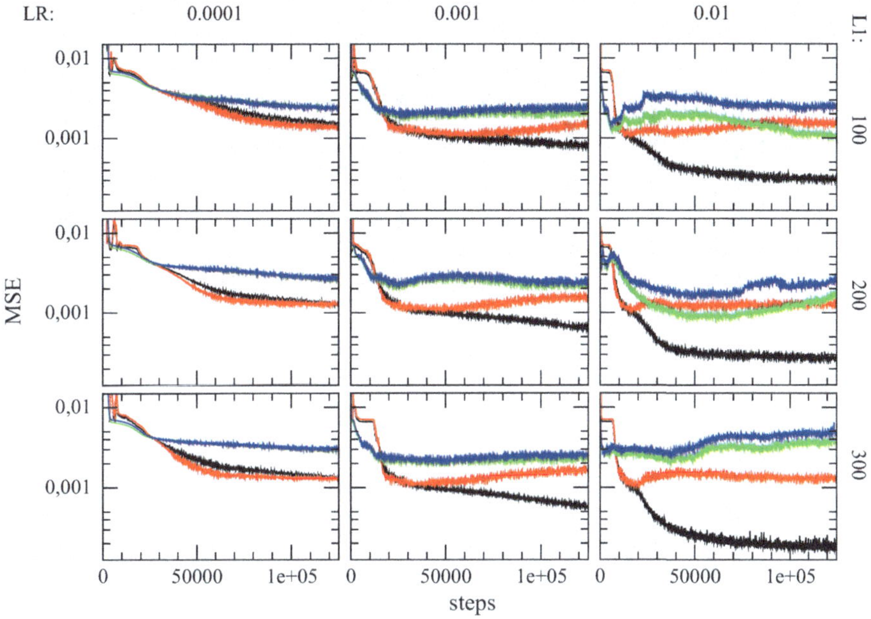

Fig. 11.6 The evolution of the MSE (mean squared error) with training for ANNs with three hidden layers (the first—L1—having 100, 200 or 300 neurons, the second 50 neurons and the third 25) with dropout probability of 15% (85% keep probability) for three learning rates (LR) of 0.0001, 0.001 and 0.01. In the case of the sigmoid activation function the training MSE is represented with black and red represents the MSE for the validation set. For the softmax activation function, the training MSE is represented in green and the validation MSE in blue

11.4 Conclusions

We have investigated the electronic properties of GNF embedded with randomly positioned atoms of boron and nitrogen by ab initio DFT simulations. Based on these results, several types of artificial neural networks were trained to reproduce the DFT gap energy. The network architectures and internal parameters were optimized. Out of the tested ANNs, the two and three-layer versions with softmax activation function indicated a good ability to generalize and predict the energy gaps of novel GNF configurations. Such an ANN based prediction of properties is suitable for batch computation on a large number of systems, for which DFT alone would prove too computationally expensive. Another advantage of this ML method is that it takes as input a fixed size distribution, which does not bind it to samples with a specific number of atoms.

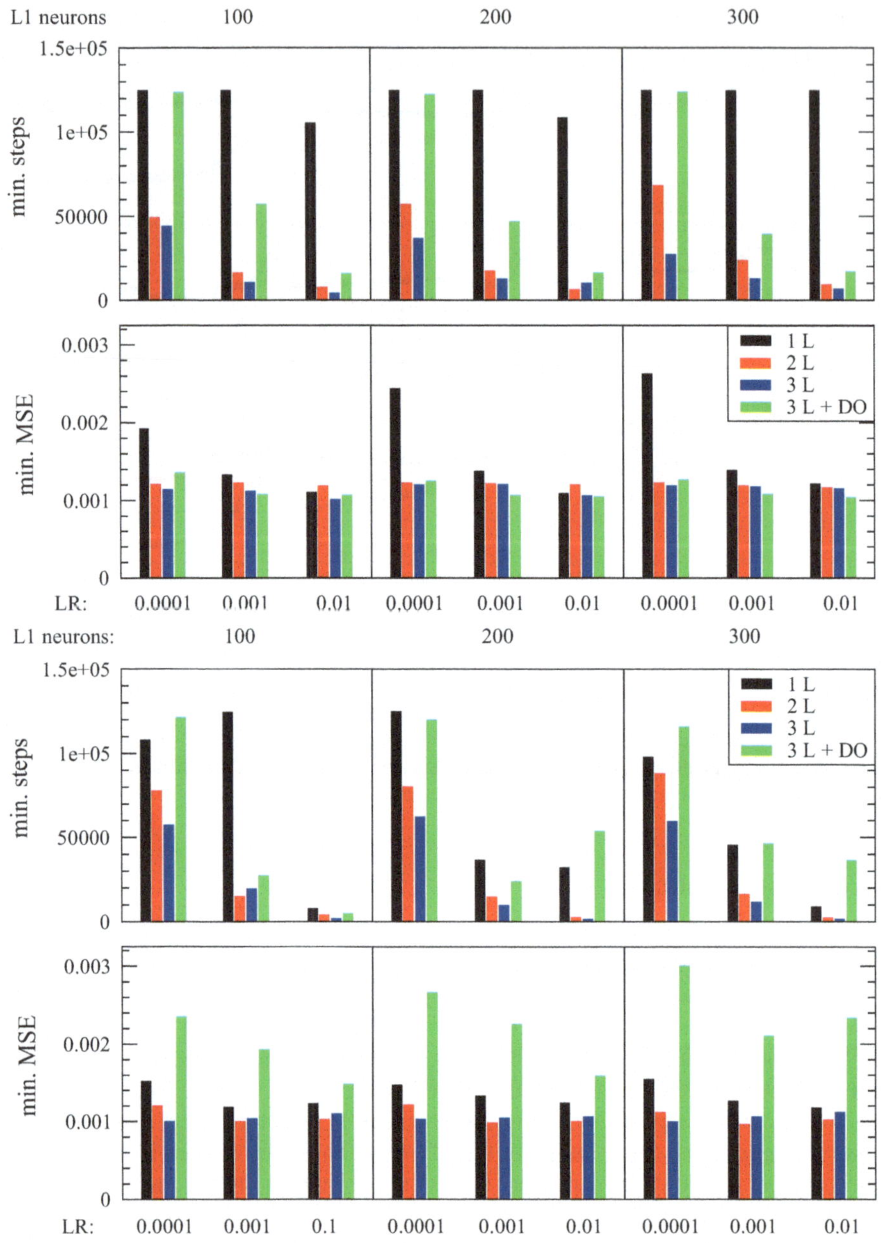

Fig. 11.7 The histogram of the lowest MSEs for validation, and the corresponding number of training steps, in the case of the sigmoid activation function (top) and softmax function (bottom). The colors designate the type of ANN: single hidden layer (1 L), two hidden layer (2 L), three hidden layers (3 L) and three hidden layers with dropout (3 L + DO)

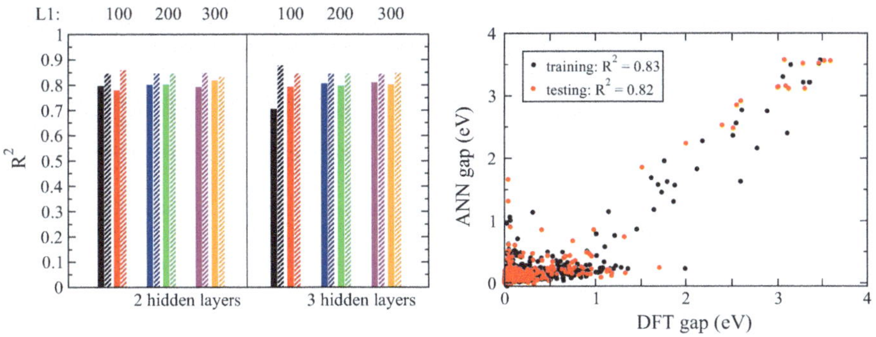

Fig. 11.8 Left: R^2 coefficient for the best performing algorithms with the softmax activation function: two hidden layers and 0.001 learning rate (LR) (and the number on neurons on the first layer: black—100, blue—200, magenta—300) or 0.01 LR (red—100, green—200, orange—300). The filled bars correspond to the test R^2 coefficient and the dashed bars to the R^2 training coefficient. Right: energy gap predicted by ANN (softmax, two hidden layers of 300 and 50 neurons and 0.01 learning rate) versus DFT results

Acknowledgements This work was supported by the Romanian Ministry of Research and Innovation under the project PN19060205/2019 and by the Romania-JINR cooperation project.

References

1. J. Carrasquilla, R.G. Melko, Machine learning phases of matter. Nat. Phys. **13**, 431 (2017)
2. A. Tanaka, A. Tomiya, Detection of phase transition via convolutional neural networks. J. Phys. Soc. Japan **86**, 063001 (2017)
3. T. Ohtsuki, T. Ohtsuki, Deep learning the quantum phase transitions in random two-dimensional electron systems. J. Phys. Soc. Japan **85**, 123706 (2016)
4. L. Wang, Discovering phase transitions with unsupervised learning. Phys. Rev. B **94**, 195105 (2016)
5. C. Dietz, T. Kretz, M.H. Thoma, Machine-learning approach for local classification of crystalline structures in multiphase systems. Phys. Rev. E **96**, 011301 (2017)
6. Y. Liu, T. Zhao, J. Wangwei, S. Shi, Materials discovery and design using machine learning. J. Materiomics **3**, 159–177 (2017)
7. T. Yamashita, N. Sato, H. Kino, T. Miyake, K. Tsuda, T. Oguchi, Crystal structure prediction accelerated by Bayesian optimization. Phys. Rev. Mater. **2**, 013803 (2018)
8. Q. Zhou, P. Tang, S. Liu, J. Pan, Q. Yan, S.-C. Zhang, Learning atoms for materials discovery. Proc. Natl. Acad. Sci. **115**, E6411–E6417 (2018)
9. B. Kolb, L.C. Lentz, A.M. Kolpak, Discovering charge density functionals and structure-property relationships with prophet: A general framework for coupling machine learning and first-principles methods. Sci. Rep. **7**, 1192 (2017)
10. B. Meredig, A. Agrawal, S. Kirklin, J.E. Saal, J.W. Doak, A. Thompson, K. Zhang, A. Choudhary, C. Wolverton, Combinatorial screening for new materials in unconstrained composition space with machine learning. Phys. Rev. B **89**, 094104 (2014)
11. A. Lopez-Bezanilla, O.A. von Lilienfeld, Modeling electronic quantum transport with machine learning. Phys. Rev. B **89**, 235411 (2014)

12. A.P. Bartók, S. De, C. Poelking, N. Bernstein, J.R. Kermode, G. Csányi, M. Ceriotti, Machine learning unifies the modeling of materials and molecules. Sci. Adv. **3** (2017)
13. C.W. Rosenbrock, E.R. Homer, G. Csányi, G.L.W. Hart, Discovering the building blocks of atomic systems using machine learning application: To grain boundaries. npj Comput. Mater. **3**, 29 (2017)
14. K. Yao, J.E. Herr, S.N. Brown, J. Parkhill, Intrinsic bond energies from a bonds-in-molecules neural network. J. Phys. Chem. Lett. **8**, 2689–2694 (2017)
15. G. Hegde, R.C. Bowen, Machine-learned approximations to density functional theory hamiltonians. Sci. Rep. **7**, 42669 (2017)
16. X. Qu, D.A.R.S. Latino, J. Aires-de Sousa, A big data approach to the ultra-fast prediction of DFT-calculated bond energies. J. Cheminformatics **5**, 34 (2013)
17. J.C. Snyder, M. Rupp, K. Hansen, K.-R. Müller, K. Burke, Finding density functionals with machine learning. Phys. Rev. Lett. **108**, 253002 (2012)
18. L. Li, J.C. Snyder, I.M. Pelaschier, J. Huang, U.-N. Niranjan, P. Duncan, M. Rupp, K.-R. Müller, K. Burke, Understanding machine-learned density functionals. Int. J. Quant. Chem. **116**, 819–833 (2016). https://doi.org/10.1002/qua.25040
19. A. Chandrasekaran, D. Kamal, R. Batra, C. Kim, L. Chen, R. Ramprasad, Solving the electronic structure problem with machine learning. npj Comput. Mater. **5**, 22 (2019)
20. G.R. Schleder, A.C.M. Padilha, C.M. Acosta, M. Costa, A. Fazzio, From DFT to machine learning: Recent approaches to materials science–A review. J. Phys.: Mater. **2**, 032001 (2019)
21. F.A. Faber, L. Hutchison, B. Huang, J. Gilmer, S.S. Schoenholz, G.E. Dahl, O. Vinyals, S. Kearnes, P.F. Riley, O.A. von Lilienfeld, Prediction errors of molecular machine learning models lower than hybrid DFT error. J. Chem. Theory Comput. **13**, 5255–5264 (2017)
22. D. Dragoni, T.D. Daff, G. Csányi, N. Marzari, Achieving DFT accuracy with a machine-learning interatomic potential: Thermomechanics and defects in BCC ferromagnetic iron. Phys. Rev. Mater. **2**, 013808 (2018)
23. F. Brockherde, L. Vogt, L. Li, M.E. Tuckerman, K. Burke, K.-R. Müller, Bypassing the Kohn-Sham equations with machine learning. Nat. Commun. **8**, 872 (2017)
24. T.L. Jacobsen, M.S. Jørgensen, B. Hammer, On-the-fly machine learning of atomic potential in density functional theory structure optimization. Phys. Rev. Lett. **120**, 026102 (2018)
25. O. SchÃitt, J. VandeVondele, Machine learning adaptive basis sets for efficient large scale density functional theory simulation. J. Chem. Theory Comput. **14**, 4168–4175 (2018)
26. A. Grisafi, A. Fabrizio, B. Meyer, D.M. Wilkins, C. Corminboeuf, M. Ceriotti, Transferable machine-learning model of the electron density. ACS Central Sci. **5**, 57–64 (2019)
27. J. Nelson, R. Tiwari, S. Sanvito, Machine learning density functional theory for the Hubbard model. Phys. Rev. B **99**, 075132 (2019)
28. X. Feng, S. Kwon, J.Y. Park, M. Salmeron, Superlubric sliding of graphene nanoflakes on graphene. ACS Nano **7**, 1718–1724 (2013)
29. J. Zhang, E. Osloub, F. Siddiqui, W. Zhang, T. Ragab, C. Basaran, Anisotropy of graphene nanoflake diamond interface frictional properties. Materials **12**, 1425 (2019)
30. C. Mansilla Wettstein, F.P. Bonafé, M.B. Oviedo, C.G. Sánchez, Optical properties of graphene nanoflakes: Shape matters. J. Chem. Phys. **144**, 224305 (2016)
31. C. Cocchi, D. Prezzi, A. Ruini, M.J. Caldas, E. Molinari, Electronics and optics of graphene nanoflakes: Edge functionalization and structural distortions. J. Phys. Chem. C **116**, 17328–17335 (2012)
32. A.M. Silva, M.S. Pires, V.N. Freire, E.L. Albuquerque, D.L. Azevedo, E.W.S. Caetano, Graphene nanoflakes: Thermal stability, infrared signatures, and potential applications in the field of spintronics and optical nanodevices. J. Phys. Chem. C **114**, 17472–17485 (2010)
33. S.K. Singh, M. Neek-Amal, F.M. Peeters, Electronic properties of graphene nano-flakes: Energy gap, permanent dipole, termination effect, and Raman spectroscopy. J. Chem. Phys. **140**(7), 074304 (2014)
34. A. Kuc, T. Heine, G. Seifert, Structural and electronic properties of graphene nanoflakes. Phys. Rev. B **81**, 085430 (2010)

35. W.L. Wang, S. Meng, E. Kaxiras, Graphene nanoflakes with large spin. Nano Lett. **8**, 241–245 (2008)
36. V. Castagnola, W. Zhao, L. Boselli, M.C.L. Giudice, F. Meder, E. Polo, K.R. Paton, C. Backes, J.N. Coleman, K.A. Dawson, Biological recognition of graphene nanoflakes. Nat. Commun. **9**, 1577 (2018)
37. S. Meenakshi, S.J. Sophia, K. Pandian, High surface graphene nanoflakes as sensitive sensing platform for simultaneous electrochemical detection of metronidazole and chloramphenicol. Mater. Sci. Eng.: C **90**, 407–419 (2018)
38. L. Liu, M. Zhou, L. Jin, L. Li, Y. Mo, S. Guoshi, X. Li, H. Zhu, Y. Tian, Recent advances in friction and lubrication of graphene and other 2D materials: Mechanisms and applications. Friction **7**, 199 (2019)
39. J. Wu, W. Pisula, K. Müllen, Graphenes as potential material for electronics. Chem. Rev. **107**, 718–747 (2007)
40. L. Zhi, K. Mullen, A bottom-up approach from molecular nanographenes to unconventional carbon materials. J. Mater. Chem. **18**, 1472–1484 (2008)
41. S. Neubeck, L.A. Ponomarenko, F. Freitag, A.J.M. Giesbers, U. Zeitler, S.V. Morozov, P. Blake, A.K. Geim, K.S. Novoselov, From one electron to one hole: Quasiparticle counting in graphene quantum dots determined by electrochemical and plasma etching. Small **6**, 1469–1473 (2010)
42. S. Mutyala, J. Mathiyarasu, Preparation of graphene nanoflakes and its application for detection of hydrazine. Sens. Actuators B: Chem. **210**, 692–699 (2015)
43. C. Berger, Z. Song, X. Li, W. Xiaosong, N. Brown, C. Naud, D. Mayou, T. Li, J. Hass, A.N. Marchenkov, E.H. Conrad, P.N. First, W.A. de Heer, Electronic confinement and coherence in patterned epitaxial graphene. Science **312**, 1191–1196 (2006)
44. G.A. Nemnes, T.L. Mitran, A. Manolescu, Gap prediction in hybrid graphene—Hexagonal boron nitride nanoflakes using artificial neural networks. J. Nanomater. **2019**, 6960787 (2019)
45. J.M. Soler, E. Artacho, J.D. Gale, A. García, J. Junquera, P. Ordejón, D. Sánchez-Portal, The SIESTA method forab initioorder-nmaterials simulation. J. Phys.: Condens. Matter **14**, 2745–2779 (2002)
46. D.M. Ceperley, B.J. Alder, Ground state of the electron gas by a stochastic method. Phys. Rev. Lett. **45**, 566–569 (1980)
47. N. Troullier, J.L. Martins, Efficient pseudopotentials for plane-wave calculations. Phys. Rev. B **43**, 1993–2006 (1991)
48. M. Abadi, P. Barham, J. Chen, Z. Chen, A. Davis, J. Dean, M. Devin, S. Ghemawat, G. Irving, M. Isard, M. Kudlur, J. Levenberg, R. Monga, S. Moore, D.G. Murray, B. Steiner, P.A. Tucker, V. Vasudevan, P. Warden, M. Wicke, Y. Yu, X. Zhang, Tensorflow: A system for large-scale machine learning. CoRR abs/1605.08695 (2016)
49. D.P. Kingma, J. Ba, Adam: A Method for Stochastic Optimization (2014). http://arxiv.org/abs/1412.6980

Chapter 12
Hydrogen in Silicon: Evidence of Independent Monomeric States

V. V. Voronkov

Abstract The data on hydrogen in saturated/quenched samples and in samples exposed to plasma have been revisited. It is concluded that the monomeric hydrogen in intrinsic silicon is represented mostly by two neutral species: H_b (presumably a ground state of tetrahedral hydrogen) and H_s (a slow monomer in a different interstitial position). At high *T* these species are in equilibrium, with a concentration ratio close to 1. At lower *T* (at least at $T \leq 500$ °C) they become independent one of the other. This conclusion differs from a conventional notion that considers bond-centred H^+(BC) ions to be dominant in intrinsic Si. In p-Si, boron is passivated not only by H^+(BC) ions (denoted $H^+(1)$) but also by another kind of H^+ denoted $H^+(2)$. A presence of several independent species (H_b, H_s, $H^+(1)$ and $H^+(2)$) gives rise to a rich variety of hydrogen depth profiles in plasma-exposed silicon; these profiles are well reproduced by simulations.

12.1 Introduction

Hydrogen impurity plays an important role in silicon materials—by passivating dopants and other defects [1–6]. In *p*-type and near-intrinsic Si, under equilibrium conditions, the dominant monomeric state of hydrogen is thought [7] to be H^+(BC)—a positive ion in a Bond-Centred position. The acceptors (normally, boron B^-) are passivated by trapping H^+ which creates neutral defects HB. The data on hydrogen in silicon were collected years ago, but they are still waiting for a proper analysis, in view of a growing interest to hydrogen as a powerful and promising tool to control the defects in Si.

Beside H^+(BC), there is another relevant state of hydrogen: the neutral atom H^0 in tetrahedral (*T*) location [7]. This one has a far lower concentration but a very high diffusivity, and the hydrogen transport in intrinsic Si is controlled by this minor species H^0 [8] rather than by the dominant one, H^+. In *p*-type Si the hydrogen

V. V. Voronkov (✉)
Global Wafers, via Nazionale 59, 39012 Merano, Italy
e-mail: vvoronkov@gw-semi.com

E. V. Levchenko et al. (eds.), *Theory and Simulation in Physics for Materials Applications*, Springer Series in Materials Science 296,
https://doi.org/10.1007/978-3-030-37790-8_12

transport is controlled by H^+ but H^0 is still important since formation of hydrogen dimers [9] proceeds by a pairing reaction $H^0 + H^+$.

Most of the data on hydrogen states in samples saturated with hydrogen at high T (around 1200 °C, up to a concentration of about 10^{16} cm^{-3}) and quenched [10–13] can be well understood [9] using the above simple notion. It turns out that during even a very fast cooling (like 500 K/s) the equilibrium relation between the monomeric states (H^+ and H^0) and the dimeric state (H_2) is supported down to about 400 °C, and the H_2 state finally becomes dominant. These dimers have been labelled H_{2A}; the subscript A is used to distinguish them from other dimeric forms observed under different experimental conditions [9]. At $T < 400$ °C, the dominant equilibrium states of hydrogen are H_{2A} and HB. The equilibrium fraction of HB increases upon lowering T, but the transition from H_{2A} to HB is slow, and the quenched-in fraction of HB remains small. If the quenched samples are then annealed at low T like 160 or 175 °C, the transition from H_{2A} to HB goes on resulting [10, 11] in an essential reduction in [H_{2A}] and a corresponding increase in [HB]. Both concentrations are monitored by IR spectroscopy. The final (equilibrium) concentration ratio of the two low-T major states, [HB]/[H_{2A}], is strongly increased at a higher boron concentration N_B. At $N_B = 10^{17}$ cm^{-3}, almost all H_{2A} are expected to be replaced with HB [9] which is indeed the case: by annealing at 175 °C, [HB] increases up to the total (SIMS-measured) hydrogen concentration [13].

There are however other phenomena—observed both in saturated/quenched samples and in samples exposed to hydrogen plasma—that disagree with the scheme involving only H^+(BC), H^0(T) and HB single-hydrogen states. The aim of this paper is to discuss these pieces of evidence—in favour of other relevant monomeric species of hydrogen. In particular the concentration profiles in plasma-exposed samples definitely show existence of two independent H^+ ions; both passivate boron forming two structurally different HB defects.

12.2 Hydrogen Loss from Saturated/Quenched Samples

Using a radioactive tritium isotope of hydrogen for saturation at around 1200 °C, it was possible to monitor out-diffusion of hydrogen (tritium) into the ambient during annealing in a range of 400–500 °C [14]. The dominant state of hydrogen, in this T range, for the total concentration $\sim 10^{16}$ cm^{-3}, is expected [9] to be H_{2A}—a species of a relatively low diffusivity D_{2A}. The hydrogen transport is then mediated by minor but highly mobile atomic species H^+ and H^0 present in equilibrium with H_{2A}. The dominant contribution is by H^0 since the high-purity float zone material of this study is intrinsic under annealing. The apparent diffusivity is thus expected to be much higher than D_{2A} but much smaller than the diffusivity D_a of atomic hydrogen [15] extrapolated from high T:

$$D_a = \left(0.0094\,\text{cm}^2/\text{s}\right)\exp(-0.48\,\text{eV}/\text{kT}). \tag{12.1}$$

However, the loss of tritium occurs in a more complicated way [14]—in two stages as illustrated in Fig. 12.1. There is a fast stage of loss that well corresponds [9] to the expected effective diffusivity of H_{2A} mediated by H^0. This fast stage is followed by a slow stage corresponding to a much smaller diffusivity:

$$D_s = \left(2.3 \times 10^{-5}\text{cm}^2/\text{s}\right)\exp(-0.53\,\text{eV}/\text{kT}). \tag{12.2}$$

This expression refers to the tritium isotope. For other isotopes of hydrogen one should take into account that the diffusivity is inversely proportional to the square root of the isotope mass.

The slow stage of the tritium loss clearly shows that, apart form a "normal" hydrogen community (composed of mutually equilibrated H_{2A}, H^+(BC) and H^0(T)) there is an independent hydrogen species denoted H_s—of a low diffusivity $D_s \ll D_a$—which is yet larger than D_{2A}. It was speculated [9] that D_s refers to "fast dimers" labelled H_{2B}. These species may be formed, along with H_{2A}, during a quench but—unlike H_{2A}—the H_{2B} dimers do not dissociate in the temperature range of 400–500 °C.

Another possible attribution of the H_s species is to a monomeric hydrogen that is abundant at high T and resides in a position different from a tetrahedral site (for instance, at the Anti-Bonding site, AB). The two monomeric subsystems, H^+(BC)/ H^0(T) and H_s, exist independent one of the other at $T \leq 500$ °C (in a range of tritium loss experiments) implying a high barrier for the reconfiguration between them. At well higher T, they are likely to coexist in an equilibrium ratio which implies that H_s amounts to about 50% of the total atomic hydrogen—since the fraction of the slow component in Fig. 12.1 is about 0.5.

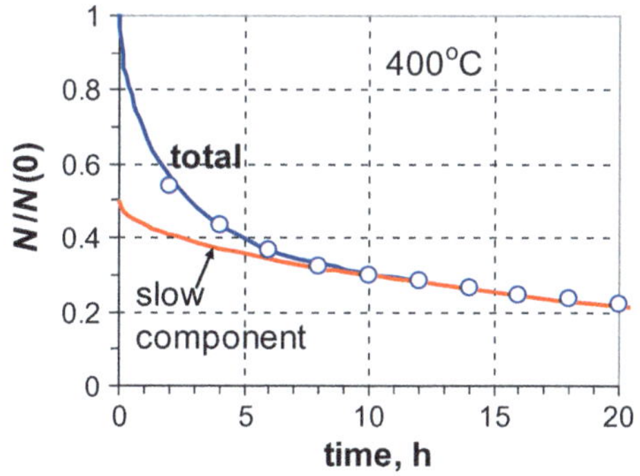

Fig. 12.1 Amount of tritium isotope in a sample (normalized by the starting value $N(0)$) in dependence of annealing time at 400 °C, after [14]. The red line is the calculated contribution of a slowly diffusing species H_s, for a uniform initial depth profile. The total calculated amount, for both the fast and the slow components, is shown by the blue line

The latter attribution of the H_s species—to an additional monomeric state—seems more natural in view of data discussed in the next section.

The as-quenched state of hydrogen in p-Si includes two hydrogen communities. The first one—originating from $H^+(BC)/H^0(T)$—is represented by H_{2A} and HB. The second one originates from the H_s monomer. Upon subsequent annealing, the total concentration of hydrogen in the H_{2A} + HB sub-community remains close to 10^{16} cm^{-3} at $N_B = 10^{16}$ cm^{-3} [11]. However at a higher doping level, $N_B = 10^{17}$ cm^{-3}, the achieved concentration of HB was considerable larger, 1.6×10^{16} cm^{-3} [13]. This difference can be understood if, at lower N_B, the H_s–related hydrogen component remains "hidden"—non-detected optically. At higher N_B, the H_s component converts into H_{2A} or HB if this reaction is enhanced by boron or by holes (or by both). In this case the concentration of HB tends to the total hydrogen concentration.

12.3 Depth Profiles of Hydrogen in Plasma-Exposed Lightly Doped n-Si

Hydrogen depth profiles produced by exposure of lightly doped n-Si samples to remote deuterium plasma at 150 °C [3] are composed of two distinct parts labelled *S* (the near-surface one) and *B* (the bulk one); an example is shown in Fig. 12.2. The shape of each part well corresponds to in-diffusion of some monomer—limited by an irreversible pairing into dimers [8, 16]. The diffusion-limited formation rate of dimers is

$$G = 4\pi r D C^2, \tag{12.3}$$

where C is the monomer concentration, D is its diffusivity and r is the pairing radius (expected to be about 0.3 nm). A steady-state profile $C(z)$ is described by an equation:

$$D\mathrm{d}^2C/\mathrm{d}z^2 = 2G. \tag{12.4}$$

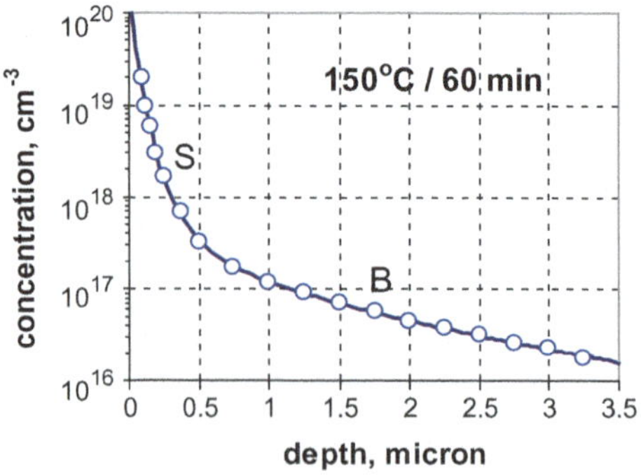

Fig. 12.2 Deuterium depth profile after exposure to a remote plasma at 150 °C for 60 min, after [3]. The donor dopant concentration is 2×10^{14} cm^{-3}, the intrinsic electron concentration is 2×10^{13} cm^{-3} at 150 °C

With a variable $S(z) = D\,C$ used instead of $C(z)$ the profile equation becomes

$$d^2S/dz^2 = 8\pi(r/D)S^2, \tag{12.5}$$

involving only one bulk parameter, D/r. The solution is

$$S(z) = S(0)/(1 + z/L)^2, \tag{12.6}$$

where the depth scale L equals $[(D/r)/(4\pi\ S(0)/3)]^{1/2}$. The total hydrogen concentration C_{tot} is mostly due to dimers, and it is close to $2\ G\ t$. The hydrogen profile is of a convex shape, and it depends, apart of the D/r ratio, on the boundary value $S(0)$ of the $D\ C$ product.

For the S-part, the deduced value of D/r is about 4×10^{-4} cm/s, and $S(0) = 1.1 \times 10^6\ \mathrm{cm}^{-1}\ \mathrm{s}^{-1}$. With $r = 0.3$ nm, the diffusivity of a species responsible for the S-component is $D = 1.2 \times 10^{-11}\ \mathrm{cm}^2/\mathrm{s}$. This value is close to the diffusivity of the H_s hydrogen species extrapolated to 150 °C by (12.2): $D_s = 1.4 \times 10^{-11}\ \mathrm{cm}^2/\mathrm{s}$ (for the deuterium isotope). The S-part is thus attributed to the H_s atomic species supplied from the plasma ambient; they in-diffuse producing dimers. If these dimers were H_{2A}, they would diffuse by a distance $(2\ D_{2A}\ t)^{1/2}$ that amounts to about 0.3 μm. The hydrogen concentration would be then peaked at a depth $z \approx 0.3$ μm due to out-diffusion of H_{2A} to the sample surface. The absence of this feature suggests that the dimers produced by H_s are different from H_{2A}—and much slower.

The parameter D/r, for the S-component, was found [8] to be the same for two available profiles [3] of essentially different doping level—showing that the in-diffusing H_s species is neutral. Accordingly, the diffusivity D_s by (12.2) refers to the neutral H_s species.

For the bulk part of the profile (marked B in Fig. 12.2) the deduced ratio D/r is about 0.6 cm/s, and $S(0) = 1.5 \times 10^6\ \mathrm{cm}^{-1}\ \mathrm{s}^{-1}$. The value of D/r is independent of the doping level of n-Si—suggesting in-diffusion of a neutral species. The diffusivity D_b for the bulk component (at $r = 0.3$ nm) is $1.8 \times 10^{-8}\ \mathrm{cm}^2/\mathrm{s}$—very similar to the atomic diffusivity by (12.1) which is $1.3 \times 10^{-8}\ \mathrm{cm}^2/\mathrm{s}$ at 150 °C for the deuterium isotope.

At this point we are forced to deviate from the conventional notion that assumes H^+(BC) to be the dominant monomeric species in intrinsic Si. It looks that actually the dominant species are neutral. In the beginning of studying hydrogen in Si it was believed that a small hydrogen atom should reside in a tetrahedral interstice, and stay neutral due to a large ionization energy. This old natural idea can be now revived if we assume that the H^0(T) state discussed above (of a low concentration and a high mobility) is actually an excited state of tetrahedral hydrogen while the ground state has a much higher concentration—higher than $[H^+(BC)]$ in intrinsic Si. The ground and the excited states of tetrahedral hydrogen will be marked by the symbols H_g and H_e, respectively.

The monomer H_b of a diffusivity D_b—responsible for the bulk component in Fig. 12.2—is then interpreted as H_g species.

The high-temperature atomic diffusivity of hydrogen [15] defined by (12.1) is thus controlled by $H_b = H_g$ neutral species, and there are two major comparable contributions into the monomeric concentration: that by H_b and that by H_s. The high-T averaged atomic diffusivity is then expressed as

$$D_a = D_b C_b / (C_b + C_s). \qquad (12.7)$$

A $D_s C_s$ product is omitted here since $D_s \ll D_b$. The average diffusivity by (12.7) is roughly half as low as D_b.

Within this modified treatment, the processes occurring during a quench from high T should be re-considered. The H_{2A} dimers are formed by a simple pairing of two neutral dominant tetrahedral species: $H_g + H_g$—the same reaction that is responsible for the B-part in Fig. 12.2. By 400 °C, the major states of hydrogen are H_{2A} and H_s. Subsequent redistribution from H_{2A} to HB occurs through the dissociation of H_{2A} into the two mutually equilibrated species—either H^+(BC) + H_e (as was assumed previously) or H^+(BC) + H_g. The dissociation is followed by trapping of H^+ by boron. Of the two dissociation paths, the latter one—that produces the major neutral species H_g—seems to be more likely.

12.4 Boron Passivation by H^+(BC) Ions

In p-type Si (normally, boron-doped) exposed to plasma, the boron acceptors are passivated by in-diffusing ions H^+(BC) which are trapped by B^- species producing neutral HB defects. For exposure at 150 °C, the bulk parts of concentration depth profiles of hydrogen are well fitted [8] considering diffusion and drift of H^+ free ions which coexist in equilibrium with trapped ions, HB; the mass action law for this equilibrium reads:

$$[H^+][B^-]/[HB] = K, \qquad (12.8)$$

where K is the equilibrium dissociation constant. The depth profile of HB is controlled by only two parameters: (1) the surface value of the diffusivity-concentration product, D^+ $[H^+]$, and (2) the material parameter D^+K. The deduced value of D^+K is around 6×10^4 cm^{-1} s^{-1} at 150 °C. With the known diffusivity D^+ [4] the dissociation constant $K \approx 3 \times 10^{13}$ cm^{-3} which corresponds to an HB binding energy close to 0.75 eV.

12.4.1 A Lower Doping Level: An Involvement of H_e Neutral Species

A representative example [17] is shown in Fig. 12.3, for a doping level $N_B = 10^{17}$ cm^{-3}. The profile (induced by an exposure at 150 °C for 60 min) includes a quasi-plateau portion (marked *P*) where boron is almost completely passivated, and the hydrogen concentration is close to the total boron concentration N_B. In the tail portion (marked *T*) a degree of passivation is small; the effective hydrogen diffusivity is controlled by a small fraction of mobile ions H^+ in the H^+ + HB community; this fraction is K/N_B according to (12.8). Hence the effective diffusivity within the tail region is D^+K/N_B. The tail shape, if well resolved, allows to extract reliably the D^+K parameter; in the present example, the best-fit value is 5.5×10^4 cm^{-1} s^{-1}.

The profile of Fig. 12.3 contains also a wide portion (to a depth of about 1.5 μm) where the concentration is well larger than N_B. This part (marked *B*) is similar to the bulk component of Fig. 12.2. A natural explanation for this excess in the hydrogen concentration is that some neutral species (denoted H*) in-diffuse and react with HB (the major single-hydrogen species) to produce dimers.

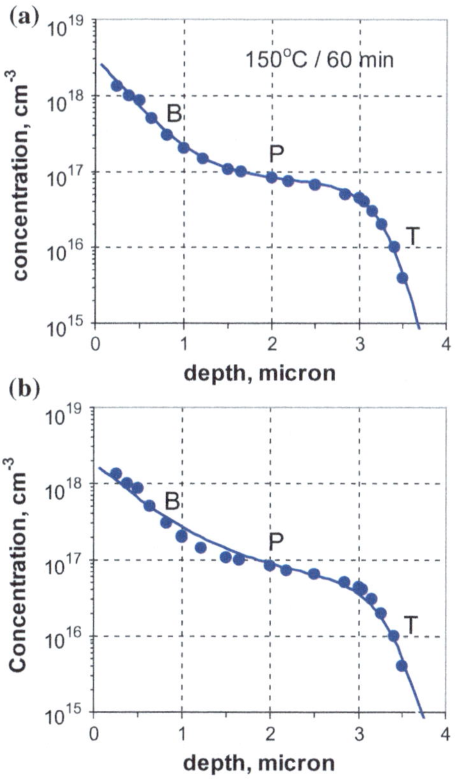

Fig. 12.3 Deuterium depth profile after exposure to a remote plasma at 150 °C for 60 min, after [17]; the boron doping concentration is 10^{17} cm^{-3}. The solid curve is fitted **a** for independent H^+ and H*, **b** for equilibrated H^+ and H*. The in-diffusing neutral species H* is likely to be H_e, an excited state of tetrahedral hydrogen

$$H^* + HB \rightarrow H_2 + B^- + h^+. \qquad (12.9)$$

The production rate of H_2 dimers is

$$G = \beta D^* C^*[HB], \qquad (12.10)$$

where C^* is the concentration of in-diffusing neutral H* monomers, D^* is their diffusivity and β is the pairing kinetic coefficient—that would be equal to $4\pi r \approx 4 \times 10^{-7}$ cm for a diffusion-limited pairing. The in-diffusing H* species may be either H_g or H_e as discussed below.

The shape of the *B*-part in Fig. 12.3 strongly depends on whether the H^+ and H* species are equilibrated or exist independently, without a transition between them (which implies a large reconfiguration barrier). In the latter case, the depth profile of H*—decoupled form that of H^+—is described by a separate diffusion equation:

$$\partial C^*/\partial t = D^* \partial^2 C^*/\partial z^2 - G. \qquad (12.11)$$

The diffusion/drift equation for H^+ [8] is now modified by including a loss of HB at a rate *G*. A best-fit computed profile of C_{tot} = [HB] + 2 [H_2] is shown in Fig. 12.3a by the solid curve. It gives a remarkably good reproduction of the whole experimental profile including the *B*-part. The best-fit value of the pairing coefficient β is 10^{-8} cm—much smaller than the above-mentioned diffusion-limited value. The best-fit surface value of the *DC* product for H* and H^+ is 4.5×10^5 and 8×10^5 cm^{-1} s^{-1}, respectively.

On the other hand, within a model of equilibrated H* and H^+, the ratio $[H^+]/C^*$ is proportional to the hole concentration *p*, and the product D^* C^* in (12.10) is expressed as $D^+[H^+]$ p^*/p where p^* is the characteristic hole concentration for the relative contributions of H^+ and H* into hydrogen transport. The computed concentration depth profiles, for various assumed values of p^*, give a poor fit; an example is shown in Fig. 12.3b for $p^* = 10^{14}$ cm^{-3} and $\beta = 1.5 \times 10^{-7}$ cm. The H* and H^+ species are thus concluded to be independent monomeric states—at low *T*, around 150 °C. If H^+ is equilibrated with H_g, the independent H* species should be H_e. Otherwise (if H^+ is equilibrated with H_e) the H* species would be H_g (identical to H_b). The former case (H* = H_e) seems to be more likely.

The dimers produced by reaction (12.9) have been treated as immobile species different from H_{2A}. If they are actually the mobile H_{2A} dimers, the best-fit pairing coefficient β will depend on the assumed dimeric diffusivity D_{2A} which is somewhat scattered [18]. For a lower number, $D_{2A} = 10^{-13}$ cm^2/s, the best-fit profile (achieved with $\beta = 1.4 \times 10^{-8}$ cm) is indistinguishable from that shown in Fig. 12.3a. It cannot be thus decided, by these data, whether the H_2 dimers are identical to H_{2A} or not. The dimeric diffusion will be manifested more definitely at a higher *T* (for larger D_{2A}) and a longer time—when the diffusion scale $(2\ D_{2A}\ t)^{1/2}$ becomes larger. A concentration profile reported [19] for 185 °C/2 h exposure (Fig. 12.4) clearly shows that the dimers do diffuse: otherwise a concentration profile would be exponential (the red curve 1). The profile computed with the best-fit parameters (the

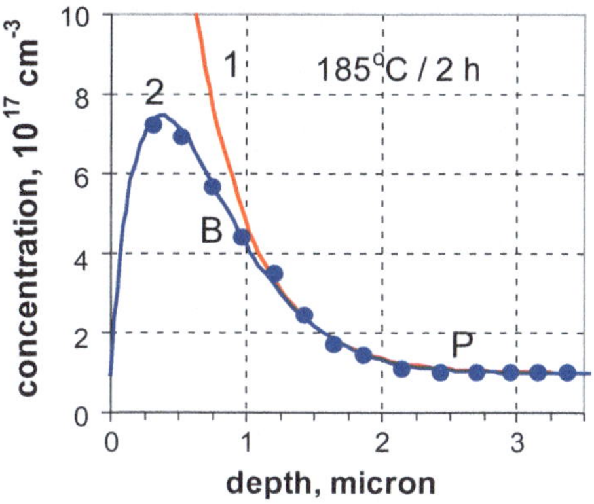

Fig. 12.4 Deuterium depth profile after exposure to a remote plasma at 185 °C for 2 h, after [19]; the boron doping concentration is 10^{17} cm^{-3}. The solid curve 2 is fitted assuming that the *B*-part is due to mobile dimers H_{2A} produced by H* + HB

dimeric diffusivity $D_{2A} = 5 \times 10^{-13}$ cm^2/s, the pairing coefficient $\beta = 2 \times 10^{-8}$ cm, and the surface value of $D^*C^* = 3.2 \times 10^5$ cm^{-1} s^{-1}) is shown by the curve 2 in Fig. 12.4. It well reproduces the experimental profile, including a near-surface part around $z = 0.5$ μm where the concentration of produced dimers (H_{2A}) is reduced by their out-diffusion to the surface.

12.4.2 A Higher Doping Level

Another example of boron passivation in plasma-exposed sample [20, 21] is shown in Fig. 12.5. A shape is very similar to that shown in Fig. 12.3, and the tail-based number for D^+K is 6×10^4 cm^{-1} s^{-1}, close to the value found above at a lower doping level. A penetration scale of H*—limited by the pairing reaction (12.9)—is equal to $(\beta N_B)^{-1/2}$, by (12.11) and (12.10) where the concentration [HB] is close to N_B. The scale becomes much shorter for a higher N_B (here, about 0.04 μm)—but not short enough to reproduce the near-surface part of the profile in Fig. 12.5, even without dimeric diffusion which actually occurs to still a larger distance of 0.2 μm. Hence the near-surface part in Fig. 12.5 cannot be attributed to the pairing reaction (12.9). It is likely that now a similar pairing reaction occurs—but involving H_s instead of H*; accordingly the near-surface part in Fig. 12.5 is marked *S*. The dimers H_2 produced by this reaction seem to be practically immobile at 150 °C. They may correspond to a so called H_2* configuration [22] composed of two oppositely charged hydrogen ions located around a Si atom along ⟨111⟩ axis.

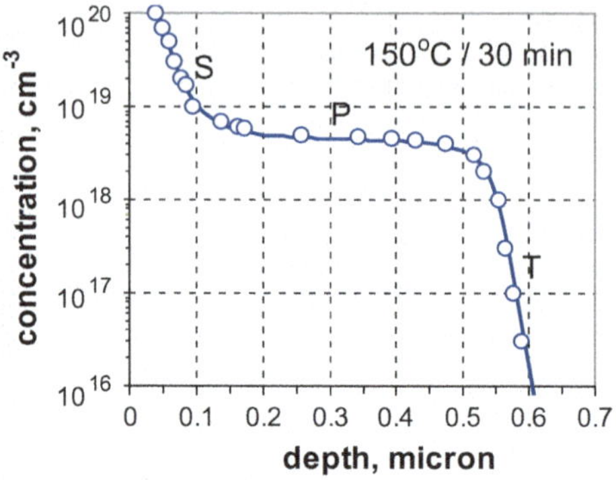

Fig. 12.5 Deuterium depth profile after exposure to a remote plasma at 150 °C for 30 min, after [20, 21]. The boron doping concentration is 5×10^{18} cm^{-3}

12.4.3 A Depth Profile of Holes: An Effect of Boron Compensation

For the previous case of a high doping level, also the hole profile $p(z)$ is available along with the concentration profile $C_{\mathrm{tot}}(z)$ [21]; the former is shown in Fig. 12.6 by circles. Both $p(z)$ and $C_{\mathrm{tot}}(z)$ profiles are computed simultaneously; the hole profile affects the drift of H^+, and the hole concentration is expressed through that of hydrogen species. Under an assumption of non-compensated boron-doped material, $p = N_B - [\mathrm{HB}] - [\mathrm{H}^+]$ where a contribution of H^+ is negligible [8]. The computed $p(z)$—shown by the curve 1 in Fig. 12.6—strongly deviates from the measured profile. A hypothesis to account for this discrepancy [23] was a simultaneous in-diffusion of two independent kinds of passivating H^+ ions. One of them (of a larger $D^+K = 6 \times 10^4$ cm^{-1} s^{-1}) is responsible for the tail part, and the other (of a smaller $D^+K \approx 10^4$ cm^{-1} s^{-1})—for the rest of the profile. Presently, a much

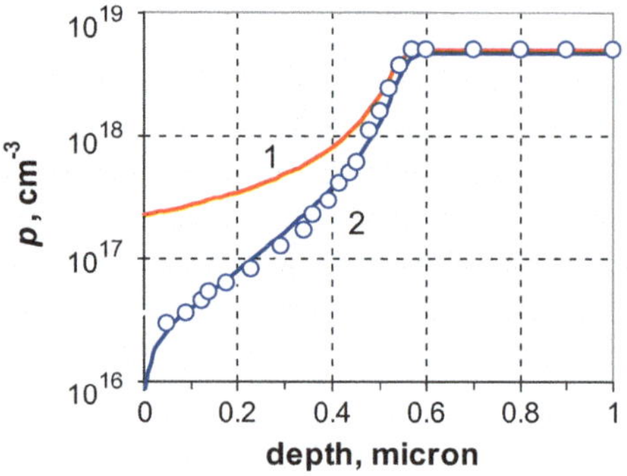

Fig. 12.6 Depth profile of holes in plasma-exposed sample—the same as that shown in Fig. 12.5. The circles are based on the spreading resistance data of [21]. The curve 1 is computed for non-compensated sample. The curve 2 corresponds to the concentration of compensating donors of 3×10^{17} cm^{-3} (a compensation degree is 6%)

simpler explanation has been found: a slight compensation of boron with donors, of a concentration N_d. In this case,

$$p = N_B - [HB] - N_d. \tag{12.12}$$

Although $N_d \ll N_B$, a contribution of donors is essential in a region of a strong passivation, where $[B^-] = N_B - [HB]$ is much smaller than N_B. The hole profile calculated with the best-fit value of $N_d = 3 \times 10^{17}$ cm^{-3} is shown by the curve 2 in Fig. 12.6, and it gives a reasonably good fit to the experimental profile—remarkably better than a fit by two kinds of H^+ [23]. The computed concentration profile is also affected by compensating donors; the solid curve in Fig. 12.5 has been computed using (12.12) with the same value of N_d. The best-fit diffusivity-concentration products used to calculate the profiles in Figs. 12.5 and 12.6 is 9.3×10^5 and 8×10^5 cm^{-1} s^{-1} for the H^+ and H_s species, respectively.

A second kind of H^+ ions is not thus required to reproduce the hole profile in Fig. 12.6. Yet a second kind of passivating H^+ ions—beside H^+(BC)—does exist as discussed in the subsequent sections.

12.5 Boron Passivation by H⁺ Ions Different from H⁺(BC)

The concentration profile for a boron-doped sample ($N_B = 1.3 \times 10^{18}$ cm^{-3}) after exposure to a deuterium plasma at 200 °C for 5 min [24] is shown in Fig. 12.7. It has a shape very similar to that shown in Figs. 12.3 or 12.5. The D^+K at 200 °C—calculated with the known D^+ and the known temperature dependence of K [8] is large—about 2×10^6 cm^{-1} s^{-1}. A profile computed even with a somewhat smaller value, 10^6 cm^{-1} s^{-1} (the curve 1 in Fig. 12.7) is inconsistent with the experimental profile that shows a very steep tail (*T*) and a strong boron passivation in the plateau region (*P*). The experimental profile is reproduced only using a much smaller value for the D^+K product, about 10^4 cm^{-1} s^{-1} (the curve 2 in Fig. 12.7).

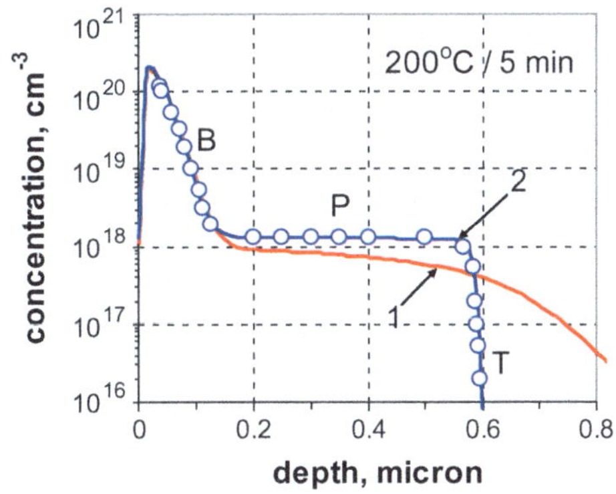

Fig. 12.7 Deuterium depth profile after exposure to a remote plasma at 200 °C for 5 min, after [24]. The boron doping concentration is 1.3×10^{18} cm^{-3}. The curve 1 (red) is a profile computed for $D^+K = 10^6$ cm^{-1} s^{-1} (for the H^+(BC) ions). The curve 2 (blue) is for $D^+K = 10^4$ cm^{-1} s^{-1}

A similarly small value of D^+K was also found [9] for H^+ ions that passivate boron-doped substrate ($N_B = 10^{17}$ cm^{-3}) separated form the surface with an n-type implanted region, for the same exposure temperature of 200 °C.

A remarkably small value of D^+K in the above examples clearly show that the boron passivation is not by H^+(BC) ions (of a large D^+K) but by some other kind of H^+, of a different lattice location. From now on, we denote these two independent kinds of passivating ions by H^+(1) and H^+(2). The first one is H^+(BC). The location of the second one, H^+(2), is yet unknown.

The near-surface part of the profile in Fig. 12.7 is attributed, as before, to the pairing reaction (12.9) or to a similar reaction involving H_s instead of H*. In the latter case the near-surface part should be marked S instead of B. A difference from the previous cases is that H* (or H_s) now reacts with a different kind of the HB defect—that formed by trapping H^+(2) by boron and denoted HB(2). The previously considered HB defect—formed by trapping H^+(1)—is denoted HB(1). The pairing kinetic constant is accordingly denoted β(2) (or β_s(2)); it is deduced to be 4×10^{-7} cm—close to the expected diffusion-limited value. The diffusivity-concentration product is 6×10^6 and 5×10^6 cm^{-1} s^{-1} for H^+(2) and H* (or H_s), respectively. The produced H_2 dimers have been allowed to diffuse, to get a somewhat better fit; the fitted diffusivity $D_2 = 1.7 \times 10^{-14}$ cm^2/s is essentially smaller than that of H_{2A} (which is about 2×10^{-12} cm^2/s). The produced dimers are therefore slow species, probably of H_2* configuration.

12.6 Two Kinds of H^+ Ions Present Simultaneously

The profiles for 150 °C exposure have been well fitted assuming only one kind of passivating ions, H^+(1), while the profiles for 200 °C—also assuming one (but different) kind, H^+(2). It can be expected, that in other examples both kinds of H^+ are essential. The equilibrium (12.8) for passivation holds for both ions, the concentration [HB] for each kind is proportional to $[H^+]/K$. With comparable diffusivity-concentration products $S^+ = D^+ [H^+]$ for the two kinds of ions, the [HB] (for each kind) is proportional to $S^+/(D^+K)$. The H^+(2) ions, due to a much smaller D^+K, are the major passivating species penetrating down to some depth z_2. The H^+(1) species, within this depth z_2, meet a very low concentration of remaining boron traps B^- since the majority of boron is converted into HB(2) defects. Hence the H^+(1) ions easily penetrate through the z_2 region and spread further into the sample bulk, where they passivate boron into HB(1) defects—down to some depth z_1 well larger than z_2. This qualitative picture will be supported below by the computed profiles of HB(1) and HB(2).

A separation of the hydrogen-passivated depth into the two regions—a near-surface region populated with HB(2) and a deeper region populated with HB (1)—will affect the production of dimers by interaction of H* with HB by the pairing reaction (12.9). The pairing coefficient β can be much larger for an interaction of H* with HB(2) than for an interaction with HB(1). Then the concentration

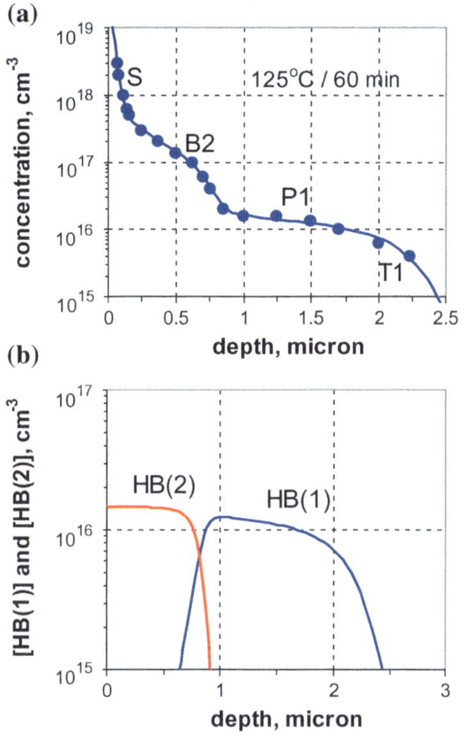

Fig. 12.8 Depth profiles of **a** the total hydrogen (deuterium) concentration, **b** of the two kinds of HB defects. The circles show the measured concentration after a plasma-exposure at 125 °C for 60 min [25]; the boron doping level is 1.5×10^{16} cm^{-3}. The solid curves are computed assuming in-diffusion of two kinds of passivating H^+ ions and production of dimers by pairing of in-diffusing neutral species H* with both kinds of HB, by reaction (12.9)

of produced dimers, for a gradually decreasing $C^*(z)$ profile, will be large within the HB(2)-populated region z_2, and will be sharply reduced at $z > z_2$. Such a steep reduction in the total concentration was indeed reported [25]; Fig. 12.8a shows such a profile for a doping level $N_B = 1.5 \times 10^{16}$ and exposure at 125 °C for 1 h. The kinetic equations for simultaneous propagation of two ions, $H^+(1)$ and $H^+(2)$, were given in [23]. They should be modified only by adding a loss term due to pairing with the H* neutral species. A loss of $H^+(1)$/HB(1) is equal to $G(1)$ specified by (12.10) that involves HB(1) and the coefficient β(1), while a loss of $H^+(2)$/HB(2)—by a similar equation involving HB(2) and the (much larger) coefficient β(2). The best-fit profile is shown by the solid curve in Fig. 12.8a. The combination D^+K for $H^+(1)$, at 125 °C, is 5×10^3 cm^{-1} s^{-1}; the adopted value of D^+K for $H^+(2)$ is 100 times smaller (the computed curve is not sensitive to a precise value of this parameter). The adjusted kinetic coefficients β(1) and β(2) are 2.5×10^{-9} and 3.5×10^{-8} cm, respectively. The surface values of the diffusivity-concentration products for $H^+(1)$, $H^+(2)$ and H* are 1.3×10^4, 1.65×10^4 and 1.6×10^5 cm^{-1} s^{-1}, respectively. The label B2 in Fig. 12.8a marks a region of dimers produced by reaction (12.9) of H* with HB(2). The labels P1 and T1 mark profile parts controlled by $H^+(1)$. The near-surface component marked S is narrow, falling into a range of plasma-damaged layer [17]; it was formally described as self-pairing of H_s.

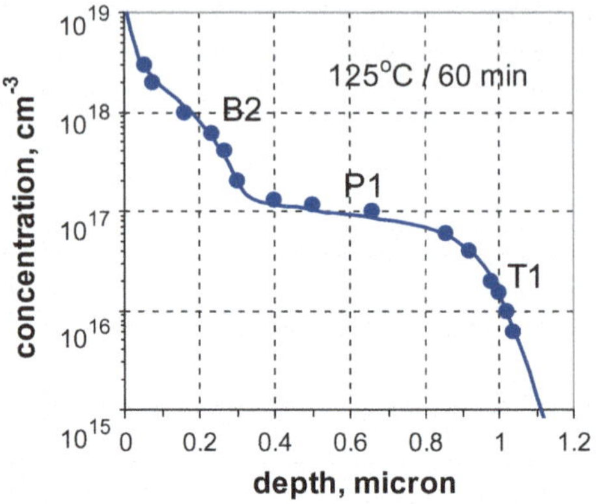

Fig. 12.9 Deuterium depth profiles after a plasma-exposure at 125 °C for 60 min [25]; the boron doping level is 1×10^{17} cm^{-3}. The solid curve is computed assuming in-diffusion of two kinds of passivating H^+ ions and a pairing reaction (12.9)

The computed individual depth profiles of HB(1) and HB(2) are shown in Fig. 12.8b. They follow an expected behaviour outlined above: the HB(2) species populate a near-surface region while HB(1)—a deeper part of hydrogen-penetrated depth.

A steep reduction in $C_{tot}(z)$—similar to that shown in Fig. 12.8a—was also found [25] at a higher doping level, $N_B = 10^{17}$ cm^{-3} (at the same T and duration). This profile is shown in Fig. 12.9. The solid curve was computed using the same material parameters (D^+K and β for the two kinds of H^+/HB) and the following boundary values of the diffusivity-concentration product: 1.25×10^4, 2.3×10^4 and 1.4×10^5 cm^{-1} s^{-1}—for $H^+(1)$, $H^+(2)$ and H*, respectively.

12.7 Summary

The conventional notion of hydrogen monomeric states in Si is that the dominant form, in *p*-type and intrinsic silicon, is the H^+(BC) ion residing in Bond-Centred position. However, a close look at the experimental data on hydrogen species manifested in saturated/quenched samples and in plasma-exposed samples suggests an essential modification of this notion.

It appears that the dominant forms of hydrogen in intrinsic Si are represented by two neutral species: H_b (probably, a tetrahedral atom in the ground state, of a moderate diffusivity) and H_s (a species of a low diffusivity, in a non-tetrahedral location—for instance H^0(AB), an atom in the Anti-Bonding site). In *p*-type silicon, there are generally two independent positive ions, $H^+(1)$ and $H^+(2)$; each passivates boron by forming its own HB defect: HB(1) and HB(2), respectively. The $H^+(1)$ corresponds to a H^+(BC), a conventional preferred site for a positive ion. The $H^+(2)$ is of not yet defined location. The two kinds of H^+ strongly differ regarding the value of D^+K parameter where D^+ is the ion diffusivity and K is the equilibrium dissociation constant of the corresponding HB defect: this parameter is about 100 times smaller for the $H^+(2)$ ion. The total hydrogen concentration in p-Si includes, beside that of HB(1) and HB(2), also a contribution of dimers H_2 produced by a pairing reactions of neutral species with both HB(1) and HB(2). For some reactions, these dimers are concluded to be H_{2A}, moderately mobile species. For other reactions, the produced dimers are of a much lower mobility. The dimeric contribution is dominant within a penetration depth of in-diffusing neutral species that is limited by their trapping and inversely proportional to $N_B^{1/2}$.

In some examples, the plasma-induced hydrogen profiles can be well simulated assuming only one kind of H^+—either only $H^+(1)$ or only $H^+(2)$. In other examples, a presence of both kinds is clearly manifested. In this case the hydrogen-passivated depth is composed of a near-surface region dominated by HB(2) and of a deeper region dominated by HB(1). The dimeric component originates mostly from HB(2), and for this reason there is a sharp reduction in $[H_2]$ at the boundary between a near-surface HB(2)-region and a deeper HB(1)-region.

Existence of several kinds of neutral and positive monomers—and of at least two kinds of produced dimers—leads to a rich variety of concentration depth profiles, dependent on the boundary conditions for these monomers.

References

1. J.I. Pankove, R.O. Wance, J.E. Berkeyheiser, Appl. Phys. Lett. **45**, 1100 (1984)
2. S.J. Pearton, J.W. Corbett, M. Stavola, *Hydrogen in Crystalline Semiconductors* (Springer, Berlin, 1991)
3. C. Herring, N.M. Johnson, Semiconductors Semimetals **34**, 225 (1991)
4. C. Herring, N.M. Johnson, C.G. Van de Walle, Phys. Rev. B **64**, 125209 (2001)
5. S. Wilking, A. Herguth, G. Hahn, J. Appl. Phys. **113**, 194503 (2013)
6. N. Nampalli, B.J. Hallam, C.E. Chan, M.D. Abbott, S.R. Wenham, IEEE J. Photovoltaics **5**, 1580 (2015)
7. S.K. Estreicher, M. Stavola , J. Weber Silicon, *Germanium, and Their Alloys: Defects, Impurities and Nanocrystals*, eds by G. Kissinger, S. Pizzini (CRC Press, 2014) (Ch. 7)
8. V.V. Voronkov, R. Falster, Phys. Status Solidi A **214**(7), 1700287 (2017)
9. V.V. Voronkov, R. Falster, Phys. Status Solidi B **254**(6), 1600779 (2017)
10. R.E. Pritchard, M.A. Ashwin, J.H. Tucker, R.C. Newman, Phys. Rev. B **57**, R15048 (1998)
11. R.E. Pritchard, J.H. Tucker, R.C. Newman, E.C. Lightowlers, Semicond. Sci. Technol. **14**, 77 (1999)
12. S.A. McQuaid, M.J. Binns, R.C. Newman, E.C. Lightowlers, J.B. Clegg, Appl. Phys. Lett. **62**, 1612 (1003)
13. M.J. Binns, R.C. Newman, S.A. McQuaid, E.C. Lightowlers, Mater. Sci. Forum **143–147**, 861 (1994)
14. T. Ichimiya, A. Furuichi, Int. J. Appl. Radiat. Isot. **19**, 573 (1968)
15. A. Van Wieringen, N. Warmoltz, Physica **22**, 849 (1956)
16. J.T. Boremstein, J.W. Corbett, S.J. Pearton, J. Appl. Phys. **73**, 2751 (1993)
17. N.M. Johnson, F.A. Ponce, R.A. Street, R.J. Nemanich, Phys. Rev. B **35**, 4166 (1987)
18. N.M. Johnson, C. Herring, Phys. Rev. B **43**, 14297 (1991)
19. R. Rizk, P. de Mierry, D. Ballutaud, M. Aucouturier, D. Mathiot, Phys. Rev. B **44**, 6141 (1991)
20. N.M. Johnson, M.D. Moyer, Appl. Phys. Lett. **46**, 787 (1985)

21. N.M. Johnson, Phys. Rev. B **31**, 5525 (1985)
22. K.J. Chang, D.J. Chadi, Phys. Rev. B **40**, 11644 (1989)
23. V.V. Voronkov, Adv. Mat. Sci. Eng. **2018**, 238543 (2018)
24. B.Y. Tong, X.W. Wu, G.R. Yang, S.K. Wong, Canad. J. Phys. **67**, 379 (1989)
25. S.J. Pearton, Mater. Sci. Eng. B **23**, 130 (1994)

Chapter 13
Architecture and Function of Biohybrid Solar Cell and Solar-to-Fuel Nanodevices

Silvio Osella, Joanna Kargul, Miriam Izzo and Bartosz Trzaskowski

Abstract In recent years, an immense research effort has been devoted to the generation of hybrid materials which change the electronic properties of one constituent by changing the optoelectronic properties of the other one. The most appealing and commonly used approach to design such materials relies on combining organic materials or metals with biological systems like redox-active proteins. Such hybrid systems can be used e.g. as bio-sensors, bio-fuel cells, biohybrid photoelectrochemical cells and nanosctuctured photoelectronic devices. Although experimental efforts have already resulted in the generation of a number of biohybrid materials, the main bottleneck of this technology is the formation of a stable and efficient (in terms of electronic communication) interface between the biological and the organic/metal counterparts. In particular, the efficiency of the final devices is usually very low due to two main problems related with the interfacing of such different materials: charge recombination at the interface and the high possibility of losing the function of the biological component, which leads to the inactivation of the entire device. In this chapter, we explore the power of computation to answer pressing questions for a rational design of the different components of the biohybrid interface.

13.1 Introduction

The need to find new sources of energy production for our everyday consumption is one of the most pressing issues of our age. To overcome the dependence on fossil fuels and to assess renewable energy resources, innovative paths must be followed,

S. Osella (✉) · B. Trzaskowski
Chemical and Biological Systems Simulation Lab, Center of New Technologies, University of Warsaw, Banacha 2C, 02-097 Warsaw, Poland
e-mail: s.osella@cent.uw.edu.pl

J. Kargul · M. Izzo
Solar Fuels Lab, Center of New Technologies, University of Warsaw, Banacha 2C, 02-097 Warsaw, Poland

E. V. Levchenko et al. (eds.), *Theory and Simulation in Physics for Materials Applications*, Springer Series in Materials Science 296,
https://doi.org/10.1007/978-3-030-37790-8_13

to go beyond the conventional technologies already available. Moreover, considering the climate change problem and environmental pollution as result of production and consumption of energy from fossil fuels, the need of green and renewable energy sources is a milestone which needs to be reached in the next 2 decades. To counteract these problems, the European Union (EU) set a 'Renewable Energy Directive' with a target of 32% of consumption of energy deriving from renewable sources, 40% reduction of greenhouse gas emission and at least 32.5% increase in energy efficiency by 2030. Within this action, the EU countries have committed to reach their own national renewables targets ranging from 10% in Malta to 49% in Sweden (ref. EU directive). In addition, each of them is required to have at least 10% of the transport fuels generated by renewable sources by 2020. In the frame of this ambitious goals, the need of new efficient and green sources of energy is a priority for material science, with much effort already dedicated to Organic Photovoltaics (OPV), Organic Field-Effect Transistors (OFET), as well as a new class of solid-state fuel cells for transport. Yet, the most promising devices had reached only 10–12% external efficiency to date, which is not enough for a widespread use of such systems. Additionally, many of the components of these organic electronic devices are constituted of non-abundant and expensive elements, such as indium or ruthenium, precluding them from being viable.

A different approach to this problem is to take advantage and mimic what nature was able to perfect upon millions of years of evolution. Yet, one of the biggest challenges in modern science is to replicate in an artificial way the most basic processes performed by nature, such as photosynthesis. This fundamental process has been found to be extremely complicated and very difficult to be reproduced in the laboratory settings using artificial, man-made systems/devices [1–3]. This is due to the many components needed from light-driven charge separation to the transport of charge between different photosynthetic reaction centres to the final production of reducing equivalents, subsequently used for solar fuel synthesis. Although almost all of these steps are nowadays fully understood, mimicking this complex photosynthetic machinery in the full solar-to-fuel artificial devices that use water as the sole source of reducing equivalents is still prohibitive at an industrial scale. In particular, in the last five years research effort was focused on the creation of an 'artificial leaf' device able to produce clean energy from absorption of light, with some encouraging results [4]. Although in its infancy, this relatively new research approach has the potential of taking advantage of two well-known branches of research: photobiology and nanomaterial science. In fact, by combining in a synergetic way the optical properties of light harvesting proteins and the electronic properties of nanomaterials, it might be possible to create a new class of biohybrid working devices with enhanced impact on sustainability.

This chapter overviews the artificial photosynthesis process and the key components for building an efficient biohybrid interface. The main part of the chapter focuses on how state-of-the-art computational methods play an important role in designing and building an efficient interface and in answering some fundamental questions not easily accessible from experimental science. It also discusses the main challenges that the computational approach faces when describing direct energy and electron transfer (ET) in such biohybrid interfaces. Finally, it examines the future perspectives for biohybrid devices.

13.2 Biohybrid Interfaces in Photosynthesis

Biohybrid devices are the meeting point of two disciplines that only recently started to interact, namely biology and material science. The potential of such collaboration is to take advantage of the best of the two fields and design and build devices that mimic the photosynthetic apparatus of plants in a laboratory setting. In nature, photosynthesis occurs through a highly complex apparatus in which the light is absorbed and converted into electrons via two large membrane proteins, photosystem I and photosystem II (PSI and PSII) [5, 6] with a multi-steps proton-coupled electron transfer chain reaction that has been perfected during evolution to efficiently convert solar energy into the biological hydrogen-storage compound, reduced nicotinamide adenine dinucleotide phosphate (NADPH). Next, ATP is generated by the ATP synthase enzyme, powered by the proton gradient generated by the processes which take place across the photosynthetic membrane. Both products, ATP and NADPH, are then used for subsequent fixation and reduction of atmospheric CO_2 into carbohydrates. The natural photosynthetic apparatus is depicted in Fig. 13.1. As a result of the sequence of reactions that take place within the photosynthetic process, the plant or alga converts solar energy into chemical

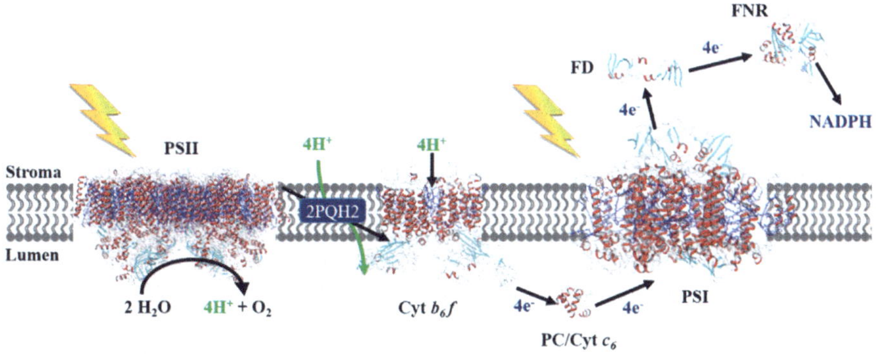

Fig. 13.1 Natural photosynthetic apparatus. The photosynthetic electron transport and subsequent generation of reducing equivalents and NADPH is performed by the concerted action of three transmembrane proteins: PSII (PDB: 3WU2), cytochrome *b₆f* (PDB: 1VF5) and PSI (PDB: 1JB0) complexes. Thanks to the presence of a mobile electron carriers which can either be plastocyanin (PC) in higher plants and green algae or cytochrome c6 (PDB: 1CYI) in algae and cyanobacteria, the electrons are transferred to PSI and subsequently to ferredoxin (Fd, PDB: 3AB5) and ferredoxin: NADP$^+$ reductase (FNR, PDB: 1FNB) to obtain the final NADPH product. The ATP forming enzyme is not shown for clarity

energy. The key to success relies on: (1) physical separation of the light collection (protein antenna) and the catalytic centre; (2) absorption of two photons for generation of sufficient potential for water oxidation; (3) spatial separation of oxidative and reductive chemistry and relevant products; (3) multielectron character of water oxidation and proton/CO_2 reduction; (4) efficient management of proton-coupled electron transfer for both catalytic reactions, and last but not least (5), use of self-assembled and self-renewable molecular catalysts embedded in responsive matrix.

In order to perform all the needed oxidative and reductive reactions, a potential gradient is required. In fact, every cofactor present is organized in such a way that the photoactivated electrons move down an energy gradient, in the so-called 'Z-scheme' of light reactions [7]. Natural photosystems operate as nearly perfect photoelectrical/photovoltaic molecular nanomachines, exhibiting internal quantum efficiency close to 100%. All the inbound redox active cofactors, photon capturing pigments and metals of the water oxidation catalyst (WOC) of PSII are spatially organized to maximize forward proton-coupled electron transfer (PCET) and minimize wasteful back reactions. The water oxidation reaction occurs through the tightly synchronized interaction of the highly oxidizing species $P680^+$ ($E_o = +1.25$ eV), formed upon absorption of red photons in the reaction center (RC) of PSII, with the WOC present in the vicinity of P680, the Mn_4O_5Ca cluster of the oxygen evolving complex (OEC). The process is underpinned by the light-driven extraction of four electrons from two substrate water molecules, thus requiring absorption of four consecutive red photons by P680 RC. The by-product of this reaction is molecular oxygen. PSI also absorbs four red photons to build sufficient potential ($E_o = -1.3$ eV) for subsequent reductive chemistry associated with CO_2 conversion into glucose (Fig. 13.2).

Recent reports have outlined the successful interaction between photoactive light-harvesting proteins (LHP) and synthetic semiconductors [8, 9]. In these hybrid systems, the tunable energy levels of the semiconductor and the flexibility of the protein at the interface allow for switchable energy and/or electron transport. The possibility of finely tune the band gap of semiconductors, the ability of absorbing light on different portion of the solar spectra and the presence of the required driving force to complete chemical reactions at the surface, make them a material of choice for biomolecular/semiconductor or biohybrid interfaces, as they function in a similar matter to the reaction centers of natural LHP. This new composite material can be applied to a vast range of applications, when either energy or electron transfer is considered as the primary mechanism, such as photochemical cells, optical biosensors and nanostructured photoelectronic devices (Fig. 13.3).

Given the fact that the natural photosynthetic apparatus is a system of high complexity, essential simplifications must be made when trying to create an artificial conductive interface in a laboratory. This usually means a complete removal of the membrane and a general architecture of the system based on a multi-layer stacked device. The easiest design for a hybrid interface is to directly assemble the LHP (i.e. PSI) over the semiconducting material. In this way, the orientation of the protein plays a crucial role in determining the direction of electron transfer either

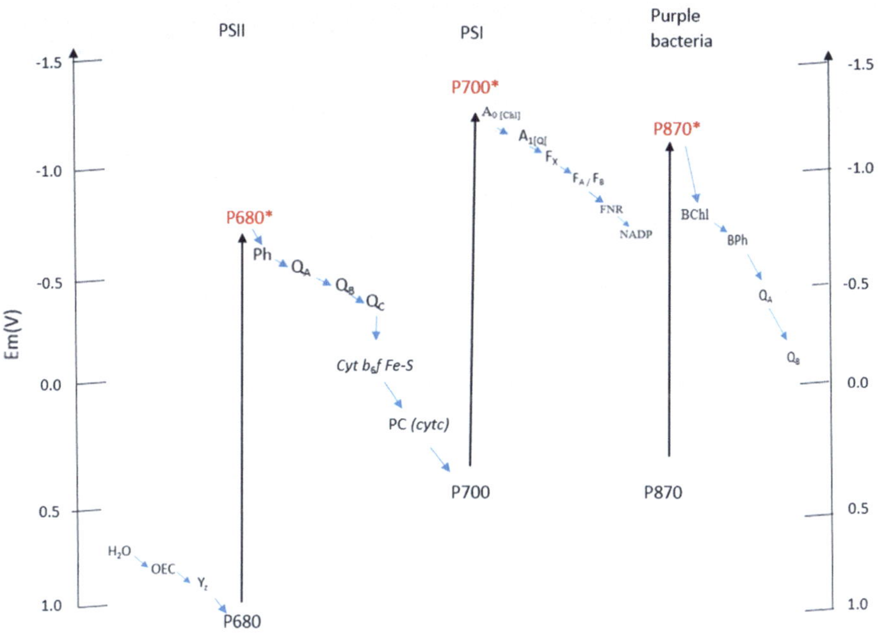

Fig. 13.2 Energetics of the components of the photosynthetic electron transfer pathways. Each primary donor and ET cofactors are organized in the highly conserved pathways in which photoactivated electrons move down the energy gradient. In oxygenic photosynthesis PSII and PSI work in tandem within the "Z-scheme" of light reactions to drive cyclic photo-oxidation of two water substrates upon absorption of four quanta of light by PSII's P680 RC. For evolution of each O_2 molecule, eight quanta of light are required to be absorbed by both photosystems present in the same membranes. In anoxygenic photosynthesis, bacterial Type II and Type I RCs that are evolutionary related to PSII and PSI RCs, respectively, drive oxidative and reductive chemistry, using other substrates than water. These bacterial RCs do not exist together in the same cell in contrast to PSII and PSI. Reproduced from [7]

from or to the electrode, leading to either cathodic or anodic currents. To work as a photoanode, the conduction band of the semiconductor should be lower in energy compared to the excited energy level of LHP and the redox potential negative enough to supply an electron to the protein. On the other hand, cathodic electron transfer occurs when the valence band of the semiconductor is lying above the reduction potential of the photosynthetic reaction center (RC) and its redox potential is more positive than the excited level of the RC [10, 11]. Yet, this strategy lacks of fine control over the orientation of the LHP and its flexibility is strongly inhibited and a strong charge recombination at the interface is observed, which is related to both the strength of the protein-metal interaction and the nature of the metal (or semiconductor) used. To enhance the flexibility of the protein and to have better control over its orientation, multi-layered devices have been considered.

Recently, Nocera's group has effectively built a so-called 'artificial leaf' which is composed of many different layers of different earth-abundant materials and

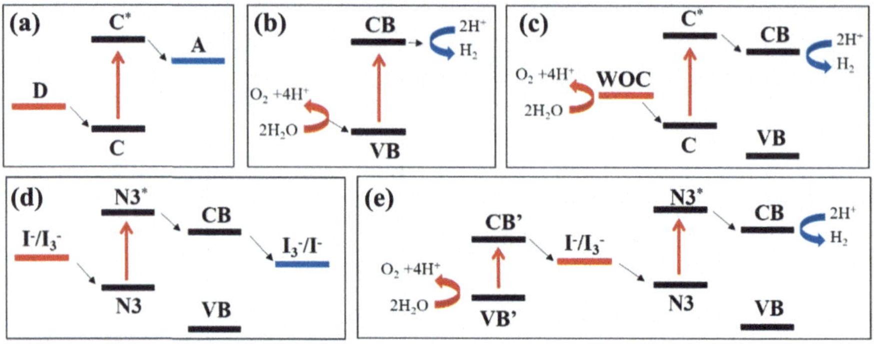

Fig. 13.3 Different photo-induced electron transfer mechanisms in common artificial systems which can be used for applications of biohybrid systems. **a** Molecular triad, **b** Semiconductor photocatalyst, **c** DSSC, **d** DSSC coupled to a water–oxidation catalyst, **e** tandem water–splitting cell. **D** refers to electron donor; **C** and **C*** ground state and excited chromophore; **A** electron acceptor; **CB** and **VB** conduction and valence band; **N3** and **N3*** ground and excited state Ru-based N3 dye; **WOC** water oxidation catalyst; **CB′** and **VB′** conduction and valence band of narrow bandgap semiconductor

operates under simple conditions and at 1 sun (AM 1.5, 100 mW cm^{-2}) illumination [12]. This cell was nanoengineered by interfacing triple junction amorphous Si to the Co-based oxygen evolving complex (Co-OEC, similar in 3D structure to $CaMn_4O_5$ catalyst that is present in the PSII water-splitting enzyme), and a NiMoZn alloy cathode for the hydrogen evolving reduction (HER) to form a sandwich-type structure (Fig. 13.4). The photosynthetic RC has been replaced here with the Si junctions which absorb light and converts it into a series of charge separation states, generating sufficient potential for water splitting. Water-derived electrons are then used for the reductive chemistry, yielding molecular hydrogen. Within this configuration, a broad light absorption spectrum can be measured, extending to the near IR part of the solar spectrum and ensuring strong photon absorption. Moreover, this simple junction cell produces 8 mA cm^{-2} of current at an open circuit voltage of 1.8 V with an overall efficiency of 6.2%, which is comparable with the full organic P3HT-PCBM bulk heterojunction solar cells of about 5–6% [13].

Despite the high efficiency, the system built by Nocera is not a full hybrid device, since it considers only the catalytic complex of the LHP as active component for the OPV process. To consider a real biohybrid interface, the whole LHP or at least its RC, responsible for the charge separation and charge transfer, should be considered. A promising biohybrid all-solid dye-sensitized solar cell (DSSC) has been recently reported, being one of the first examples of integration of PSI-based devices with an improved efficiency [14]. The structure of the device is relatively simple, consisting in a layer of hematite-coated fluorinated tin oxide (FTO) on which an oriented multilayer of robust PSI has been adsorbed with the reducing side close to the electrode surface. The PSI multilayers were covalently cross-linked, to

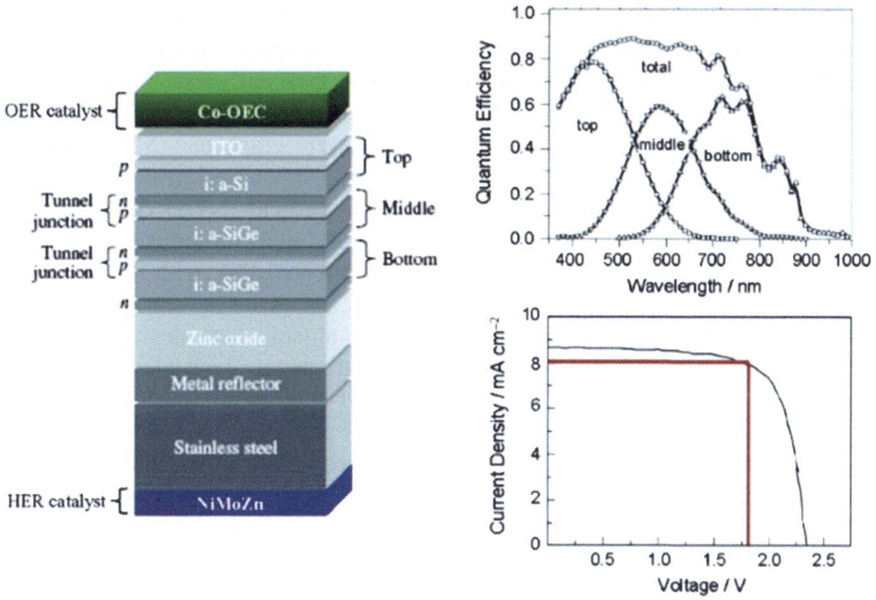

Fig. 13.4 The artificial leaf. The *p*-side of a triple junction Si solar cell is coated with an ITO protective layer and Co-OEC. The *n*-side is coated with a NiMoZn HER catalyst. The Si absorption (*top right*) and current/voltage (*bottom right*) properties are shown. The overall solar cell efficiency is 6.2%. Reprinted with permission from Account Chem. Res. 45, 767–776 (2012). Copyright (2019) American Chemical Society

give structural integrity to the final device. This simple design led to an overall current density of 56.9 μA cm^{-2} and an open circuit voltage of 321 mV, which translates into an efficiency of 0.17%. This value might seem extremely small when compared with the previous device of Nocera, but it is twice more efficient than previously reported device consisting of hybrid PSI/semiconductor materials [15]. In addition, this is one of the first attempts to incorporate the full LHP into a working solid-state DSSC. Considering the complexity of the protein itself and the inherent difficulties in orienting the different layers, it is an interesting proof of concept. The enhanced light absorption and charge separation intrinsic for hematite have been considered as the probable cause for the high efficiency and long-term stability (up to 90 days) of the device, proving the possibility of creating a truly green biohybrid cell for solar-to-fuel conversion.

The fabrication of a full bioelectrode has been reported very recently [16]. It consists of highly-oriented PSI on a single layer graphene electrode involving a 'sandwich' stacking structure with different biological and organic molecular linkers. To assure a high orientation of the PSI, the protein should be wired to the functionalized graphene surface and genetically engineered to incorporate affinity tags (or metal-binding peptides) into their structure. Another strategy is to create intermolecular interaction between the LHP and natural electron donors, such as cytochrome c_{553}, which promotes the electron transfer to the electrode and can be used as the biological conductive interface between graphene and LHP. Yet, with this strategy an additional degree of complexity is added, since now also the cytochrome should have a specific orientation over the electrode. The final layered device consists of a functionalized single layer graphene (SLG) with a self-assembled monolayer (SAM) of organic molecules composed of π-π stacked pyrene derivatives functionalized with nitrilotriacetic acid (NTA) chelating Ni^{2+} cations. Through this direct chelation, the His-tagged cytochrome is bound to the SAM-SLG electrode, and on the top of this, the oriented PSI (with its donor side pointing towards the electrode) is incorporated by means of electrostatic and hydrophobic interactions. The final structure of the all-solid-state device is reported in Fig. 13.5a. From fluorescence and photoelectrochemical characterization the authors of the study conclude that not only the PSI in close proximity to SLG take part in the electron transfer, but also the other PSI present in the multilayer play a role for the transfer of charge. Moreover, a measured photocurrent of 370 nA cm^{-2} at high negative bias of −300 mV indicates that for this particular bioelectrode configuration the electrons flow from the SLG to the PSI layers through the SAM interface due to the cytochrome-driven orientation of the LHP, which exposes the hole-generating side towards the electrode. In addition, in devices where PSI is randomly oriented, the photocurrent was reduced over five fold, thus defining the key role played by the proper orientation of the protein for improvement of the final efficiency of the device.

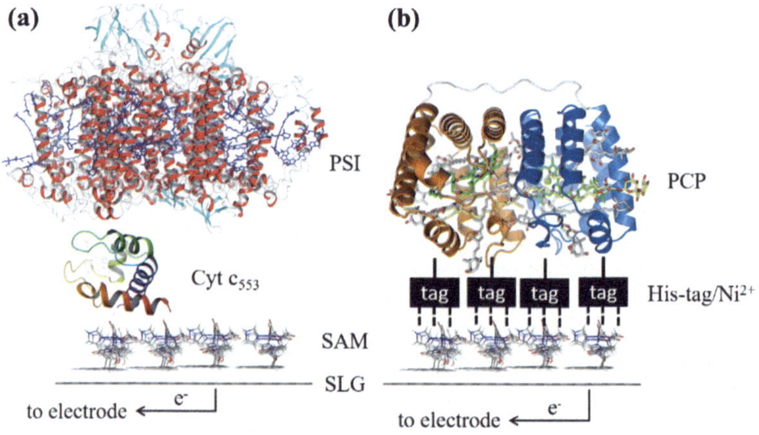

Fig. 13.5 Overview of two rationally designed biohybrid architectures proposed from experiments. The absorption of light triggers the charge transport to the graphene through the tag linking the (cytochrome) self-assembled monolayer (SAM), which activates the bio-electronic device

13.2.1 Mechanism of Action

The mechanism of action governing the biohybrid interface is fairly straightforward and fully understood. This is due to a strong advancement into the knowledge of the mechanism of natural light harvesting proteins such as PSI and PSII, which only recently have been fully unraveled due to advancement in structural and biophysical methods.

Briefly, in the natural photosynthetic process, a photon is absorbed by the antenna of PSII and funneled to the reaction center to produce an excited hole/electron pair; subsequently, the hole is transferred to the oxygen evolving complex which is responsible for the water splitting resulting in the release of O_2 and four protons from 2 substrates bound within the catalytic metal center. On the other hand, the electron is transferred to PSI (generating another hole upon absorption of a red photon in its reaction center) via a series of redox-active cofactors to generate reducing equivalents in the form of NADPH. The outstanding quantum yield of hole/pair production and separation in photosynthetic reaction centers, close to one, and not reached by any man-made synthetic systems, is the main advantage of the use of light harvesting proteins for solar-to-fuel nanodevice applications. The key to understand this extremely high efficiency has been discovered only recently by means of different experimental approaches [17]. One of the most important factors is the spatial separation of PSI and PSII reaction centers (i.e. separation of reducing and oxidative processes), which allows the generation of one hole/electron pair one photon at the time. Yet, when trying to reproduce this in an artificial way, one should keep in mind that the full process generates four hole/electron pairs in total, in the presence of a strongly oxidizing environment.

With the hole/electron pair production mechanism unraveled, the next step is to create a system in which these free charges are effectively transported from the photoactivated protein to the electrode to generate current. To achieve this, the light harvesting protein should be coupled with linkers which are then connected to the photoelectrode of choice. A lot of literature report different methods and linkers to have high yield electron transfer (see the review of Kargul [18] for a detailed description). In this chapter we focus on the overall transfer of electron from the LHP to the electrode, as depicted in Fig. 13.5. In fact, after absorption of light, the photon is funneled to the reaction center of the LHP and one hole/electron pair is formed. Then, the electron transfer to the photoelectrode can occur via the linker, either a small protein such as cytochrome or a self-assembled monolayer of organic conjugated molecules.

13.2.2 Competing Pathways

State-of-the-art devices for photoconversion designed up to date possess poor efficiencies (only up to 1–2%) due to the complexity of the structure and the lack of knowledge and design of efficient interfacing between the different components of the system [19–21]. To obtain an efficient electron transport pathway, rational design of the biohybrid interfaces is required. The main challenges are related to competing ET pathways which have detrimental effects on the performance of the device, and are related to: (1) the degradation of the protein induced by the formation of reactive oxygen species upon charge separation and water splitting, (2) charge recombination at the interface and within electrode and LHP modules and consequent inefficient charge transfer (CT) between the LHP and the electrodes of the nanodevice and (3) the presence of a non-physiological environment that can disrupt the protein's function [22]. Hence, in order to achieve high efficiency, there is a need for developing rational design of efficient electronic contacts between the photoactive layer and the conducting electrode, as well as to obtain uniform and dense packing [23].

Recent studies identified the link between the reduced CT and anodic electron transfer generated in a PSII/ITO electrode, which arises from the oxygen reduction that occurs inside the PSII structure, in the inner antenna subunit [24]. In turn, this pathway caused a short-circuit in charge flow, pointing to the significance of the environment when constructing the device (in the specific case, in absence of oxygen). Moreover, the final efficiency of the device is not merely due to the proximity and orientation of the LHP over the semiconductor, but is the net result of multiple competitive energy (and electron) transfer pathways.

As a result, it is obvious that the choice of the different components of the final device is of crucial importance [25]. From the biological point of view, due to the anisotropy of LHPs, the appropriate orientation upon assembly is crucial and must be assessed either experimentally or theoretically, to maximize the electron transfer. Thus, four important steps should be considered: (1) relationship between the chemical and electronic structure of the linkers and the LHP; (2) orientation of the bioactive component over the linker and photoelectrode surface; (3) creation of a dense and uniform monolayer of both linker and LHP; (4) supramolecular electron transfer process between the different interfaces (i.e. photoelectrode/SAM; SAM/LHP) to minimize charge recombination.

From the material point of view the main challenge is to find a nanomaterial that has the following characteristics [26–28]: (1) should be highly conducting or semiconducting material; (2) should have a suitable bandgap in between 1 and 2 eV; (3) should have good absorption in the range of visible/near-IR wavelengths and (4) efficient charge separation with little or negligible recombination of charges. To date, no material that fulfils all the aforementioned requirements has been found. Traditionally, the semiconductors used for to build bioelectronics materials are either noble metals (e.g. gold, silver) or inorganic metal-oxides [29–31] (see Sect. 13.3.2). Yet, emerging 'organic metals' or semimetals can be promising candidates to fulfill all the needed properties. In addition, there are two important issues which need to be considered to overcome the charge recombination bottleneck: the choice of the organic material to assure high flow of charges and the biocompatibility of the material. On one hand the organic material must not disrupt the properties and functionality of the bioactive component upon the formation of the interface in a non-physiological environment, since the final device is all-solid state, while ensuring high charge transport rate. Furthermore, the modification of the organic semiconductor can be problematic, leading to structural changes that can strongly alter the conjugated lattice needed for the charge transport (i.e. for graphene, the π-electron delocalization).

13.3 Proof-of-Concept System: Components

One of the most promising approaches towards the solution of the presence of competitive pathways is to reduce the complexity of the system while preserving the fundamental building blocks of this process. In this view, the two light harvesting complexes PSI and PSII as well as small light harvesting/redox active proteins (SLHP) such as azurine, plastocyanin and peridinin-chlorophyll-protein complexes (dubbed PCP) were used in proof-of-concept systems for the efficient harvesting of light and subsequent transport of charges [7, 32]. In fact, once the photon is absorbed by the antenna of PSI (PSII) or SLHP, it is converted into electrons with a quantum efficiency close to one, thus ensuring a high charge carrier towards the electrode. This biomimetic device should be capable of both conducting thermodynamically demanding water splitting reaction and use the water-derived protons and electrons to produce molecular hydrogen and/or carbon-based liquid fuel from reduction of CO_2 such as methanol and formate [33, 34]. This can pave the way to photosystem-based solar cells, bio-photosensors and biohybrid solar-to fuel devices. As a result, a lot of research effort has been focused on the connection between photosynthetic proteins and electrodes to obtain a truly green photosensing device in bio-optoelectronics or solar-to-fuel devices [35–38]. In order to obtain an adequate interaction between the protein and the electrode, attention must be paid to the material used for the photoelectrode, as well as the possible linker between the protein and the electrode, either molecular or biological. Figure 13.5 shows a rationally designed biohybrid interface that consists of three main components: light harvesting protein, molecular linkers in form of a self-assembled monolayer and single layer graphene as a photoelectrode.

In this section, all three components of biohybrid solar-converting devices are examined in details and their prominent characteristics as well as their functions are described. This section is not intended to be a complete review on the components, but a short introduction to the systems considered for the computational study. For exhaustive reviews on each of the components, the authors suggest recent literature [4, 7, 18, 39–41].

13.3.1 Light Harvesting Proteins (LHP)

PSI (plastocyanin-ferredoxin oxidoreductase) supercomplex consists of the core complex containing the photochemical reaction center coupled to the light-harvesting complex (LHCI) [7, 39, 42]. This large pigment-proteins supercomplex is found in the membranes of oxygenic photosynthetic organism (plants, algae and cyanobacteria) and operates with a quantum yield close to unit (for $\lambda < 680$ nm) for generation of the long-lived primary charge separation state [41]. In natural photosynthesis, the hole generated in the reaction center of photoactivated PSI is filled with water-derived electrons generated by PSII, transferred by cyt b_6f and ultimately donated to PSI by a mobile electron cofactor, cytochrome c_6 or plastocyanin.

13.3.1.1 Cyanobacterial PSI Complex

The first structure of the PSI complex was obtained from the thermophilic cyanobacterium *Synechococcus elongatus* (*S. elongatus*) at a resolution of 2.5 Å. The cyanobacterial PSI complex exist as a homotrimer of 1068 kDa [43]. Each monomer contains 12 protein subunit (9 membrane-bound subunits and 3 stromal subunits PsaE, PsaC and PsaD), and 127 cofactors including 96 chlorophylls, 2 phylloquinones, 3 Fe_4S_4 clusters, 22 carotenoids, 4 lipids, and a putative Ca^{2+} ion, accompanied by 201 water molecules [44]. The PSI multimeric protein complex consists of the reaction centre (RC) where most of the redox-active electron transfer cofactors (ETC) are present, and the peripheral antenna, which in cyanobacteria is composed of phycobilisomes (PBS) and in algae and higher plants of the LHCI comprising a variable number of the subunits with inbound pigment cofactors [18].

The core of PSI is formed by a PsaA/PsaB heterodimer, each subunit composed of 11 TMs. The ETC identity and arrangement are evolutionary conserved in terms of their location and ligation within the structure of the PSI RC [7]. They are arranged along two pseudo-symmetrical branches that join at the Fx iron-sulphur cluster located on the stromal side of the membrane, at the interface of PsaA/PsaB hetrerodimer (Fig. 13.6) [44–46].

In contrast to trimeric PSI of cyanobacteria the higher plant PSI-LHCI supercomplex exists as a monomer of 600 kDa composed of the core complex asymmetrically associated with a crescent-like LHCI antenna on the PsaJ/PsaK side (Fig. 13.7). The outer LHCI antenna is composed of a varying number of the Lhca subunits depending on the species and it undergoes rapid remodeling in response to growth conditions in order achieve the optimal size of absorption cross-section relative to photoprotection of the photochemical reaction centre. The higher plant core complex contains ten TM subunits (PsaA, PsaB, PsaF, PsaG, PsaH, PsaI, PsaJ, PsaK, PsaL, PsaR) and three peripheral subunits (PsaC, PsaD and PsaE). Among these subunits four have been found only in higher plant PSI-LHCI (PsaG, PsaH, PsaN and PsaR).

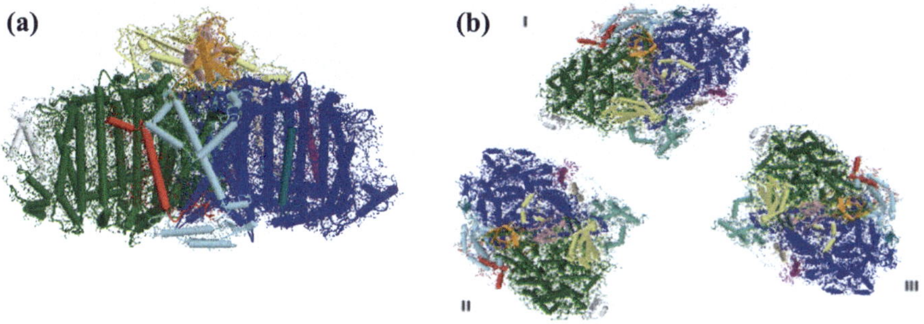

Fig. 13.6 Structure of cyanobacterial photosystem I. **a** A side view of cyanobacterial *S.elongatus* PSI structure. **b** View along the membrane normal from the stromal side. The subunits of PSI and inbound cofactors are shown in each of three monomers (I, II and III). Colors codes: PSI core subunits: green, PsaA; blue, PsaB; pink, PsaC; yellow, PsaD; orange, PsaE; cyano, PsaF; white, PsaK; brown, PsaI; red, PsaJ; tiffany, PsaL; magenta, PsaM; dark green, PsaX

Fig. 13.7 A view of the structure of *Pisum sativum* photosystem I from the side view. The structural coordinates were taken from the theoretical refinement (PDB) of original structural data (PDB 2WSC). Color codes: PSI core subunits: cyan, PsaA; light green, PsaB; light pink, PsaC; yellow, PsaD; red, PsaE; light blue, PsaF; gold, PsaG; orange, PsaH; brown, PsaI; blue, PsaJ; magenta, PsaK; tiffany, PsaL; white, PsaN; dark blue, PsaR. Lhcr proteins: dark pink and white Lhcr1; multicolor, Lhcr2; military green, Lhcr3; gray, Lhcr4

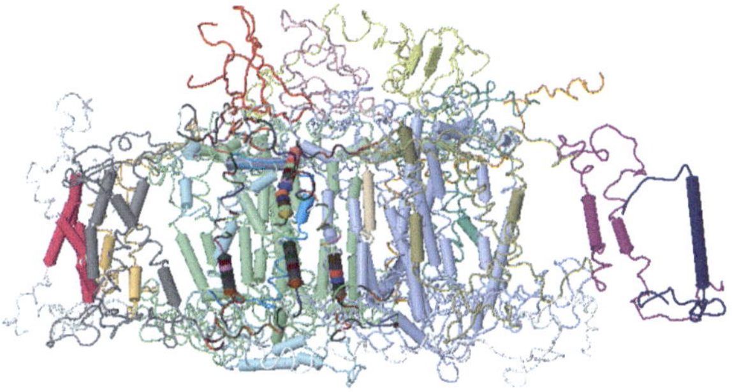

The primary charge separation occurs as the result of excitation of the chlorophyll dimer P700 (eC-A1/eC-B1 the primary electron donors). The electron passes through the ETC consisting of 2 pairs of Chla molecules (eC-A2/eC-B2 and eC-A3/eC-B3) and a pair of phylloquinones (Q_K-A/Q_K-B secondary electron acceptor) [47]. The electron from either of the phylloquinones is subsequently intercepted at the F_x cluster, and subsequently transferred to two additional iron-sulphur clusters, F_A and F_B, which are bound within the stromal PsaC (Fig. 13.8).

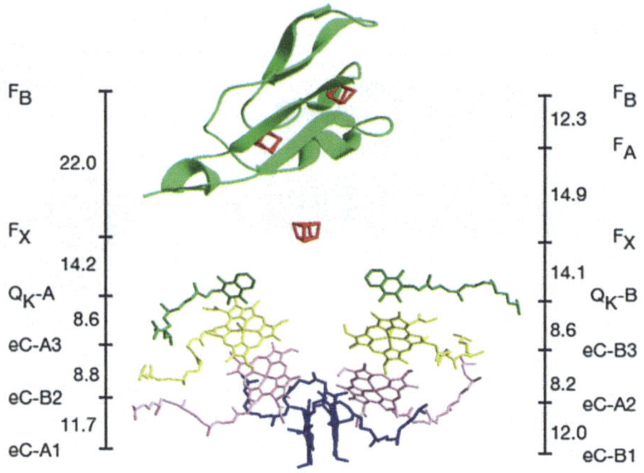

Fig. 13.8 Arrangement of electron transfer cofactors present in the reaction centre of photosystem I. View parallel to the membrane plane. The chlorophyll 'special pair' is shown in blue. The additional two Chl pairs are arranged in two branches A and B. They are termed as eC-A2 and eC-B2 (pink) and eC-A3/eCB3 (yellow). The phylloquinines are labelled QK-A and QK-B (green). The first of the three Fe_4S_4 clusters (shown in red), F_X is located on the stromal side of the membrane at the interface of the PsaA/PsaB heterodimer. The two additional Fe_4S_4 clusters, F_A and F_B are bound within the stromal PsaC subunit (green). The distances between the cofactors are shown in Å. (Based on [44])

13.3.1.2 Highly Robust PSI-LHCI Supercomplex from an Extremophilic Microalga *Cyanidioschyzon Merolae*

Various types of PSI complexes and their counterparts from anoxygenic photosynthetic bacteria have been used in solar converting devices including dimeric PSII, bacterial Type I RCs and PSI. However, the most promising devices have utilized the highly robust PSI components from extremophilic microalgae including thermophilic cyanobacteria and thermo-acidophilic red algae. A prominent example is PSI-LHCI supercomplex from a red thermo-acidophilic microalga *Cyanidioschyzon merolae*. The red algal (Rodophyta) photosynthetic apparatus is unique, as it represents an evolutionary link between prokaryotic cyanobacteria and eukaryotic phototrophs [48, 49]. In fact, PSII from *Cyaniodioschyzon merolae* contains cyanobacterial-like light harvesting system (phycobilisomes) and plant-like LHCI composed of Chl *a*-containing Lhcr subunits (Lhcr1-3) that are assymmetrically associated with PSI core complex on the PsaK and PsaJ side [50]. The fact that the photosynthetic apparatus of *C. merolae* comprises two different types of light-harvesting antenna systems, phycobilisomes and chlorophyll *a*-binding Lhcr proteins, expand light-harvesting capacity of the photosynthetic apparatus for this alga thriving in shallow volcanic hot springs [49]. The PSI-LHCI supercomplex from *C. merolae* is monomeric, with molecular mass of 580 kDa [51, 52]. Recently, both X-ray and cryo-EM structures of *C. merolae* PSI-LHCI supercomplex have been reported at a resolution of 3.82–3.63 Å providing an insight into energy transfer pathways of this complex [53, 54]. The cryo-EM analysis identified two isoforms of *C. merolae* PSI-LHCI supercomplex (Fig. 13.9), one with three Lhcr antenna subunits (PSI-3Lhcr) and the other one with five Lhcrs (PSI-5Lhcr) attached to the PSI core, whereas the X-ray crystallography identified the isoform with 3 Lhcr subunits only.

Investigation of *C. merolae* PSI-LHCI structure and function under varying light conditions revealed even more complex and dynamically adjusting arrangement of the LHCI structure in this extremophilic alga, with 4–8 Lhcr subunits present in the PSI-LHCI supercomplex depending on the intensity of growth illumination [51]. The *C. merolae* PSI core complex is evolutionary conserved and contains most of the core subunits of the higher plant counterpart (PsaA-PsaO), with the notable exception of the PsaG, PsaH and PsaN subunits whose genes are missing in the *C. merolae* genome [18]. Interestingly, the cyanobacerial-like PsaM subunit is also present in the structure of the red algal PSI core complex. Within the protein matrix

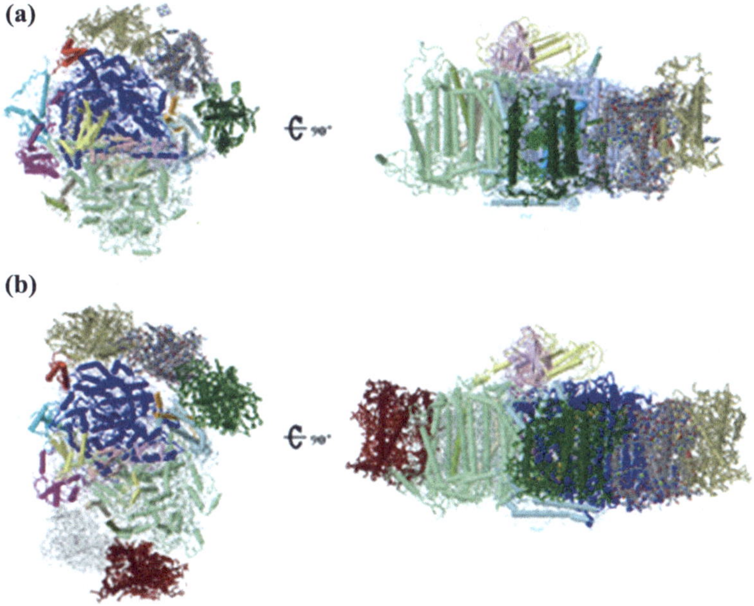

Fig. 13.9 Two isomers of the PSI-LHCI supercomplex from *C. merolae*. **a** Structure of PSI-3Lhcr viewed along the membrane normal from the stromal side (*Left*) and its side view (*Right*). **b** Structure of PSI-5Lhcr viewed along the membrane normal from the stromal side (*Left*) and its side view (*Right*). Color codes: PSI core subunits: blue, PsaA; light green, PsaB; pink, PsaC; yellow, PsaD; violet, PsaE; light blue, PsaF; red, PsaK; brown, PsaI; orange, PsaJ; magenta, PsaL; green, PsaM; cyano, PsaO. Lhcr proteins: hot green, Lhcr1; grey, Lhcr2; gold, Lhcr3; white, Lhcr1*; hot red, Lhcr2*. Cofactors: green, Chls *a* of the PSI core complex; yellow and orange, Chls *a* of Lhcrs; blue, carotenoids of the PSI core complex; purple, Zeas of Lhcrs; red, cofactors of the electron transfer chain (Chls *a*, phylloquinones, and Fe_4S_4 clusters). Based on Proc. Natl. Acad. Sci. U S A. 115, 4423–4428 (2018)

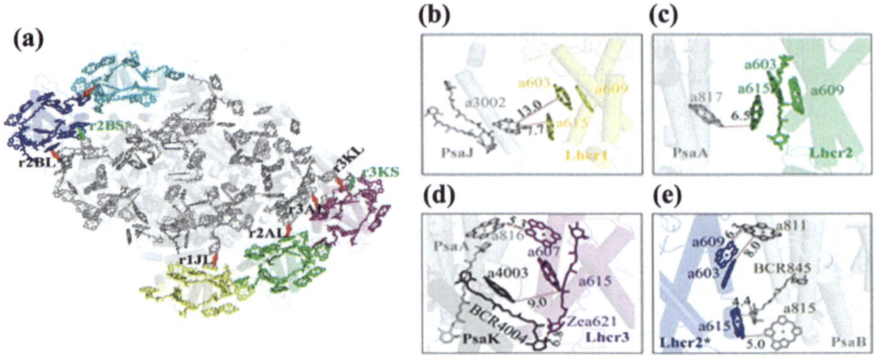

Fig. 13.10 Possible energy transfer pathways from LHCR to the PSI core. **a** The energy transfer pathways (red arrows) from Lhcrs to the PSI core. **b** Luminal side pathway r1JL from Lhcr1 to the core. **c** Luminal side pathway r2AL from Lhcr2 to the core. **d** Pathways r3KL, r3AL, and r3KS from Lhcr3 to the core. **e** Pathways r2KL* and r2BS* from Lhcr2* to the core. Reprinted with permission from Proc. Natl. Acad. Sci. U S A. 115, 4423–4428 (2018)

more than 200 cofactors are bound including 158 Chl *a*molecules, 21 β-carotenes (BCRs), 25 zeaxanthins (Zeas), 2 phosphatidyglycerols, 1 digalactosyldiaclycerol, 3 Fe_4S_4 clusters and 2 phylloquinones [53, 54].

The mechanism of the energy transfer pathways (EET) in PSI has been the subject of intensive research [55–58]. The structural analysis of the *C. merolae* PSI-LHCI supercomplex brings in significant developments concerning the possible EET pathways present within this extremophilic complex compared to higher plants. According to recent structural investigation there is closer edge-to edge distance between the pigments in Lhcrs and those of the PSI core complex, resulting in the presence of some unique EETs in this extremophilic PSI complex (see Fig. 13.10).

13.3.2 Photoelectrodes

From the material point of view, *n*-type semiconductors are often used to oxidize water to molecular oxygen by a photoactive process which generates holes in the semiconductor lattice. Due to the multi-electron nature of the process, often metal oxides are used. The main advantages of this family of materials is their cost-effectiveness, abundance, easy manipulation of the raw material to obtain a structured material and high stability in corrosive environments.

13.3.2.1 Metal Oxides

One of the most common oxide used as photoelectrode is titanium dioxide (TiO_2). Due to its excellent stability upon absorption of light, it has been used for a long time as protection layer for unstable electrodes [59]. Moreover, being a semiconductor, it can be directly used for water oxidation reaction [60]. Despite its wide usage, the absorption of TiO_2 occurs in the UV region of the solar spectrum, strongly decreasing its efficiency. Thus, in order to shift the absorption to longer wavelengths, titanium dioxide can be doped with different non-metals such as nitrogen, phosphorus, carbon, boron as well as metal cations (i.e. Fe, Cu, Au) which effectively enhance the light absorption in the visible range due to the decrease of the band gap of the doped oxide. In recent years, more complex composite materials have been suggested, such as CuO/TiO_2, SnO_2/TiO_2, ZnO/TiO_2 and SiO_2/TiO_2 interfaces, in which the band gap of the resulting photoelectrode is finely tuned by the interaction at the interface of the oxides [61–64].

Another widely used metal oxide is hematite (α-Fe_2O_3) which possesses a narrow band gap of around 2 eV and strong absorption over 600 nm wavelength [65]. Hematite is very stable in ambient condition and is inexpensive due to its high abundance. Despite its appealing properties, the strong charge recombination is a major push back for this material. Doping can be considered to enhance its photoactivity, the electronic conductivity and at the same time decrease the charge recombination [66]. A more interesting approach is to build nanostructures with hematite such as nanotubes, nanowires or nanorods which lead to a decrease diffusion length of the photogenerated holes and their efficient transport within the hematite lattice [67–69]. More complex interfaces, with composite materials of different oxides such as Fe_2O_3/Cu_2O coupled with titanium proved to be successful for improvement of hematite photocatalytic activity [70].

The two materials reported here are the most commonly used for this type of applications, though it is worth mentioning that other promising metal oxides are often reported as valid alternative to hematite and titanium dioxide. In particular, we would like to mention, among the *n*-type semiconductors, tungsten oxide (WO_3) and zinc oxide (ZnO) which are reported to have strong desirable electronic properties to be efficient photoelectrodes [71]. A yet different approach to design a photoelectrode is to combine the oxide to well-known metal catalysts to finely tune their opto-electronic properties to enhance the efficiency of water splitting. Among the many different classes of catalysts commonly used, ruthenium-based compounds show very strong activity in water oxidation as well as high robustness [72]. Iridium-based catalysts have been also extensively used, but (as to less extent Ru) its high cost and scarcity of raw metal on earth's crust are strong detrimental factors for a widespread usage [73].

13.3.2.2 Single Layer Graphene (SLG)

Despite the wild usage of inorganic oxide as photoelecrodes, only recently the attention started to shift towards organic semiconductors, which might provide better opto-electronic properties at a cheaper production cost as well as easy manipulation. The most intriguing class of materials are 2D materials, which came to attention with the discovery of graphene in 2004 [74]. Among the vast variety of 2D materials available, graphene still remains the semi-metal of choice due to its outstanding and unique electronic properties. Graphene is an infinite, flat monolayer of sp2-hybridized carbon atoms bonded together to form a 2D honeycomb lattice, with extended π-electron delocalization along the whole sheet [75]. It can be considered as a semi-metal and due to symmetry considerations, the conduction band and the valence band intersect at the K points within the first Brillouin zone [76]. The unique property of graphene is to have linear dispersion energy near the Dirac point K (and not quadratic as commonly observed), that can be described by the Dirac equation and results in massless fermions [77] (see upper panel in Fig. 13.11). This feature makes the Fermi level to be located at the intersection between the top of the valence and the bottom of the conduction bands [78]. The linear relationship defines two cones in the k space that meet only at the K point, creating a conical intersection and hence leading to a zero band-gap material with perfect electron-hole symmetry. Such a characteristic favors ambipolar charge transport, since the charge carrier density can be continuously tuned for electrons or holes, with mobilities measured for exfoliated graphene exceeding 50,000 $cm^2\ V^{-1}\ s^{-1}$ in ambient conditions [79]. Additionally, in absence of charged impurities and ripples, mobilities of 200,000 $cm^2\ V^{-1}\ s^{-1}$ for a graphene sheet have been predicted [80]. In this view, graphene already possesses a number of desirable properties to be a good candidate as nanomaterial for biohybrid electronics [76, 80]. In spite of the impressive charge mobility, the main drawback of graphene is the lack of band-gap, which limits its use in any electronic devices where a band-gap, even very small, is needed. To overcome such a problem, the band gap can be opened by cutting the graphene sheet along specific directions, in strips with nanometric sized width (<10 nm) which are referred to as graphene nanoribbons (GNR), see Fig. 13.11.

An alternative strategy consists of functionalizing the graphene layer with atoms or functional groups, by means of either covalent or non-covalent interactions. The first strategy has the secondary effect of doping the graphene layer, shifting the Fermi level in the valence or conducting band and making this new material a p-type or n-type semiconductor, respectively. Besides substitution with heteroatoms, doped graphene layers can be generated by chemically grafting electron-rich or electro-poor molecules [81, 82]. Though this approach offers the advantage of forming stable, hybrid structures, the honeycomb structure of graphene becomes locally disrupted and as consequence its electronic and optical properties are altered. On the other hand, in the case of physisorption, the non-covalent

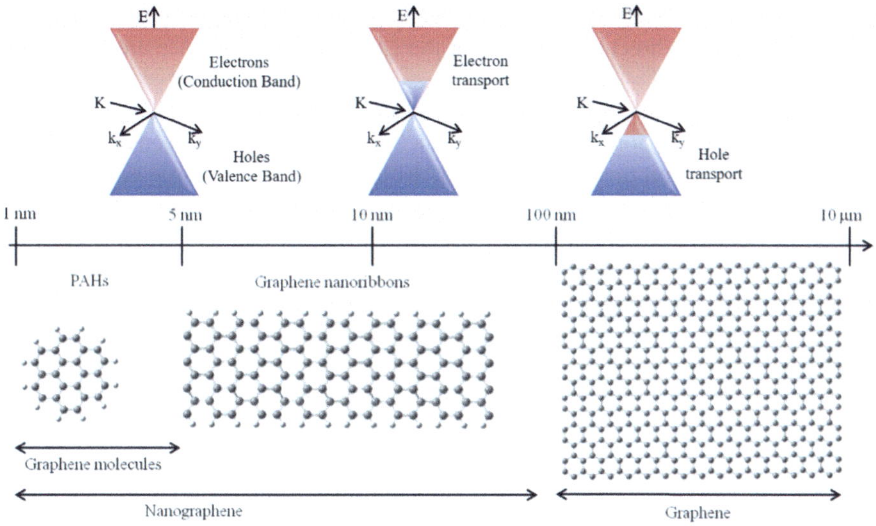

Fig. 13.11 Electronic peculiarity of graphene. (*top*) Representation of the linear dispersion of the valence and conduction bands of graphene at the *K* point of the fist Brillouin zone and the ambipolar character of charge transport. (*bottom*) Length scales in graphene objects: graphene molecules, within a range of 1–5 nm; GNRs defined as graphene strips with a width <10 nm and a length/width ratio >10. Fragments with a dimension less than 100 nm are defined as nanographene, while graphene exceeds 100 nm in both dimensions

interactions between graphene and small molecules preserve the π-conjugated structure of the former, but the reversibility of the interaction can lead to the desorption of molecules from the surface [83].

13.3.3 Self-Assembled Monolayers (SAM)

To secure the proper connectivity and orientation between the photoelectrode and the LHP, a functionalization of both components is required. This can be achieved by chemically modify the LHP (i.e. with biological linker) to orient the protein on the metal surface, optimize the charge injection and improve the total photoconversion by increasing the optical cross-section of the LHP. On the one hand, non-covalent interactions via π-π stacking hydrophobic or electrostatic interactions only mildly perturbs the structure of the electrode. One of the most widely used anchoring group is pyrene and recent studies showed its versatility over different kind of metal substrates [84, 85]. On the other hand, covalent functionalization enables strong and robust bonding and can offer control over the degree of functionalization via selective immobilization of the LHP in an oriented way by covalent linkage, affinity tags or genetically modified proteins [86]. The most common targets for direct conjugation of the LHP and metal surface are lysine and cysteine residues [87].

To functionalize the metal substrate, a self-assembled monolayer (SAM) of organic molecules is commonly used, which can function as effective interlayer between the metal and the protein [88]. SAM are stable, highly ordered and dense supramolecular arrays of molecules that can be deposited on specific surfaces either through electrostatic or chemical bonding interaction. Prototypical molecules that self-assemble on a surface are formed by three distinct parts: a head (or anchoring group) made of a functional group (like thiol, amine, cyanide, carboxylic acid) or aromatic core (i.e. pyrene derivatives), which binds the molecule to the surface; the backbone of the molecule—in the typical range in between 0.5 and 5 nm—and the tail, which can bring different functionalities with respect to the backbone. The purpose of depositing SAMs on a metal surface relies on their ability to influence the electronic properties (such as interface dipole, band structure, absorption band and band binding) at the interface between the (bio)organic material and the metal (or metal oxide) surface [89]. The control over the energy level alignments of the different components/moieties enables a fine tuning of the electronic properties between different units in a multicomponent material, leading to the modulation of fundamental optoelectronic processes such as charge generation, transfer and recombination, as well as energy transfer, which are the governing mechanisms of a working device [90]. In particular, the charge in injection barrier depends on the alignment between the energy levels of the molecule and the Fermi energy of the metal. The presence of a monolayer at the interface tunes the barrier to increase or decrease the work function (WF) of the electrode and favor the hole or electron injection, respectively (see Sect. 13.4.3). The key parameters to consider are the dipole of the molecules forming the assembly and the bond dipole arising upon the formation of the bond between the electrode and the anchoring group of the molecule for chemisorption, and the charge transfer when the SAM is physisorbed; if the dipole points towards the substrate it lowers the WF of the electrode and thus the electron barrier for injection, while if it points away from the substrate it increases the WF of the electrode and lowers the hole barrier for injection [91]. To obtain a well-packed SAM, the molecules must find a balance between intermolecular repulsion (i.e. van der Waals forces) and the density of packing [92]. The more the molecules are densely packed and perpendicular to the surface, the larger the influence of the molecular dipole on the WF of the metal, and *vice versa*. As a consequence, slight variations in the geometry of the molecules forming the SAM can lead to significant changes in electronic properties at the interface [93–95].

Different routes are normally considered for the functionalization of the photoelectrode with SAM; briefly, the SAM can be entrapped in polymer films, anchored on a metal or metal oxide nanoparticle or chemisorbed on a gold surface through the creation of gold-sulphur bonds. When considering carbon based materials as photoelectrodes, two different paths can be followed. The first is covalent immobilization and can be reached by mean of either diazonium ion reduction through a radical bond formation, or alkyne-azide 'click' reaction. The second is non-covalent

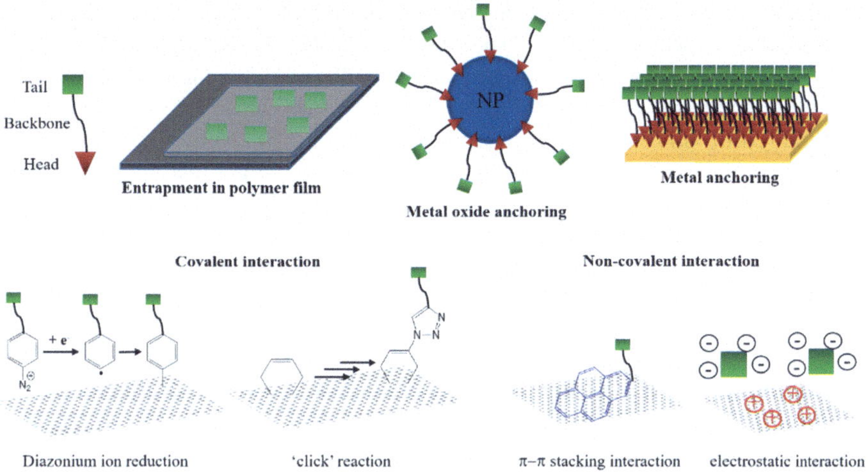

Fig. 13.12 Functionalization of electrodes with SAM. Schematic representation of the structure of a SAM forming molecule is commonly divided into anchoring head, backbone and functional tale. The SAM can then be incorporated into polymer films, or coated around metal (oxide) nanoparticle of planar metal electrodes. In the particular case of graphene electrode, covalent coverage can be achieved either via diazonium ion reduction or 'click' reaction, while the non-covalent ones are mostly through π-π stacking or electrostatic interactions

interaction, by either π-π staking with i.e. pyrene and porphirine derivatives or ion pairing electrostatic interactions (Fig. 13.12). Through either paths, the presence of a SAM alters the electronic properties of the metal, enhancing the transport of charges from the protein to the electrode or vice versa from the electrode to the protein, to increase the water oxidation yield [96].

An exciting variant of simple SAMs is the possibility to use functionalized molecules that are photochromic [97]. These class of molecules can undergo reversible photo-triggered isomerization between (at least) two (meta)stable states.

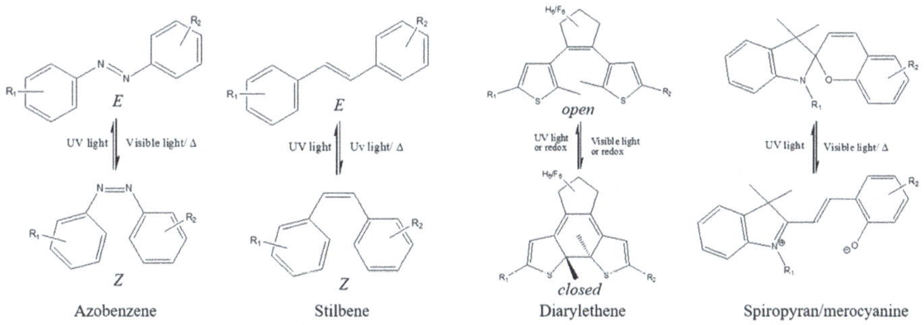

Fig. 13.13 Chemical structures of common photochromic molecules. The photoswitch can occur either by light, redox reactions or heat

Depending on the process involved, photochromic molecules can be divided into: (1) pericyclic reactions such as spiropyrans/oxazines and diarylethenes; (2) *E/Z* isomerizations such as azobenzenes and stilbenes; (3) photo-induced bond cleavages such as perchlorotoluene and triarylmethanes; (4) intramolecular hydrogen/group transfer such as polycyclic quinones; (5) electron transfer (redox) such as viologens. Among all photochromic molecules, the most widely used are azobenzenes, spiropyranes, diarylethenes and stilbenes (Fig. 13.13).

Thus, the incorporation of such photochromic molecules into SAM will lead to strong control of local variation of optical, mechanic and electrostatic environments via light input, in order to enhance the transfer of charge at the electrode/protein interface. If such moieties can be included in the design of new SAMs, the final device will not only be activated by absorption of light from the protein, but can have novel enhanced functions. The type of the chosen SAM and photochromic molecules and the chemical nature of the linker will define the charge transfer and separation processes between them, thus influencing the efficiency of the interface. In addition, changes in packing density and orientation of the SAM can be also altered by light, resulting in additional changes in conformation of the structure to be used as molecular machines or sensors [98].

13.4 Computation at Work: A Case Study

From computational point of view, the main interest lies in the definition of the different interfaces and in the description of the electron and energy transfer from one interface to the other. In the simple model proposed in Fig. 13.5, at least three interfaces can be obtained, namely LHP/linker, LHP/SLG and linker/SLG.

The most detrimental competitive pathway is the charge recombination process at the different interfaces, which can strongly hamper the efficiency of the device. Therefore, insights into the process are vital to a rational design and optimization of charge transport, and computation can lead to a strong understanding into it.

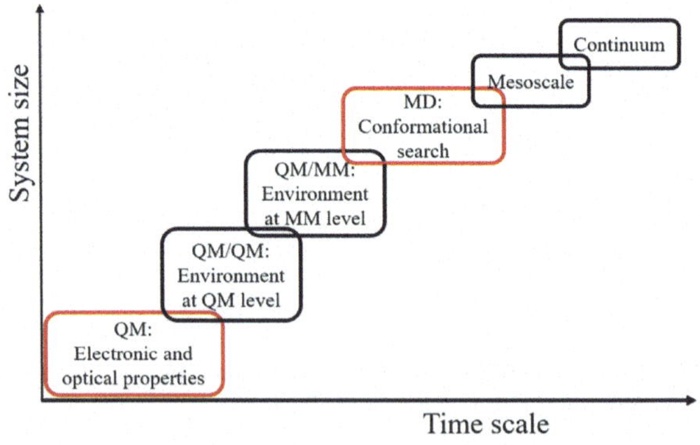

Fig. 13.14 Multiscale approach to computational problems. Increasing the time scale and system size lead to a loss of accuracy in the description of the details of the investigated systems. In this chapter, we will focus on the two blocks highlighted in red: QM and MD

13.4.1 Multiscale Approach

The study of this new interface is a high demanding task for computational calculations, yet with the implementation of high-performance clusters it is nowadays possible to study such large and complex systems using a multiscale approach (Fig. 13.14).

Different computational methods and approaches are considered in this chapter, such as *ab initio* calculations (QM), classical molecular mechanics (MD) and docking, as well as kinetic methods for the computation of the charge transport. Moreover, the presence of a non-physiologic environment is an additional challenge for the description of the biological part of the system. A prototypical system can be schematized by dividing the full system in three different blocks: (1) small photoactive protein; (2) biological and/or molecular linkers and (3) photoelectrode in the form of graphene monolayer. Considering the high complexity of the LHP PSI and PSII (consisting of more than 200,000 heavy atoms), we instead consider model systems named as small LHP (SLHP), which inherently retain the extremely high absorption quantum yield of the parent proteins, but at an affordable computational cost.

13.4.2 Docking and Molecular Dynamics: Conformational Search

The first approach to study the proposed system considers the use of docking of the protein/SLG interface in order to obtain reliable and stable starting structures, followed by molecular mechanics (MM) and molecular dynamics (MD) simulations to gain more insight into the stability of the interface in time. The orientation of the protein on graphene can be mapped using protein-protein protocols, which allow to obtain orientations of one system versus the other for relatively large system using relatively simple algorithms. While most of available software is optimized for proteins, some (FTDock) are also well suited for protein-surface problems [99]. This program discretizes the two molecules onto orthogonal grids and performs a global scan of translational and rotational space with the scoring method being primarily a surface complementarity score between the two grids.

For our interface, we consider the interaction between the cytochrome c_{553} (cyt c_{553}) and SLG as archetype of SLHP/SLG interactions. The cytochrome c_{553} from *Cyanidioschyzon merolae* structure was obtained using homology modelling, as implemented in Prime v.3.9 [100], considering the cytochrome c_6 protein from alga *porphyra yezoensis* as the template (PDB code: 1GDV). Due to the presence of graphene as substrate, the FTDock program was used, in which both cyt and graphene surface were considered as rigid bodies. From the docking analysis a set of stable cyt c_{553}/SLG interfaces was obtained, from which four with the highest docking score were selected, and considered as starting configurations for MD simulations. The differences in the starting conformations (from docking) are mostly related to the different heme-SLG distance and orientation, which in turn results from different interactions of the amino acids of the protein with graphene surface (i.e. different orientations of the protein) ranging from a minimum distance of 0.54 nm to a maximum of 1.53 nm (considered as minimum distance between the SLG and the heavy atoms of the heme structure).

These four model interfaces have then been used as starting configurations for MD simulations performed using Gromacs 2016.3 software [101] with the CHARMM27 force field [102]. The parameters for cyt c_{553} have been modified to account for the heme-His, heme-Cys and heme-Met interactions [103]. The interface was then solvated and charge neutralized with Na^+ and Cl^- ions, and a 100 ns dynamic was performed in the NTV ensemble at room temperature and periodic boundary condition applied [104]. To assure homogeneity, the SLG was considered frozen. Since the starting configurations were not optimized, a certain equilibration time was required and only the second half of the MD, from 50 to 100 ns was considered for the analysis. The systems were considered equilibrated when the drift in total energy was found to be less than 10 kJ/mol. The final structures are reported in Fig. 13.15.

Interestingly, two of the interfaces (system 1 and 3) preserve a very close orientation as obtained from docking while systems 2 and 4 collapse into the same position, with a negligible root means square displacement (RMSD). The RMSD for all the interfaces have values between 0.09 and 0.11 nm. A short 50 ns MD simulation of the cyt in water without SLG shows RMSD value for cyt c_{553} of 0.10 nm, thus ensuring the integrity and function of the protein upon absorption. Despite these differences, the cyt c_{553}–SLG distance for all interface is constant at 0.25 nm, suggesting a strong interaction between the protein and the graphene surface, as also confirmed by the interaction energy analysis with values in the

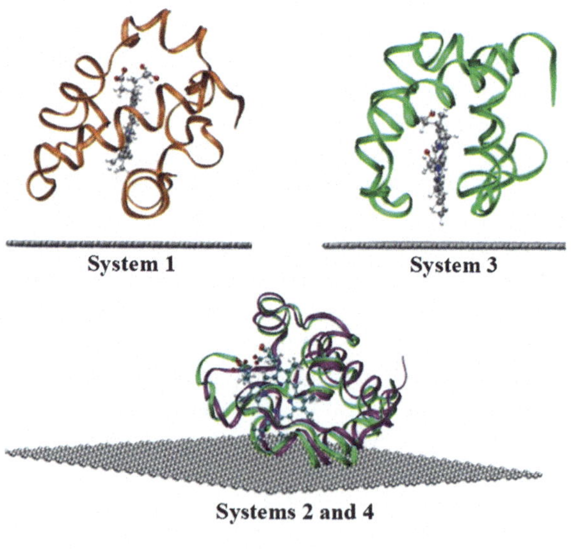

Fig. 13.15 Comparison of structures after 100 ns of MD simulations. For the sake of comparison, System 2 and 4 are overlapped to better present their similarity in structure. For clarity, water is omitted. Reprinted from J. Phys. Chem. C 122, 29405–29413 (2018) (https://pubs.acs.org/doi/10.1021/acs.jpcc.8b10517; further permissions related to the material excerpted should be directed to the ACS)

Table 13.1 Geometric analysis of the average distances from the cyt c_{553} components (protein, heme and Fe(II) to SLG as well as cyt c_{553}–SLG interactions energies obtained from 100 ns MD simulations

	System 1	System 2 and 4	System 3
cyt c_{553}—SLG distance	0.26	0.25	0.25
heme—SLG distance	1.11	0.60	0.28
Fe(II)—SLG distance	1.53	1.22	0.97
Interaction Energy cyt c_{553}—SLG	−120	−110	−160

All distances are reported in nm, while the energies are in kJ/mol

−110/−160 kJ/mol range. Yet, a closer look at the protein components shows deeper differences in the relative distance from graphene, especially when the heme group (reflected over the Fe(II)–SLG distance) is considered; the distances range from a short value of 0.28 (0.97) nm for System 3 to a maximum value of 1.11 (1.53) nm for System 1 and with intermediate values around 0.60 (1.20) nm for Systems 2 and 4. The values of the geometrical analysis are summarized in Table 13.1.

As first result we can state that the heme-SLG distance is only one of the many factors influencing the interactions of the interface, since also specific hydrophobic interactions of the apolar groups of the cyt c_{553} which are facing the graphene layer stabilize selected orientations of the protein on surface. Another major contribution arises from the relative orientation of the heme group with respect to graphene. To determine this structural parameter, we computed the tilting angle, defined as the angle between the plane of the heme molecule and the graphene layer plane (see insert in Fig. 13.16).

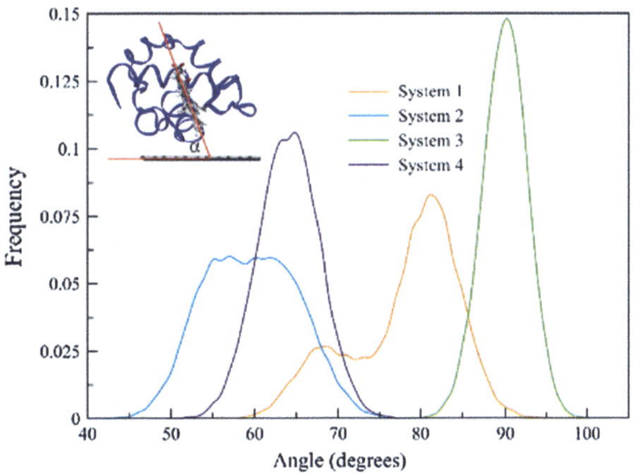

Fig. 13.16 Heme–SLG tilt angle analyses. The insert shows how the heme–SLG tilt angle has been calculated. The angle analysis has been performed for the 50–100 ns window. Reprinted from J. Phys. Chem. C 122, 29405–29413 (2018) (https://pubs.acs.org/doi/10.1021/acs.jpcc.8b10517; further permissions related to the material excerpted should be directed to the ACS)

A broad range of angles has been obtained for the four interfaces studied. While for system 3 the heme is perpendicular to SLG, with angle of approximately 90° and a very narrow distribution, the other three interfaces are more flexible. In particular, system 2 present a plateau of stable tilting angles in the 55–62° window and system 4 a smaller window in between 63 and 65°. Finally, the high degree of rotation of the heme due to the greater distance from graphene results into two distinct orientations of the heme on graphene for system 1: the first at 68°, and the second, more pronounced, at 81°. To summarize, with MD simulations is possible to harness not only geometrical information on the most probable thermodynamically stable orientation of the protein on the surface, but also make a step forward into the prediction on the behavior of the different components of the interfaces on their electron transfer abilities.

13.4.3 Quantum Mechanics: Electronic Properties

Due to the relatively large size of the studied system and the need to calculate electronic properties at a relatively accurate level, density functional methods are a perfect choice for this part of the investigations. Density Functional Theory (DFT) is probably the most used *ab initio* quantum-mechanical method in computational chemistry. Rather than based on wavefunctions, DFT instead relies on the electron density $\rho(r)$. The electron probability density $\rho(r)$ is defined as the probability to find any of the N electrons of a system within a volume element, and vanishes at the infinity [105–108]. The main advantage in using $\rho(r)$ is that, unlike the wavefunction, it is an observable that can be measured experimentally e.g. by X-ray diffraction.

Ab initio calculations at the DFT level of theory are performed to study the electronic and transport properties of SAM/SLG interfaces. For biomimetic devices, two different strategies can be followed, namely physisorption of the SAM via π-π interaction using pyrene derivatives, or chemisorption using 'click' reaction (common adsorbed molecules are reported in Fig. 13.17).

In this section we focus on a pyrene derivative connected to a nitrilotriacetic acid (NTA) via a saturated backbone. The NTA moiety is used to chelate metal cations (M^{2+}) to bind the SAM to the light harvesting protein which has been genetically modified to include a polyhistidine-tag (six histidine residues at the end of the protein chain, commonly used in protein purification), which ensures a stable link between LHP and the SAM. To efficiently reproduce the link at an affordable computational cost, the His6-tag is commonly replaced with the parent imidazole molecules, which complete the coordination of the metal-NTA molecule.

Since one of the most common competitive pathway to the efficient transfer of charge at the interface is the charge recombination, computation focuses on the analysis which shed light to this problem, namely the work function study of the flow of charges at the interface. As mentioned in Sect. 13.3.3, the charge injection from electrodes to organic semiconductors depends on the alignment between the

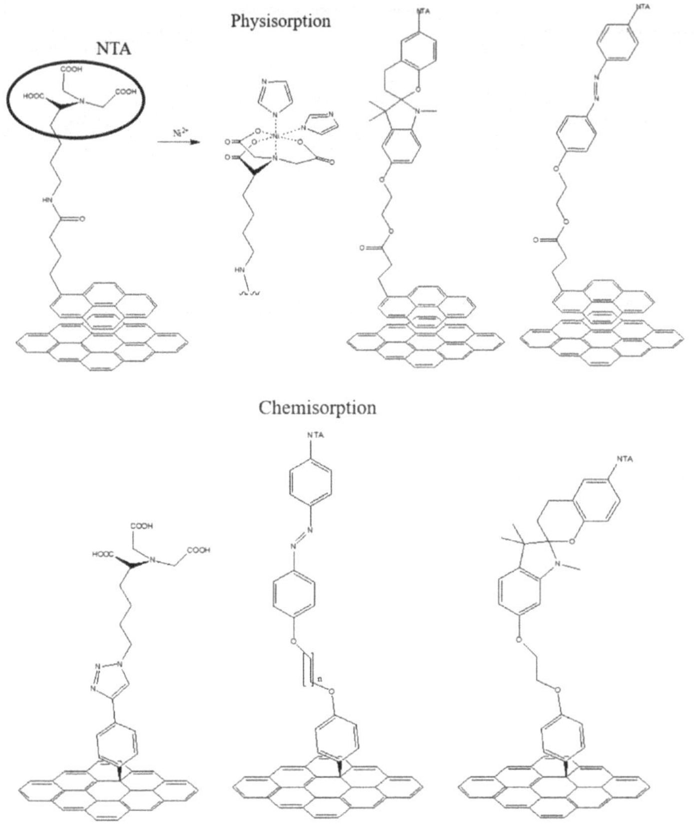

Fig. 13.17 Commonly used self-assembled monolayer (SAM) structures together with their chemical modifications. The NTA tail brings coordination to a metal dication which, in turn, coordinates the His-tag connected to the protein

energy levels of the molecule and the Fermi energy of the photoelectrode. A large mismatch between the Fermi level of the metal and the HOMO (LUMO) frontier orbital of the organic material leads to a large barrier for hole (electron) injection, as depicted in Fig. 13.18. The presence of a monolayer at the interface tunes the barrier to increase or decrease the WF of the electrode and favor the hole or electron injection, respectively. The work function is defined as the minimum work needed to extract one electron from the surface of a metal to a point in the vacuum just outside the surface. The key parameters to consider are the dipole of the molecules (oriented from negative to positive charges) forming the assembly and the bond dipole arising upon the formation of the bond between the electrode and the anchoring group of the molecule; if the dipole points towards the substrate it lowers the WF of the electrode and thus the electron barrier for injection, while if it points

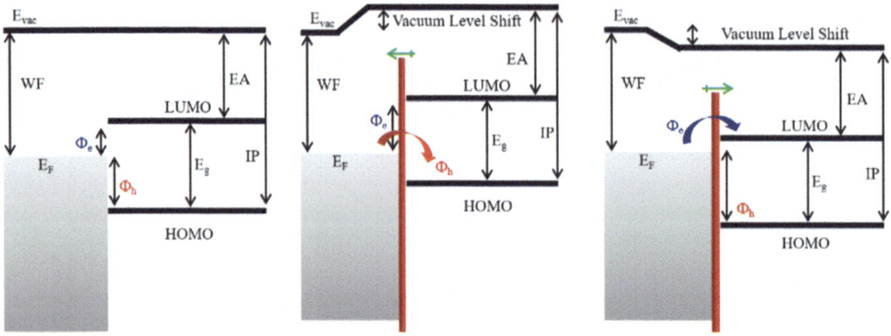

Fig. 13.18 Schematic level alignment between the frontier orbitals of the molecules and the work function of the metal at the organic-metallic interface. (*left*) Electron (Φ_e, blue) and hole (Φ_h, red) injection barriers when the SAM is not present. The interfacial dipole created upon formation of SAM lowers the barrier for holes (*centre*) or electrons (*right*) depending on the dipole direction (± indicate the direction of the interface dipole generated by the SAM). The red (blue) arrow indicates the flow of holes (electrons) from the metal to the organic substrate. E_F refers to the Fermi energy of the metal, E_g to the energy gap between the frontier orbitals, EA to the electron affinity and IP to the ionization potential

away from the substrate it increases the WF of the electrode and lowers the hole barrier for injection, see Fig. 13.18.

The pyrene-NTA-M^{2+} system considered here is chelated with three different metal centers: Co^{2+} (pyrNTA-Co-IM), Ni^{2+} (pyrNTA-Ni-IM) and Cu^{2+} (pyrNTA-Cu-IM), physisorbed on SLG. The choice of the metal center is not trivial and random, but pondered due to biocompatibility and easy of construction of such assemblies from experiments. The most common metal center used for direct binding of His6-tag engineered proteins is Ni^{2+}. Nevertheless, to enhance the direct electron transfer (DET) within nanostructured bioelectronic devices, other metal centers, such as the neighboring metals in the periodic table like cobalt and copper, can be considered. Both Co^{2+} and Cu^{2+} cations present a radical character which might, in turn, strongly change the final DET output at the interface [109, 110].

The geometry optimization of the full, periodic interfaces has been performed at the DFT level of theory, with the PWscf package of the Quantum Espresso suite of programs [111]. The PBE functional coupled with the vdw-DF2 term [112] for the exchange and correlation was used to account for van der Waals interactions together with ultrasoft pseudopotentials [113] with a cut-off of 50 and 200 Ry for the expansion of the wave function and density, respectively. Since periodic boundary conditions (PBC) conditions are used, the final structure should be neutral; thus, to assure a null net charge, one carboxyl group of the NTA moiety has been protonated. As a consequence, the geometry of the system is now distorted and does not reflect the expected octahedral coordination, but a square planar geometry. Although this might not be the lowest energy conformation of the SAM, all three interfaces have been built in the same way, thus assuring consistency for the calculations. To consider the radical nature of the systems when Co and Cu are present, spin magnetization was included during the calculations. After optimization at the DFT level of theory with the PBE functional, the distance between pyrene and graphene was measured to be between 3.7 and 3.8 Å. The obtained electronic density was used to plot the electrostatic potential along the axis normal to the interface (i.e. z-axis) to be able to compute the WF shift by comparing the converged potential on the bare side and on the SAM-covered side of the surface. Two main contributions are responsible of the total shift in the work function ($\Delta\Phi$): the first arises from the dipole moment of the molecular backbone, named hereafter molecular contribution ($\Delta\Phi_{SAM}$) and the second from the interfacial electronic reorganization upon physisorption of the SAM on graphene, namely the charge transfer contribution ($\Delta\Phi_{CT}$):

$$\Delta\Phi = \Delta\Phi_{SAM} + \Delta\Phi_{CT} \tag{13.1}$$

$\Delta\Phi_{SAM}$ has been estimated by computing the electrostatic potential profile across the molecules without graphene, while keeping the coordinates of the system frozen. The $\Delta\Phi_{CT}$ contribution is then calculated by subtracting $\Delta\Phi_{SAM}$ from $\Delta\Phi$. Only when a dipole moment is present in the SAM the shift in the work function is observed. In fact, in the Helmholtz model, the molecular contribution $\Delta\Phi_{SAM}$ is directly proportional to the molecular dipole along the axis normal to the surface, $\mu_{\perp}$, and inversely proportional to the surface area per adsorbed molecule A:

$$\Delta\Phi_{SAM} = e\mu\perp/\varepsilon_0 A \tag{13.2}$$

with ε_0 the vacuum permittivity [114, 115]. As a consequence, if the SAM dipole is pointing towards the metal surface the shift in the work function is positive, while if the dipole points away from the surface the shift in the work function is negative.

The plane average potential profiles of the three interfaces are depicted in Fig. 13.19. The work function for the whole interface is equal to 5.92 eV, 5.01 eV and 5.74 eV for SLG/pyrNTA-Co-IM, SLG/pyrNTA-Ni-IM and SLG/pyrNTA-Cu-IM, respectively. These show an increase of 1.34 eV, 0.44 eV and 1.16 eV with respect to the SLG work function of 4.58 eV (in excellent agreement with experimental measurements of 4.6 eV [116]) when the SAM molecules are absent.

To quantify the contribution from the molecular dipole moment versus the supramolecular effect arising from the transfer of charges at the interface, the $\Delta\Phi_{SAM}$ and $\Delta\Phi_{CT}$ components were separately investigated and summarized in Table 13.2. The shifts arising from the free SAM have values of 0.85 eV, 1.05 eV and 0.70 eV for pyrNTA-Co-IM, pyrNTA-Ni-IM and pyrNTA-Cu-IM, respectively. The supramolecular effect arising from the CT contribution is equal to 0.36 eV, −0.52 eV and 0.31 eV for SLG/pyrNTA-Co-IM, SLG/pyrNTA-Ni-IM and SLG/pyrNTA-Cu-IM interfaces, respectively. Here, two main conclusions can be drawn. First, the $\Delta\Phi_{SAM}$ contribution prevails when Co^{2+} or Cu^{2+} cations are considered, despite the smaller dipole moment calculated (see below). Second, while for Co^{2+} and Cu^{2+} the $\Delta\Phi_{SAM}$ and the $\Delta\Phi_{CT}$ contribution act to enhance the

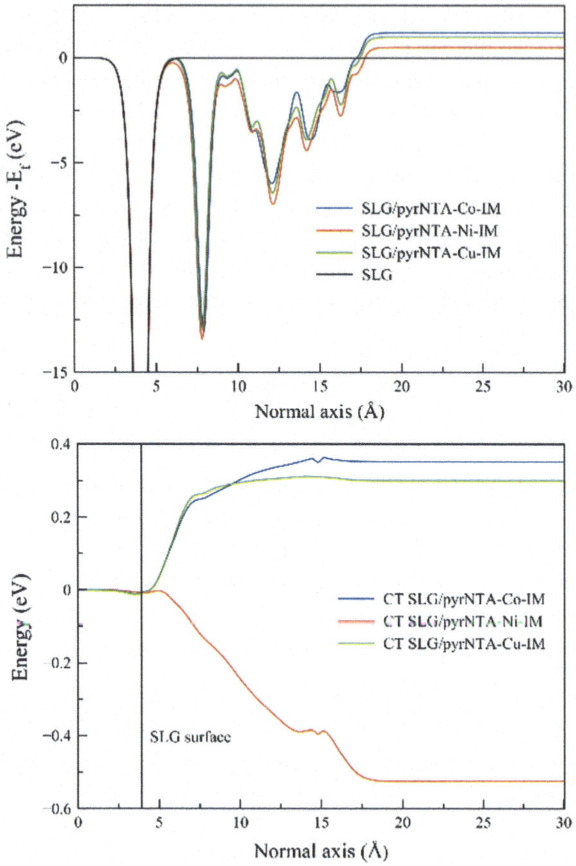

Fig. 13.19 Plane averaged potential of the three interfaces. The bare graphene monolayer potential is shown in black, the graphene surface covered by SAM in blue, red and green for Co, Ni and Cu coordination, respectively (*top*). The evolution of the CT is reported (*bottom*). From left to right in each plot, we move from the graphene surface to the SAM contribution, away from the surface. Adapted with permission from J. Phys. Chem. C 123, 8623–8632 (2019). Copyright (2019) American Chemical Society

total work function shift, the presence of Ni^{2+} alters the trend with an opposite interaction, which is the main factor responsible of the smaller total calculated work function shift.

The dipole moment of the molecule along the direction perpendicular to the SLG surface is responsible for the $\Delta\Phi_{SAM}$ contribution. Surprisingly, for the SLG/pyrNTA-Ni-IM its value is quite small, −1.83 D, while rises up to −4.23 D and −3.55 D for SLG/pyrNTA-Co-IM and SLG/pyrNTA-Cu-IM interfaces, respectively. This strong difference (up to 2.4 D) can be attributed to the different nature of the M^{2+} cations, which is the key to quantify the $\Delta\Phi_{SAM}$ to the total work function. In fact, although all three metal centers considered in the study are the first row transition metals and neighbors in the periodic table, Co^{2+} and Cu^{2+} are formally radicals, which character strongly alter the electronic properties of the final assembly.

The CT contribution $\Delta\Phi_{CT}$, which is substantial, is not confined to the SLG/SAM interface ligated with various M^{2+} cations (i.e. interaction between graphene and pyrene) but increases until the end of the molecular backbone (Fig. 13.19).

Table 13.2 Work function analysis casted down among the three components, $\Delta\Phi$ $\Delta\Phi_{SAM}$ and $\Delta\Phi_{CT}$

Interface	$\Delta\Phi$ (eV)	$\Delta\Phi_{SAM}$ (eV)	$\Delta\Phi_{CT}$ (eV)	Dipole moment (D)
SLG/pyrNTA-Co-IM	1.21	0.85	0.36	−4.23
SLG/pyrNTA-Ni-IM	0.53	1.05	−0.52	1.83
SLG/pyrNTA-Cu-IM	1.01	0.70	0.31	−3.55

The dipole moments of the SAM are calculated along the *z*-axis

The plot illustrates that the CT has a similar evolution when the Co^{2+} and Cu^{2+} metal cations are considered, while it is negative with Ni^{2+} cation is present. This translates into a different direction of charge flow for the three interfaces which can be quantified by considering the unbalance of charge between the two components of the interface in their ground state. The CT contribution translates into a (partial) transfer of charges form one fragment to the other, once the interface is created. Bader charge analysis allows to calculate the excess/depletion of charge at the interface as:

$$\Delta\rho(z) = \rho_{SLG/SAM}(z) - [\rho_{SLG}(z) + \rho_{SAM}(z)] \tag{13.3}$$

where $\rho_{SLG/SAM}$ is the charge density of the full system and ρ_{SLG} and ρ_{SAM} the charge densities on the two non-interacting fragments. An excess of electrons of $-0.08\ |e|$ on the SAM for both SLG/pyrNTA-Co-IM and SLG/pyrNTA-Cu-IM interfaces has been computed, while the presence of Ni^{2+} cation reversed the flow, with an excess of electrons ($-0.05\ |e|$) on the graphene surface. We suggest that the strong positive shift of the work function for both systems with Co^{2+} and Cu^{2+} coordination arises from the synergic interaction of the two contributions which enhance the overall work function shift, resulting in an electron flow from graphene to the SAM. On the other hand, for the SLG/pyrNTA-Ni-IM interface these contributions counteract, resulting in the opposite direction of the electron flow, from SAM to the SLG.

To gain an even deeper insight into these difference, the analysis of the density of states (DOS) is considered. Total DOS and the projection over the fragments (SAM and SLG) as well over atomic types are presented in Fig. 13.20.

Both systems with Co^{2+} and Cu^{2+} metal ions present the states pinned at the Fermi level (−4.06 eV for SLG/pyrNTA-Co-IM and −3.99 eV for SLG/pyrNTA-Cu-IM interfaces) very close in energy to the valence band maximum level (VBM) arising from the presence of an open shell system, with energy at −3.82 eV and −3.87 eV for SLG/pyrNTA-Co-IM and SLG/pyrNTA-Cu-IM interfaces, respectively. Therefore, both levels contribute to the Fermi energy, while for the Conduction Band Minimum (CBM), lying at −2.12 eV and −2.04 eV for SLG/pyrNTA-Co-IM and SLG/pyrNTA-Cu-IM, respectively, there is a sole contribution from the graphene surface. Moreover, the LUMO of the SAM is located at much higher energy with respect to the CBM, at −1.63 eV and −1.52 eV

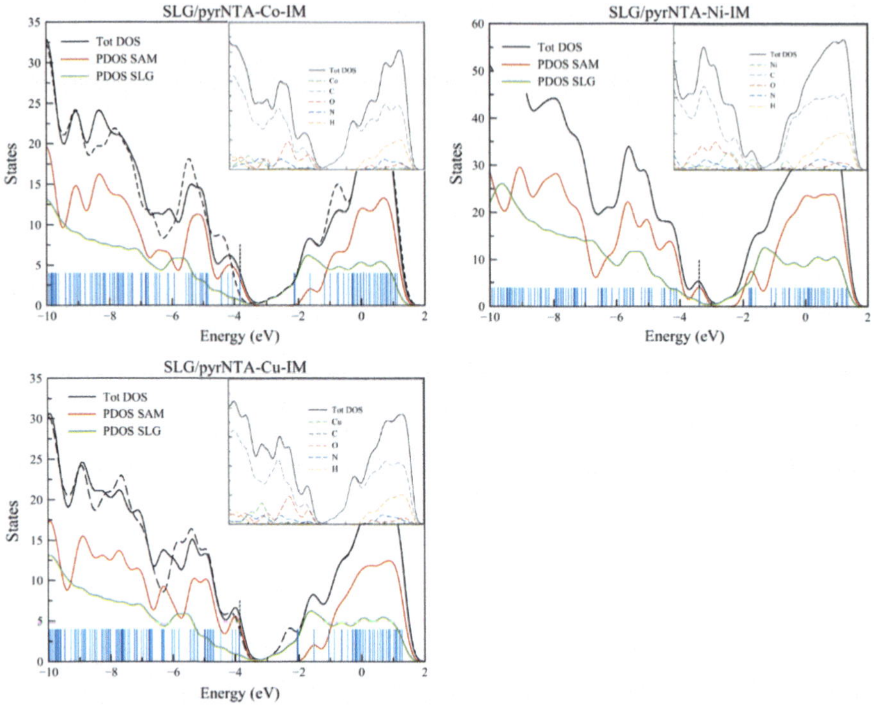

Fig. 13.20 Total density of states and projected density of states over the fragments and over each atom (inserts) for the three interfaces analyzed. Vertical blue lines indicate the eigenvalues; dotted line indicates the position of the Fermi energy level. Black dashed line refers to the total DOS calculated for the 'spin down' electrons for the two open shell systems. Reprinted with permission from J. Phys. Chem. C 123, 8623–8632 (2019). Copyright (2019) American Chemical Society

for pyrNTA-Co-IM and pyrNTA-Cu-IM, respectively. From the PDOS, we observe that the peak at the Fermi level is a sum of different contributions from carbon and oxygen atoms, while the CBM is purely that of carbon DOS. The strong different localization of this frontier levels is the main responsible for the high CT observed. A different scenario occurs when the SLG/pyrNTA-Ni-IM interface is considered (Fig. 13.20), where the contribution at the Fermi level is due to the SAM only, while for CBM the contribution is mixed, arising from both SAM and SLG. In complete opposition with respect to the previous interfaces, the PDOS analysis reveals that the Ni^{2+} cation is the main element responsible for both the VBM and CBM peaks, together with a strong contribution of the carbon atoms.

The computational data presented here are in full agreement with reported electrochemical data obtained on the same working device, created considered a SLG on a fluorinated tin oxide (FTO) electrode on which the pyrNTA-M^{2+} monolayer was deposited with and without imidazole coordination. Measurement of cyclic voltammetry, differential pulse voltammetry and photochronoamperometric experiments were performed at different potentials for each electrode, ranging from −300 to +100 mV [110]. All these electrochemical data confirm the DFT findings of the different WF shift affecting the flow of charges at the interfaces depending on the metal center considered. Interestingly, from these experimental measurements was found that the imidazole modification of the pyrNTA films resulted in a strong suppression of the photocurrents for all the electrodes. At an open circuit potential (OCP), for which direct comparison with DFT results can be considered, the cathodic photocurrent decreases following the trend: SLG/pyrNTA-Co-IM > SLG/pyrNTA-Cu-IM > SLG/pyrNTA-Ni-IM (Fig. 13.21), with highest currents values of 4.8 nA cm^{-2} for the samples with Co^{2+}, the lowest ones of 2.1 nA cm^{-2} for the assemblies with Ni^{2+} and an intermediate situation of 3.1 nA cm^{-2} when Cu^{2+} is present.

Likewise, the highest cathodic photocurrents with value of 43 nA cm^{-2} at −300 mV were recorded for SAM with Co^{2+} and the lowest ones, of 17.2 nA cm^{-2} when Cu was present. On the other hand, the highest anodic photocurrent densities were recorded for the samples with Ni^{2+} with high value of 18.1 nA cm^{-2}, confirming that Ni^{2+} cation promotes the charge transfer from SAM to graphene. Despite this finding, the redox behavior of the SAM is more complex when Ni^{2+} cation is present, as it seems to promote generation of also cathodic currents at a negative bias of at least −200 mV (see Fig. 13.21). In addition, an analogous behavior is observed for the pyrNTA-Cu-IM and pyrNTACo-IM interfaces, in which the measured anodic current is enhanced up to 8.8 and 7.2 nA cm^{-2}, which is in apparent contrast with the charge flow direction. Both behaviors can be explained by (1) the unbalance between the DET and the overpotential applied, in

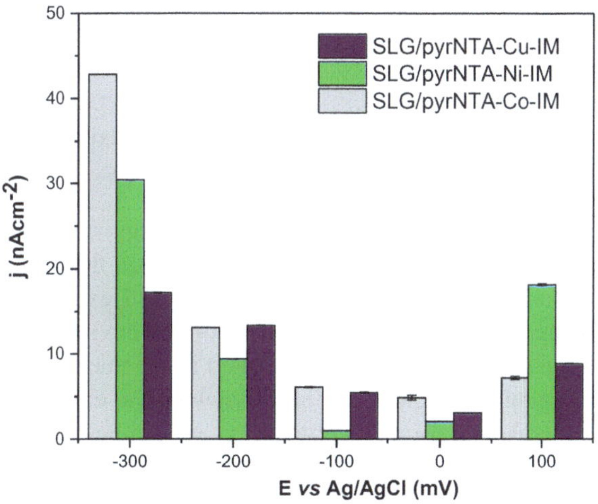

Fig. 13.21 Photochronoamperometric analysis of SLG/pyrNTA-M-IM assemblies. Photocurrent densities generation at respective electrodes as a function of external potential. The average of the current densities values was estimated from two independent measurements ($n = 2$). Adapted with permission from J. Phys. Chem. C 123, 8623–8632 (2019). Copyright (2019) American Chemical Society

which the last one overturns the (small) ground state charge transfer and (2) the radical nature of the Co and Cu cations, which can strongly affect the charge flow at the interface [109, 110].

In conclusion, quantum mechanics (QM) is a powerful method to assess different directionalities of the charge flow at various interfaces and to pin the subtle changes in energy level alignment due to the formation of the interfaces between (in the current case) SLG and complex organic molecules. With this rational design of interfaces, computation becomes a powerful way to assess a priori the efficiency of devices and the effect that small changes in the molecular structure of the SAM (such as metal bounding atoms, backbone conjugation or change in functional groups) can have on the charge transfer of complex interfaces. Moreover, with a deep understanding of the mechanism governing the charge flow and energy alignment of the frontier molecular orbital involved in the process, computation can answer to the pressing question on how to suppress the charge recombination detrimental pathway for efficient injection of charges into the photoelectrode.

13.4.4 Quantum Mechanics: Transport Properties

Considering the complexity of the biohybrid interface and the long distance the charge has to travel from the protein to reach the electrode (higher than 4 nm), the electron transfer can be safely considered via hopping mechanism. In this regime, the charge carrier is localized at individual sites and proceeds by a sequence of non-coherent (*hopping* motion) transfer events. To compute the charge transfer abilities of the interfaces and the hopping rate, the semi-classical Marcus-type equation can be considered [117, 118]:

$$R_{\text{hop}} = \frac{|V|^2}{\hbar}\sqrt{\frac{\pi}{\lambda k_B T}}\exp\left[-\frac{(\Delta E + \lambda)^2}{4\lambda k_B T}\right] \tag{13.4}$$

where λ denotes the reorganization energy, ΔE the change in energy associated to the hop and V the transfer integral between the initial and final states. This equation is valid when the energy of internal and external phonons (vibrations) are small compared to the thermal energy of the environment. For a fixed temperature, the large transfer rate can be attributed to the maximal transfer integral and the minimal reorganization energy. Yet, in view of the complexity of the system, both transfer integral and reorganization energy should be considered. The reorganization energy λ results from the distortion the system experiences when an electron is transferred from one site to another. This quantity is normally divided into two contributions $\lambda = \lambda_{\text{inner}} + \lambda_{\text{out}}$ in which the first term refers to intra-molecular distortion occurring when a molecule become charged; the second term arises from the change in molecular energy due to the inter-molecular displacement and polarization due to solvent effects. The latter can be of importance in aqueous solutions due to strong solvation effects, while in solid-state devices the former is dominant. The transfer integral V characterizes the strength of the electronic coupling between two adjacent molecules, which can be written as [119]:

$$V = \frac{J_{AB} - \frac{1}{2}(e_A + e_B)S}{1 - S^2} \tag{13.5}$$

with S being the overlap matrix and e_A and e_B the diagonal elements of a Hamiltonian defined by the frontier molecular orbitals of the isolated molecule A and B in the dimer representation, respectively. Namely, for hole (electron) transport, the HOMO (LUMO) should be considered. In an orthonormalized basis, these e_n (n = A, B) give access to specific site energies ε_n (n = A, B). The correction by means of these site energies is necessary in view of the pronounced polar character of the different moieties and the different nature of both monomers in the complex (i.e. SAM/SLG, linker/SAM/SLG interfaces). With other words, since the dimer is non-symmetric, the energy split between the HOMO (LUMO) and HOMO-1 (LUMO+1) in the dimer is both influenced by the electronic coupling V and the difference in site energies. Among the different computational methods developed to calculate these quantities, the projective method will be used to quantify the transfer integral V between the two non-equivalent components of the interface [120, 121]. Within the projective method, the system is divided into fragments, in which an electron or hole is localized on a fragment and can hop from one fragment to another. The fragment calculations, performed by using the DFT method, follow a procedure where the orbitals of a pair of molecules (a dimer) are projected onto a basis set defined by the orbitals of each individual molecule (the fragments). The obtained set of orthogonal molecular orbital energies of the dimer are then used to express the Fock matrix in function of the new localized basis set, which enables the formation of a block-diagonal matrix. The main advantage of this method is the possibility to analyse pairs of different molecules, since it is not necessary to assume that the energies of the HOMO of the two fragments are equal.

In addition, since the system is non-crystalline, Kinetic Monte Carlo (KMC) simulations can be performed to assess the transfer of charges of the whole interface. In this model, disorder medium possessing some density of states is randomly generated and processes of charge generation, separation, transport and injection can be simulated [122, 123]. The response of the material is then averaged over a large number of replicas, to account for a strong statistic. With this approach, a charge is places at the reaction center of the SLHP (i.e. the heme group) and hop to its nearest neighbours, within a given cut-off. The procedure is then repeated until the charge is transfer to SLG. Considering the complexity of the systems, KMC simulations at frozen morphology are performed on several snapshots extracted from the MD simulations. The transfer integrals (for every hop) is then set in the ensemble as the average of the transfer integrals distribution over time and for every pair of nearby molecules (nearer than a chosen cut-off), the transfer integral is calculated with the projective method described above.

13.4.4.1 Transfer Integral and Coupling

As example for the calculation of the transfer integral and coupling parameters for biohybrid interfaces let us consider the cyt/SLG interface described before in Sect. 13.4.2. Considering that in biological environment the electron transfer can occur up to a distance of 1.4 nm [124, 125], all the analysed structures can, in principle, be good candidates for a direct electron transfer mechanism (DET) from the cyt c_{553} to SLG or vice versa. Yet for efficient DET one should consider not only the heme-graphene distance but also the orientation of the two species involved in this process. In particular, in the case of cyt/SLG interface the orientation of the heme may be a crucial parameter in assessing whether the transfer of charge is efficient. Moreover, it is not yet clear if the transfer occurs through-linker or through-space due to the long distance of heme and the presence of the protein backbone surrounding it and being directly in contact with SLG. Yet, we can assume that the shortest pathway might be followed and thus focus only on the through-space mechanism, with a possible overall super-exchange transfer [126, 127].

The four different heme/SLG interfaces were extracted from MD simulations and graphene cut in such a way that no zig-zag edges are present. In addition, to avoid artefacts due to thermal conformational fluctuations present in the MD calculations, the geometry of the heme group has been optimized at the HSE/6-31G(d,p) level of theory and then superimposed the heavy atoms to the original position by mean of RMSD minimization, as found from MD simulations. The obtained interfaces are reported in Fig. 13.22. The calculated Fe(II)–SLG distances are 1.53, 1.24, 0.97 and 1.21 nm for the four interfaces, respectively [104].

To assess the transfer integral, the interface is separated into two distinct fragments (i.e. heme and SLG), in which an electron or hole is localized on a fragment and can hop from one fragment to another. The fragments are thus considered as a donor and an acceptor part, and the coupling is computed considering any amount of frontier molecular orbitals from the donor and acceptor sites, considering all important contributions (vertical and diagonal coupling terms). In the specific case, the frontier MOs degeneracy for system 4 was considered, while for the other three structures only the HOMO-HOMO coupling is taken into account. Results of the analysis are reported in Table 13.3.

A negligible transfer integral for both electrons and holes carrier is computed when the distance between heme and SLG is substantial, higher than 0.5 nm; in particular, system 1 presents virtually no transfer for either carriers, with values lower than 10^{-8} and 10^{-7} meV; system 2 and 4 which present similar distance but different orientation of the heme on SLG, show an increase of transfer integral of few order of magnitude, up to 7.1×10^{-4} meV for electrons and 1.6×10^{-5} meV holes; only system 3 presents a substantial transfer integral for both carrier, with values of 1.2 meV for holes and 0.25 meV for electrons, comparable with common polymers used in organic photovoltaics [128]. As expected for a through-space charge transfer, a strong correlation between the transfer integral and the distance, which decays exponentially, is found, but in addition, a correlation with respect to the tilting of the heme group and the intensity of the transfer integral can be

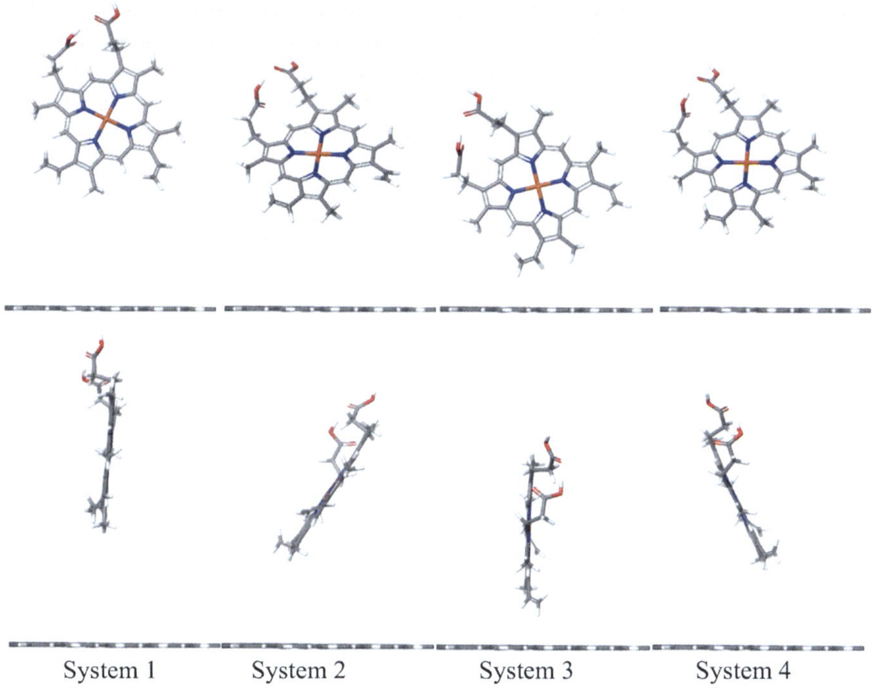

Fig. 13.22 Frontal and side views of the four different interfaces extracted from MD simulations and used for the charge transfer calculations. Reprinted from J. Phys. Chem. C 122, 29405–29413 (2018) (https://pubs.acs.org/doi/10.1021/acs.jpcc.8b10517; further permissions related to the material excerpted should be directed to the ACS)

Table 13.3 Dependence of the calculated transfer integrals for the heme/SLG interfaces for hole (*h*) and electron (*e*) transfer on the different positions

Interfaces	Carrier	$V_{heme/GNF}$ (meV)
System 1	*h*	6.4×10^{-8}
	e	1.5×10^{-7}
System 2	*h*	3.8×10^{-4}
	e	4.3×10^{-5}
System 3	*h*	1.2
	e	0.25
System 4	*h*	1.6×10^{-5}
	e	7.1×10^{-4}

All values are given in meV

observed. In details, system 1 and 3 show the same tilt angle but different distance from SLG, while for system 2 and 4 the situation is opposite, with similar distance from of about 0.6 nm from SLG but a different tilt angle, which is responsible for the substantial difference in transfer integral ability, up to one order of magnitude.

To conclude, the transfer integral analysis provides deeper insight into the mechanism of charge transfer which are complementary to the quantum mechanics analysis, showing strong dependence on both relative orientation and position of the fragments, with strong sensitivity for subtle changes that can lead to strong alteration of charge transfer ability of the interface. We can then assume that in a real device, thermodynamic effects can play a major role in determining the DET efficiency, and more complex models can be envisaged.

13.4.4.2 Charge Transport Rate

To assess the kinetic of charge separation for biohybrid interfaces, the Marcus theory can be considered, since the system is intrinsically disordered and strong fluctuations are present due to the protein. The interplay between disorder, molecule distance and orientation significantly affect the electronic properties at the interfaces, which in turn is responsible for the mechanism and nature of free charge carrier generation. Thus, neglecting these factors will lead to non-accurate results and the inability to describe the performance of real devices. As the electronic properties of the molecules depend on the environment, the various energies to take into account (such as frontier orbital and singlet energies) might be highly dependent upon the configuration of the surrounding molecules as well as the proximity to the interface. Since the charge transfer states are strongly dependent on distance and relative orientation of the surrounding molecules, disorder comes to play a crucial role [129]. Thus, a simple Marcus curve as depicted in Fig. 13.23 should be replaced with many Marcus curves, all of them retaining the unique rate that at the end has to be averaged to get experimentally relevant rates.

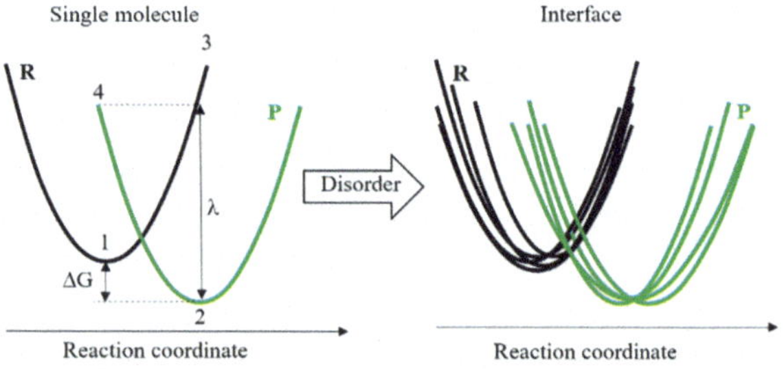

Fig. 13.23 Marcus free energies describing transition from reactant (*R*) to product (*P*) states. Numbers indicate the four-point procedure to determine λ and ΔG. The presence of static disorder creates a distribution of both reactant and product energy landscapes

The electron transport can be evaluated considering both holes and electrons carriers in different environments. As a result, Marcus relates the activation energy to the reorganization energy, seen as the energy required for the reactant to adapt its configuration to the final charge distribution in the products, and the Gibbs free energy for the reaction. The main consequence is that due to the non-linear dependence of the activation energy on the Gibbs free energy, a maximum rate can be reached for specific value of the driving force, that can be followed by and inverse region. In this region, the increase in driving force results in the decrease of the reaction rate. Yet, for the description of layers of materials deposited on a surface such as an electrode, the Marcus's approach has to be modified since now we deal with electrodes, which inherently account for a large number of electronic bands near the Fermi level and the overpotential of the metal. Thus, the expression shall be modified to include the various states which can act as donor or acceptor. In a non-adiabatic regime between the redox center and the electrode, the reaction rate can be expressed as [130, 131]:

$$k_c(\varepsilon,\eta) = \frac{|V_{AB}|^2}{\hbar}\sqrt{\frac{\pi}{\lambda k_B T}}\int\limits_{-\infty}^{+\infty} \rho(\varepsilon)\exp\left[-\frac{(\Delta E+\lambda+e\eta)^2}{4\lambda k_B T}\right] f(\varepsilon)\mathrm{d}\varepsilon \qquad (13.6)$$

where V_{AB} is the electronic coupling between the redox active group and the metal of the electrode, $\rho(\varepsilon)$ the density of the electronic states in the metal, $\Delta E = (\varepsilon_F - \varepsilon)$ the energy difference between the Fermi energy ε_F and the orbital energy ε of the transport state of the layer material, λ is the reorganization energy, η is the overpotential and $f(\varepsilon)$ the Fermi-Dirac distribution of occupied states of the metal. For the oxidation process (the inverse reaction) the $f(\varepsilon)$ term becomes $[1 - f(\varepsilon)]$. Thus, the main factors affecting the rate constant for electron transfer at the interface are the overpotential, the reorganization energy and the nature of the metal, as well as the strength of the coupling between the redox site (i.e. the heme group) and the electrode. As result, the transfer depends on the distance between the two groups and the chemical nature of the linking between them. This last term in turn affect the height of the barrier for transfer and it depends on the chemical nature of the molecule, via the decay constant β: the lower the decay constant, the higher the electron transfer. Once the abovementioned equations are fitted against experimental dependence of rate constant and the applied bias, standard rate constant can be obtained.

Recently, Plumeré developed a new protocol to study the complex mechanism governing bio-photoelectrochemical kinetics, which consider the processes generating the photocurrent and the processes competing with photocurrent generation (such as charge recombination) [132]. The model consists of both photosystems and electron mediators (which is confined in the redox film on the electrode surface) in which both outer-sphere electron transfer between the photosystem and the electron acceptor and photoenzymatic reactions are considered (Fig. 13.24). Moreover, the electrons can be transferred either to a charge carrier which diffuses to the bulk

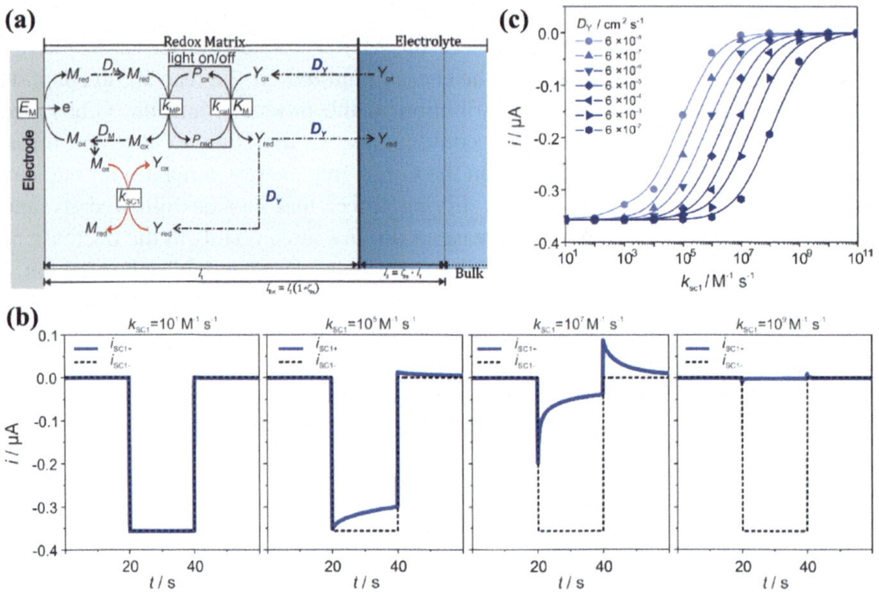

Fig. 13.24 **a** Schematic illustration of the reactions for the biophotocathode based on a photoenzyme generating a charge carrier that diffuses into the bulk of the electrolyte. The recombination pathway is limited to the reaction between the reduced charge carrier and the oxidized electron mediator. **b** Photocurrent predicted for the photocatalytic process alone (black dashed line) and for the recombination pathway associated to the photocatalytic process (blue solid line) for increasing values of kSC1 at $DY = 6 \times 10^{-6}$ cm^2 s^{-1}. **c** Photocurrent at 40 s (at the end of the illumination period) plot versus log(kSC1) for increasing values of DY. Reprinted from Faraday Discuss. (2019) Advance Article, https://doi.org/10.1039/c8fd00168e. Published by The Royal Society of Chemistry

(to generate electricity) or to be transferred to redox catalyst (for solar fuel cell applications).

This model has strong predictive power, especially when the experimental outcome is not clear. Due to the large amount of parameters used in the model, it will be possible to test the effect of each of them on the final device performance and photocurrent generation. With a simple case study, the authors were able to show the prediction power of this new method, and the accuracy in predicting both the current generation and the recombination processes with qualitative trend for photocurrent decrease and quantitative analyses of the diffusion coefficient impact on the process. The flexibility of the model allows its use for several different interfaces with increased degree of complexity, providing the presence of the catalytic center (i.e. LHP) and the electron acceptor (i.e. SAM, SLG).

13.5 Conclusions, Challenges and Perspectives

Due to the advances in computer power and implementation of new software and methods, computation is becoming one important component not only in the study of fundamental mechanisms at a molecular level, but also in the understanding of more complex environments which comprehend biological systems. Additionally, as a result of the introduction of multiscale simulations, the development of novel molecular dynamic and docking software as well as the high accuracy reached by DFT and post Hartree-Fock methods, it is nowadays possible to fully describe complex interfaces like those analysed in this chapter. The key to success of modern computation is multifaceted and relies on advances in various areas. First of all, it is possible to cast down complex problems to simpler building block, in a so called bottom-up approach, where the focus is on the constituents of a complex system; by discerning among the most important central aspects of the problem (such as the centers essential for electron transfer), it is possible to access electronic and optical properties of complex interfaces; in addition, the inclusion of the environment, which is described at a less computationally-demanding level, leads to the description of model systems which are close to the real experimental counterparts.

In a nutshell, by a rational design of complex interfaces, computation proved itself a useful approach to answer questions which are still difficult to assess experimentally, and shed light into key processes such as the electron transfer which are essential for any kind of solar-related devices. As the computational approach is more and more refined, with inclusion of more complex systems described at different level of theory (from MD to QM/MM to full ab initio methods), the model systems studied are closer to the experimental ones, and thus more reliable results can be obtained and the prediction power of computation can be exploited.

Yet, many limitations in such approach are still present. These include mainly the difficulties in merging different size and time scales of simulations and to go from the microscopy to the macroscopic modelling. This is particularly clear when trying to combine a system investigated at an atomic scale (with either Schrödinger or Newton equations) and combine it with macroscopic models which consider experimentally-accessible observables. Despite these drawbacks, the use of DFT based methods which allow the description of large systems (up to thousands of atoms) at an atomic scale, strongly improved the reliability of computation and its prediction power.

Thus, when considering computation to study complex systems such as those involved in artificial photosynthesis, one has to carefully design the computational experiment and take particular care at the multiscale approach to use. In addition, as many simplifications are needed, care must be also paid to the definition of the size of the final model interfaces, and the level of theory at which each component is described. The final aim of this approach is to accurately describe light-driven charge separation, charge transfer and concomitant oxidative and reductive chemistry of model interfaces, and to assess a priori the efficiency of the final device. Optimization of efficient directional ET is the one of the main bottlenecks that has been recently rationally addressed by generation of highly ordered architectures of photoactive modules, both by computation and by the creation of nanoengineered conductive interfaces between working modules of the photoelectrodes. In these first attempts, quantum modelling of the charge transfer, especially at the connecting interfaces, has been helpful for the rational development of different nanoengineering strategies. In our view, a strong interconnection between computation, synthesis, characterization and creation of a working devices is the key aspect to be able to move this innovative technology out of the laboratory scale. We believe that this is the direction that computation should follow: not only to answer fundamental questions but also to predict the behavior of complex interfaces in close-to-real models.

Acknowledgements SO acknowledges the National Science Centre, Poland (grant no. UMO-2018/31/D/ST4/01475) and grant (UMO-2015/19/P/ST4/03636) for the funding from the European Union's Horizon 2020 research and innovation program under the Marie Skłodowska-Curie grant agreement No. 665778. SO and MI acknowledge the DSM 2018 grant from the Centre of New Technologies (grant no. 501-D313-86-0119600-01). JK and MI are grateful for the support from the National Science Centre (grant no. UMO-2017/27/B/ST5/00472) and the Polish National Centre for Research and Development (grant no. DZP/POLTUR-1/50/2016, agreement no. 5/POLTUR-1/2016), within the framework of 1st Bilateral Polish-Turkish POLTUR program.

References

1. M.G. Walter, E.L. Warren, J.R. McKone, S.W. Boettcher, Q. Mi, E.A. Santori, N.S. Lewis, Solar water splitting cells. Chem. Rev. **110**, 6446–6473 (2010)
2. R. Cao, W. Laia, P. Du, Catalytic water oxidation at single metal sites. Energy Environ. Sci. **5**, 8134–8157 (2012)
3. R.E. Blankenship, D.M. Tiede, J. Barber, G.W. Brudvig, G. Fleming, M. Ghirardi, M.R. Gunner, W. Junge, D.M. Kramer, A. Melis et al., Comparing photosynthetic and photovoltaic efficiencies and recognizing the potential for improvement. Science **332**, 805–809 (2011)
4. J.D. Janna Olmos, J. Kargul, A quest for the artificial leaf. Int. J. Biochem. Cell. B. **66**, 37–44 (2015)
5. A. Amunts, O. Drory, N. Nelson, The structure of a plant photosystem I supercomplex at 3.4 A resolution. Nature **447**, 58–63 (2007)
6. N. Kamiya, J. Shen, Crystal structure of oxygen-evolving photosystem II from Thermosynechococcus vulcanus at 3.7-A resolution. Proc. Nat. Acad. Sci. **100**, 98–103 (2003)
7. J. Kargul, J. Barber, Structure and function of photosynthetic reaction centres, in *Molecular Solar Fuels*, ed. by T.J. Wydrzynski, W. Hillier (Royal Society of Chemistry, Cambridge, 2011), pp. 107–142
8. N.J.M. Sanghamitra, T. Ueno, Expanding coordination chemistry from protein to protein assembly. Chem. Commun. **49**, 4114–4126 (2013)
9. G.P. Whyburn, Y. Li, Y. Huang, Protein and protein assembly based material structures. J. Mater. Chem. **18**, 3755–3762 (2008)

10. J. Barber, P.D. Tran, From natural to artificial photosynthesis. J. R. Soc. Interface **10**, 20120984 (2013)
11. I. McConnell, G. Li, G.W. Brudvig, Energy conversion in natural and artificial photosynthesis. Chem. Biol. **17**, 434–447 (2010)
12. D.G. Nocera, The artificial leaf. Accounts Chem. Res. **45**, 767–776 (2012)
13. P.R. Berger, M. Kim, Polymer solar cells: P3HT:PCBM and beyond. J. Renew. Sustain. Ener. **10**, 013508 (2018)
14. K. Ocakoglu, T. Krupnik, B. van den Bosch, E. Harputlu, M.P. Gullo, J.D. Janna Olmos, S. Yidirimcan, R.K. Gupta, F. Yakuphanoglu, A. Barbieri, J.N. Reek, J. Kargul, Photosystem I-based biophotovoltaics on nanostructured hematite. Adv. Funct. Mater. **24**, 7467–7477 (2014)
15. A. Mershin, K. Matsumoto, L. Kaiser, D. Yu, M. Vaughn, Md.K. Nazeeruddin, B.D. Bruce, M. Graetzel, S. Zhang, Self-assembled photosystem-I biophotovoltaics on nanostructured TiO_2 and ZnO. Sci. Rep. **2**, 1–7 (2012)
16. M. Kiliszek, E. Harputlu, M. Szalkowski, D. Kowalska, C.G. Unlu, P. Haniewicz, M. Abram, K. Wiwatowski, J. Niedziółka-Jönsson, S. Maćkowski, K. Ocakoglu, J. Kargul, Orientation of photosystem I on graphene through cytochrome c553 leads to improvement in photocurrent generation. J. Mater. Chem. A **6**, 18615–18626 (2018)
17. M.E. El-Khoulya, E. El-Mohsnawy, S. Fukuzumi, Solar energy conversion: from natural to artificial photosynthesis. J. Photoch. Photobio. C **31**, 36–83 (2017)
18. J. Kargul, J.D. Janna Olmos, T. Krupnik, Structure and function of photosystem I and its application in biomimetic solar-to-fuel systems. J. Plant Physiol. **169**, 1639–1653 (2012)
19. L. Duan, L. Tong, Y. Xua, L. Sun, Visible light-driven water oxidation—from molecular catalysts to photoelectrochemical cells. Energy Environ. Sci. **4**, 3296–3313 (2011)
20. S.Y. Reece, J.A. Hamel, K. Sung, T.D. Jarvi, A.J. Esswein, J.J.H. Pijpers, D.G. Nocera, Wireless solar water splitting using silicon-based semiconductors and earth-abundant catalysts. Science **334**, 645–648 (2011)
21. M. Kamran, J. Delgado, V. Friebe, T. Aartsma, R. Frese, Photosynthetic protein complexes as bio-photovoltaic building blocks retaining a high internal quantum efficiency. Biomacromol **15**, 2833–2838 (2014)
22. K.S. Joya, J.L. Vallés-Pardo, Y.F. Joya, T. Eisenmayer, B. Thomas, F. Buda, H.J.M. de Groot, Molecular catalytic assemblies for electrodriven water splitting. Chem. Plus. Chem. **78**, 35–47 (2013)
23. P.N. Ciesielski, C.J. Faulkner, M.T. Irwin, J.M. Gregory, N.H. Tolk, D.E. Cliffel, G.K. Jennings, Enhanced photocurrent production by photosystem I multilayer assemblies. Adv. Funct. Mater. **20**, 4048–4054 (2010)
24. J.Z. Zhang, K.P. Sokol, N. Paul, E. Romero, R. van Grondelle, E. Reisner, Competing charge transfer pathways at the photosystem II–electrode interface. Nat. Chem. Biol. **12**, 1046–1052 (2016)
25. J.D. Janna Olmos, J. Kargul, Oxygenic photosynthesis: translation to solar fuel technologies. Acta Soc. Bot. Pol. **83**, 423–440 (2014)
26. R. van de Krol, Y.Q. Liang, J. Schoonman, Solar hydrogen production with nanostructured metal oxides. J. Mater. Chem. **18**, 2311–2320 (2008)
27. R. van de Krol, M. Graetzel (eds.), *Photoelectrochemical Hydrogen Production* (Springer, New York, 2012)
28. B. O'Regan, M. Graetzel, A low-cost, high-efficiency solar cell based on dye-sensitized colloidal TiO_2 films. Nature **353**, 737–740 (1991)
29. J. Brillet, M. Cornuz, F. Le Formal, J.H. Yum, M. Graetzel, K. Sivula, Examining architectures of photoanode–photovoltaic tandem cells for solar water splitting. J. Mater. Res. **25**, 17–24 (2010)
30. V. Shah, W. Henson, T. Chadha, G. Lakin, H. Liu, R. Blankenship, P. Biswas, Linker-free deposition and adhesion of photosystem I onto nanostructured TiO_2 for biohybrid photoelectrochemical cells. Langmuir **31**, 1675–1682 (2015)

31. N. Sekar, R.P. Ramasamy, Recent advances in photosynthetic energy conversion. J. Photochem. Photobiol., C **22**, 19 (2015)
32. R.J. Cogdell, A.T. Gardiner, H. Hashimoto, T.H.P. Brotosudarmo, A comparative look at the first few milliseconds of the light reactions of photosynthesis. Photochem. Photobiol. Sci. **7**, 1150–1158 (2008)
33. R. Angamuthu, P. Byers, M. Lutz, A.L. Spek, E. Bouwman, Electrocatalytic CO_2 conversion to oxalate by a copper complex. Science **327**, 313–315 (2010)
34. H. Li, P.H. Opgenorth, D.G. Wernick, S. Rogers, T.-Y. Wu, W. Higashide, P. Malati, Y.-X. Huo, K.M. Cho, J.C. Liao, Integrated electromicrobial conversion of CO_2 to higher alcohols. Science **335**, 1596 (2012)
35. J. Torella, C. Gagliardi, J. Chen, D. Bediako, B. Colón, J. Way, P. Silver, D.G. Nocera, Efficient solar-to-fuels production from a hybrid microbial–water-splitting catalyst system. Proc. Nat. Acad. Sci. **112**, 2337–2342 (2015)
36. H. Toporik, I. Carmeli, I. Volotsenko, M. Molotskii, Y. Rosenwaks, C. Carmeli, N. Nelson, Large photovoltages generated by plant photosystem I crystals. Adv. Mater. **24**, 2988–2991 (2012)
37. G. LeBlanc, E. Gizzie, S. Yang, D.E. Cliffel, G.K. Jennings, Photosystem I protein films at electrode surfaces for solar energy conversion. Langmuir **30**, 10990–11001 (2014)
38. T. Kothe, S. Pöller, F. Zhao, P. Fortgang, M. Rögner, W. Schuhmann, N. Plumeré, Engineered electron-transfer chain in photosystem I based photocathodes outperforms electron-transfer rates in natural photosynthesis. Chemistry **20**, 11029–11034 (2014)
39. N. Nelson, W. Junge, Structure and energy transfer in photosystems of oxygenic photosynthesis. Annu. Rev. Biochem. **84**, 659–683 (2015)
40. S. Fukuzumi, Y.M. Lee, W. Nam, Mimicry and functions of photosynthetic reaction centers. Biochem. Soc. Trans. **46**, 1279–1288 (2018)
41. J. Kargul, G. Bubak, G. Andryianau, Biophotovoltaic systems based on photosynthetic complexes. in *Encyclopedia of Interfacial Chemistry: Surface Science and Electrochemistry*, ed by K. Wandelt (Elsevier, 2018), pp. 43–63
42. X. Qin, M. Suga, T. Kuang, J.-R. Shen, Structural basis for energy transfer pathways in the plant psi-lhci supercomplex. Science **348**, 989–995 (2015)
43. E.J. Boekema, J.P. Dekker, M.G. van Heel, M. Rögner, W. Saenger, I. Witt, H.T. Witt, Evidence for a trimeric organization of the photosystem I complex from the thermophilic cyanobacterium Synechococcus sp. FEBS Lett. **217**, 283–286 (1987)
44. P. Jordan, P. Fromme, H.T. Witt, O. Klukas, W. Saenger, N. Krauß, Three-dimensional structure of cyanobacterial photosystem I at 2.5 Å resolution. Nature **411**, 909–917 (2001)
45. A. Ben-Shem, F. Frolow, N. Nelson, Crystal structure of plant photosystem I. Nature **426**, 630–635 (2003)
46. Y. Mazor, A. Borovikova, N. Nelson, The structure of plant photosystem I super-complex at 2.8 Å resolution. Elife **4**, 1–18 (2015)
47. K. Brettel, W. Leibl, Electron transfer in photosystem I. BBA-Bioenergetics **1507**, 100–114 (2001)
48. H. Ohta, T. Suzuki, M. Ueno, A. Okumura, S. Yoshihara, J.R. Shen, I. Enami, Extrinsic proteins of photosystem II: an intermediate member of PsbQ protein family in red algal PS II. Eur. J. Biochem. **270**, 4156–4163 (2003)
49. A. Busch, M. Hippler, The structure and function of eukaryotic photosystem I. Biochim. Biophys. Acta **1807**, 864–877 (2011)
50. L. Tian, Z. Liu, F. Wang, L. Shen, J. Chen, L. Chang, S. Zhao, G. Han, W. Wang, T. Kuang, X. Qin, J.R. Shen, Isolation and characterization of PSI-LHCI super-complex and their sub-complexes from a red alga Cyanidioschyzon merolae. Photosynth. Res. **133**, 201–214 (2017)
51. P. Haniewicz, M. Abram, L. Nosek, J. Kirkpatrick, E. El-Mohsnawy, J.D. Janna Olmos, R. Kouřil, J. Kargul, Molecular mechanisms of photoadaptation of photosystem I supercomplex from an evolutionary cyanobacterial/algal intermediate. Plant Physiol. **176**, 1433–1451 (2018)

52. B. Drop, M. Webber-Birungi, F. Fusetti, R. Kouřil, K.E. Redding, E.J. Boekema, R. Croce, Photosystem I of Chlamydomonas reinhardtii contains nine light-harvesting complexes (Lhca) located on one side of the core. J. Biol. Chem. **286**, 44878–44887 (2011)
53. M. Antoshvili, I. Caspy, M. Hippler, N. Nelson, Structure and function of photosystem I in Cyanidioschyzon merolae. Photosynth. Res. **139**, 499–508 (2019)
54. X. Pi, L. Tian, H.-E. Dai, X. Qin, L. Cheng, T. Kuang, S.-F. Sui, J.R. Shen, Unique organization of photosystem I–light-harvesting supercomplex revealed by cryo-EM from a red alga. Proc. Natl. Acad. Sci. **115**, 4423–4428 (2018)
55. B. Gobets, R. van Grondelle, Energy transfer and trapping in photosystem I. Biochim. Biophys. Acta **1507**, 80–99 (2001)
56. A.N. Melkozernov, J. Kargul, S. Lin, J. Barber, R.E. Blankenship, Spectral and kinetic analysis of the energy coupling in the PS I-LHC I supercomplex from the green alga Chlamydomonas reinhardtii at 77 K. Photosynth. Res. **86**, 203–215 (2005)
57. J.A. Ihalainen, P.E. Jensen, A. Haldrup, I.H.M. van Stokkum, R. van Grondelle, H.V. Scheller, J.P. Dekker, Pigment organization and energy transfer dynamics in isolated photosystem I (PSI) complexes from Arabidopsis thaliana depleted of the PSI-G, PSI-K, PSI-L, or PSI-N subunit. Biophys. J. **83**, 2190–2201 (2002)
58. J.J. Snellenburg, J.P. Dekker, R. van Grondelle, I.H.M. van Stokkum, Functional compartmental modeling of the photosystems in the thylakoid membrane at 77 K. J. Phys. Chem. B **117**, 11363–11371 (2013)
59. S. Hu, M.R. Shaner, J.A. Beardslee, M. Lichterman, B.S. Brunschwig, N.S. Lewis, Amorphous TiO_2 coatings stabilize Si, GaAs, and GaP photoanodes for efficient water oxidation. Science **344**, 1005–1009 (2014)
60. W. Xie, R. Li, Q. Xu, Enhanced photocatalytic activity of Se-doped TiO_2 under visible light irradiation. Sci. Rep. **8**, 1–10 (2018)
61. Z. Jin, X. Zhang, Y. Li, S. Li, G. Lu, 5.1% Apparent quantum efficiency for stable hydrogen generation over eosin-sensitized CuO/TiO_2 photocataly:st under visible light irradiation. Catal. Commun. **8**, 1267–1273 (2007)
62. K.-R. Wee, B.D. Sherman, M.K. Brennaman, M.V. Sheridan, A. Nayak, L. Alibabaei, T. J. Meyer, An aqueous, organic dye derivatized SnO_2/TiO_2 core/shell photoanode. J. Mater. Chem. A **4**, 2969–2975 (2016)
63. M.M. Momeni, Y. Ghayeb, Visible light-driven photoelectrochemical water splitting on ZnO–TiO2 heterogeneous nanotube photoanodes. J. Appl. Electrochem. **45**, 557–566 (2015)
64. N. Rungjaroentawon, S. Onsuratoom, S. Chavadej, Hydrogen production from water splitting under visible light irradiation using sensitized mesoporous-assembled TiO2–SiO2 mixed oxide photocatalysts. Int. J. Hydrogen. Energ. **37**, 11061–11071 (2012)
65. M. Barroso, S.R. Pendlebury, A.J. Cowan, J.R. Durrant, Charge carrier trapping, recombination and transfer in hematite (α-Fe_2O_3) water splitting photoanodes. Chem. Sci. **4**, 2724–2734 (2013)
66. K. Sivula, F. Le Formal, M. Grätzel, Solar water splitting: progress using hematite (α-Fe_2O_3) photoelectrodes. Chem. Sus. Chem. **4**, 432–449 (2011)
67. S.K. Mohapatra, S.E. John, S. Banerjee, M. Misra, Water photooxidation by smooth and ultrathin α-Fe_2O_3 nanotube arrays. Chem. Mater. **21**, 3048–3055 (2009)
68. X. Wen, S. Wang, Y. Ding, Z.L. Wang, S. Yang, Controlled growth of large-area, uniform, vertically aligned arrays of α-Fe_2O_3 nanobelts and nanowires. J. Phys. Chem. B **109**, 215–220 (2005)
69. J.Y. Kim, D.H. Youn, J.H. Kim, H.G. Kim, J.S. Lee, Nanostructure-preserved hematite thin film for efficient solar water splitting. ACS Appl. Mater. Inter. **7**, 14123–14129 (2015)
70. D. Sharma, S. Upadhyay, A. Verma, V.R. Satsangi, R. Shrivastav, S. Dass, Nanostructured Ti-Fe_2O_3/Cu_2O heterojunction photoelectrode for efficient hydrogen production. Thin Solid Films **574**, 125–131 (2015)
71. J. Kargul, M. Kiliszek (2020) Artificial Photosynthesis: current advances and challenges, in *Bioelectrochemical Interface Engineering*, eds by R.N. Krishnaraj, R. Sani (Wiley, Hoboken, NJ, 2020), pp. 271–310

72. F. Li, K. Fan, L. Wang, Q. Daniel, L. Duan, L. Sun, Immobilizing Ru(bda) catalyst on a photoanode via electrochemical polymerization for light-driven water splitting. ACS Catalysis **5**, 3786–3790 (2015)
73. J.M. Spurgeon, J.M. Velazquez, M.T. McDowell, Improving O_2 production of WO_3 photoanodes with IrO_2 in acidic aqueous electrolyte. Phys. Chem. Chem. Phys. **16**, 3623–3631 (2014)
74. K.S. Novoselov, A.K. Geim, S.V. Morozov, D. Jiang, Y. Zhang, S.V. Dubonos, I.V. Grigorieva, A.A. Firsov, Electric field effect in atomically thin carbon films. Science **306**, 666 (2004)
75. H.P. Boehm, Graphene—how a laboratory curiosity suddenly became extremely interesting. Angew. Chem. Int. Ed. **49**, 9332 (2010)
76. A.K. Geim, K.S. Novoselov, The rise of graphene. Nat Mater. **6**, 183 (2007)
77. F.D.M. Haldane, Model for a quantum hall effect without landau levels: condensed-matter realization of the "parity anomaly". Phys. Rev. Lett. **61**, 2015 (1988)
78. G.W.J. Flynn, Perspective: the dawning of the age of graphene. Chem. Phys. **135**, 050901 (2011)
79. L. Banszerus, M. Schmitz, S. Engels, J. Dauber, M. Oellers, F. Haupt, K. Watanabe, T. Taniguchi, B. Beschoten, C. Stampfer, Ultrahigh-mobility graphene devices from chemical vapor deposition on reusable copper. Sci. Adv. **1**, e1500222 (2015)
80. V.S. Morozov, K.S. Novoselov, M.I. Katsnelson, F. Schedin, D.C. Elias, J.A. Jaszczak, A.K. Geim, Giant intrinsic carrier mobilities in graphene and its bilayer. Phys. Rev. Lett. **100**, 016602 (2008)
81. K.C. Chandra Bikram, S.K. Das, K. Ohkubo, S. Fukuzumi, F. D'Souza, Ultrafast charge separation in supramolecular tetrapyrrole− graphene hybrids. Chem. Commun. **48**, 11859–11861 (2012)
82. P. Lazar, F. Karlicky, P. Jurečka, M. Kocman, E. Otyepková, K. Safářová, Otyepka M, Adsorption of small organic molecules on graphene. J. Am. Chem. Soc. **135**, 6372–6377 (2013)
83. X. Dong, D. Fu, W. Fang, T. Shi, P. Chen, L.J. Li, Doping single-layer graphene with aromatic molecules. Small **5**, 1422 (2009)
84. M. Singh, M. Holzinger, M. Tabrizian, S. Winters, N.C. Berner, S. Cosnier, G.S. Duesberg, Noncovalently functionalized monolayer graphene for sensitivity enhancement of surface plasmon resonance immunosensors. J. Am. Chem. Soc. **137**, 2800–2803 (2015)
85. Q. Wei, X. Wang, F. Zhou, A versatile macro-initiator with dual functional anchoring groups for surface-initiated atom transfer radical polymerization on various substrates. Polym. Chem. **3**, 2129–2137 (2012)
86. F. Schröper, A. Baumann, A. Offenhäusser, D. Mayer, Bidirectional immobilization of affinity-tagged cytochrome c on electrode surfaces. Chem. Commun. **46**, 5295 (2010)
87. Y. Kim, S.A. Shin, J. Lee, K.D. Yang, K.T. Nam, Hybrid system of semiconductor and photosynthetic protein. Nanotechnology **25**, 342001 (2014)
88. H. Yaghoubi, Z. Li, D. Jun, E. Lafalce, X. Jiang, R. Schlaf, J.T. Beatty, A. Takshi, Hybrid wiring of the rhodobacter sphaeroides reaction center for applications in bio-photoelectrochemical solar cells. J. Phys. Chem. C **118**, 23509 (2014)
89. H. Ma, H.-L. Yip, F. Huang, A.K.-Y. Jen, Interface engineering for organic electronics. Adv. Funct. Mater. **20**, 1371 (2010)
90. K. Asadi, Y. Wu, F. Gholamrezaie, P. Rudolf, P.W.M. Blom, Single-layer pentacene field-effect transistors using electrodes modified with self-assembled monolayers. Adv. Mater. **21**, 4109 (2009)
91. L. Wang, G.M. Rangger, L. Romaner, G. Heimel, T. Bucko, Z.-Y. Ma, Q.-K. Li, Z. Shuai, E. Zojer, Electronic structure of self-assembled monolayers on Au(111) surfaces: the impact of backbone polarizability. Adv. Funct. Mater. **19**, 3766 (2009)
92. C. Vericat, M.E. Vela, G. Benitez, P. Carrob, R.C. Salvarezza, Self-assembled monolayers of thiols and dithiols on gold: new challenges for a well-known system. Chem. Soc. Rev. **39**, 1805 (2010)

93. C.-Y. Chen, K.-Y. Wu, Y.-C. Chao, H.-W. Zan, H.-F. Meng, Y.-T. Tao, Concomitant tuning of metal work function and wetting property with mixed self-assembled monolayers. Org. Electron. **12**, 148 (2011)
94. K.-Y. Wu, S.-Y. Su, Y.-T. Tao, Continuous modulation of electrode work function with mixed self-assembled monolayers and its effect in charge injection. Langmuir **25**, 6232 (2009)
95. E. Orgiu, N. Crivillers, J. Rotzler, M. Mayor, P. Samorí, Tuning the charge injection of P3HT-based organic thin-film transistors through electrode functionalization with oligophenylene SAMs. J. Mater. Chem. **20**, 10798 (2010)
96. B. Zhang, L. Sun, Artificial photosynthesis: opportunities and challenges of molecular catalysts. Chem. Soc. Rev. **48**, 2216–2264 (2019)
97. X. Zhang, L. Hou, P. Samori, Coupling carbon nanomaterials with photochromic molecules for the generation of optically responsive materials. Nat. Commun. **7**, 11118 (2016)
98. G. Pace, V. Ferri, V. Grave, M. Elbing, C. von Hänisch, M. Zharnikov, M. Mayor, M.A. Rampi, P. Samorì, Cooperative light-induced molecular movements of highly ordered azobenzene self-assembled monolayers. Proc. Natl. Acad. Sci. **104**, 9937–9942 (2007)
99. E. Katchalski-Katzir, I. Shariv, M. Eisenstein, A.A. Friesem, C. Aflalo, I.A. Vakser, Molecular surface recognition: determination of geometric fit between proteins and their ligands by correlation techniques. Proc. Natl. Acad. Sci. **89**, 2195–2199 (1992)
100. M.P. Jacobson, R.A. Friesner, Z. Xiang, B. Honig, On the role of crystal packing forces in determining protein sidechain conformations. J. Mol. Biol. **320**, 597–608 (2002)
101. M.J. Abraham, T. Murtola, R. Schulz, S. Páll, J.C. Smith, B. Hess, E. Lindahl, GROMACS: high performance molecular simulations through multi-level parallelism from laptops to supercomputers. SoftwareX **1**, 19–25 (2015)
102. A.D. Mackerell Jr., Empirical force fields for biological macromolecules: overview and issues. J. Comput. Chem. **25**, 1584–1604 (2004)
103. K. Kaszuba, P.A. Postila, O. Cramariuc, M. Sarewicz, A. Osyczka, I. Vattulainen, T. Róg, Parameterization of the prosthetic redox centers of the bacterial cytochrome bc_1 complex for atomistic molecular dynamics simulations. Theor. Chem. Acc. **132**, 1370–1383 (2013)
104. A.M. Kowalska, B. Trzaskowski, S. Osella, Assessing the charge transfer at the cytochrome c553/grapheneinterface: a multiscale investigation. J. Phys. Chem. C **122**, 29405–29413 (2018)
105. P. Hohenberg, W. Kohn, Inhomogeneous electron gas. Phy. Rev. **136**, B864 (1964)
106. W. Kohn, L.J. Sham, Self-consistent equations including exchange and correlation effects. Phys. Rev. **140**, A1133 (1965)
107. N. Argaman, G. Makov, Density functional theory: an introduction. Am. J. Phys. **68**, 69 (2000)
108. G. Kotliar, S.Y. Savrasov, K. Haule, V.S. Oudovenko, O. Parcollet, C.A. Marianetti, Electronic structure calculations with dynamical mean-field theory. Rev. Mod. Phys. **78**, 865 (2006)
109. S. Osella, M. Kiliszek, E. Harputlu, C.G. Unlu, K. Ocakoglu, J. Kargul, B. Trzaskowski, Controlling the charge transfer flow at the graphene/pyrene–nitrilotriacetic acid interface. J. Mater. Chem. C **6**, 5046 (2018)
110. S. Osella, M. Kiliszek, E. Harputlu, C.G. Unlu, K. Ocakoglu, B. Trzaskowski, J. Kargul, Role of metal centers in tuning the electronic properties of graphene-based conductive interfaces. J. Phys. Chem. C **123**, 8623–8632 (2019)
111. P. Giannozzi, S. Baroni, N. Bonini, M. Calandra, R. Car, C. Cavazzoni, D. Ceresoli, G.L. Chiarotti, M. Cococcioni, I. Dabo et al., QUANTUM ESPRESSO: a modular and open-source software project for quantum simulations of materials. J. Phys. Condens. Matt **21**, 395502 (2009)
112. K. Lee, E.D. Murray, L. Kong, B.I. Lundqvist, D.C. Langreth, Higher-accuracy van der waals density functional. Phys. Rev. B: Condens. Matter Mater. Phys. **82**, 081101 (2010)
113. D. Vanderbilt, Soft self-consistent pseudopotentials in a generalized eigenvalue formalism. Condens. Mater. Phys. **41**, 7892–7895 (1990)

114. D. Chaen, A. Kahn, Electron energetics at surfaces and interfaces: concepts and experiments. Adv. Mater. **15**, 271–277 (2003)
115. G. Heimel, I. Salzmann, S. Duhm, N. Koch, Design of organic semiconductors from molecular electrostatics. Chem. Mater. **23**, 359 (2011)
116. Y.-J. Yu, Y. Zhao, S. Ryu, L.E. Brus, K.S. Kim, P. Kim, Tuning the graphene work function by electric field effect. Nano Lett. **9**, 3430–3434 (2009)
117. R.A. Marcus, On the theory of oxidation-reduction reactions involving electron transfer. I. J. Chem. Phys. **24**, 966–978 (1956)
118. G.R. Hutchison, M.A. Ratner, T.J. Marks, Hopping transport in conductive heterocyclic oligomers: reorganization energies and substituent effects. J. Am. Chem. Soc. **127**, 2339–2350 (2005)
119. E.F. Valeev, V. Coropceanu, D.A. da Silva Filho, S. Salman, J.-L. Bredas, Effect of electronic polarization on charge-transport parameters in molecular organic semiconductors. J. Am. Chem. Soc. **128**, 9882–9886 (2006)
120. J. Kirkpatrick, An approximate method for calculating transfer integrals based on the ZINDO Hamiltonian. Int. J. Quantum Chem. **108**, 51–56 (2008)
121. B. Baumeier, J. Kirkpatrick, D. Andrienko, Density-functional based determination of intermolecular charge transfer properties for large-scale morphologies. Phys. Chem. Chem. Phys. **12**, 11103–11113 (2010)
122. H. van Eersel, R.A.J. Janssen, M. Kemerink, Mechanism for efficient photoinduced charge separation at disordered organic heterointerfaces. Adv. Funct. Mater. **22**, 2700–2708 (2012)
123. R. Volpi, S. Kottravel, M.S. Nørby, S. Stafström, M. Linares, Effect of polarization on the mobility of C60: a kinetic monte carlo study. J. Chem. Theory Comput. **12**, 812–824 (2016)
124. H.B. Gray, J.R. Winkler, Long-range electron transfer. Proc. Natl. Acad. Sci. **102**, 3534–3539 (2005)
125. Sjulstok, E., Olsen, J.M.H., Solov'yov, I.A.: Quantifying electron transfer reactions in biological systems: what interactions play the major role? Sci. Rep. 5, 18446 (2015)
126. M.F. Lucas, D.L. Rousseau, V. Guallar, Electron transfer pathways in cytochrome c oxidase. Biochim. Biophys. Acta **1807**, 1305–1313 (2011)
127. H.R. Williamson, B.A. Dow, V.L. Davidson, Mechanisms for control of biological electron transfer reactions. Bioorg. Chem. **57**, 213–221 (2014)
128. J.-L. Brédas, D. Beljonne, V. Coropceanu, J. Cornil, Charge-transfer and energy-transfer processes in π-conjugated oligomers and polymers: a molecular picture. Chem. Rev. **104**, 4971–5004 (2004)
129. S. Difley, L.-P. Wang, S. Yeganeh, S.R. Yost, T. Van Voorhis, Electronic properties of disordered organic semiconductors via QM/MM simulations. Acc. Chem. Res. **437**, 995–1004 (2010)
130. C.E. Chidsey, Free energy and temperature dependence of electron transfer at the metal-electrolyte interface. Science **251**, 919 (1991)
131. S. Gosavi, R.A. Marcus, Nonadiabatic electron transfer at metal surfaces. J. Phys. Chem. B **104**, 2067 (2000)
132. D. Buesen, T. Hoefer, H. Zhang, N. Plumeré, A kinetic model for redox-active film based biophotoelectrodes. Faraday Discuss. (2019). https://doi.org/10.1039/c8fd00168e

Chapter 14
Mathematical Modeling of the Kinetics of Counter Diffusion During the Formation of Boron-Containing Coatings on Steels

A. S. Borsyakov, V. A. Yuryev, V. V. Ozyerelyev and E. V. Levchenko

Abstract For the first time, a mathematical model of the kinetics of the formation of diffusion boride coatings on iron has been created, allowing to calculate the increase in the dimensions of parts due to the displacement of the outer surface of the saturated sample. Models of diffusant distribution in boride phases of different chemical composition are obtained. The experimental results confirmed the validity of the developed models.

14.1 Introduction

Improving the performance characteristics of metal products for various purposes by creating boron-containing coatings has been used in practice for several decades. Electrolytic boron saturation of products increases corrosion resistance in aggressive media, wear resistance and surface hardness of parts. But at the same time, the geometrical parameters of the product change due to the increase in the surface.

It should be noted in mind that for production workers engaged in the chemical-thermal treatment of precision parts and assemblies (dies, punches, broaches, etc.), the issue of increasing the dimensions of parts when boriding them is key. So far, the record of such an increase in the geometry of the product is either

A. S. Borsyakov
Voronezh State University of Engineering Technologies, 19 Revolution Avenue, Voronezh 394036, Russian Federation

V. A. Yuryev (✉) · V. V. Ozyerelyev
Voronezh State Technical University, 14 Moscow Avenue, Voronezh 394026, Russian Federation
e-mail: yuryevva@gmail.com

E. V. Levchenko
School of Mathematical and Physical Sciences, Priority Research Centre for Computer-Assisted Research Mathematics and Its Applications (CARMA), The University of Newcastle, Callaghan, NSW 2308, Australia

E. V. Levchenko et al. (eds.), *Theory and Simulation in Physics for Materials Applications*, Springer Series in Materials Science 296,
https://doi.org/10.1007/978-3-030-37790-8_14

carried out empirically, or subsequent machining is carried out after boriding. There are no analytical analyzes of changes in the size of products during electrolysis boriding.

The aim of this work is to develop an analytical model that establishes the relationship between the parameters of the electrolytic saturation of steel products with boron and the increase in their external dimensions during the synthesis of boride coating. Such a model will make it possible to derive mathematical expressions that make it possible to determine the full thickness of the boride coating taking into account the movement of the outer surface of the saturable sample.

14.2 Physical Model of Boriding Process

The available experimental data show that the size of the parts and components is increasing during boronizing process. The boriding is associated with the diffusion saturation of the crystal lattice of iron; therefore, the observed effect of increasing the size is possible only due to the appearance of a counter diffusion flux of iron atoms toward the boron source. The phenomenon of inequality of counter-diffusion flows of atoms of different components, known as the Kirkendall effect. However, in the practice of diffusion saturation, for example, with the same boron, no solution of such problems with regard to crystal chemical reactions was undertaken.

The mechanism for the synthesis of the boride coating during the formation of the low-boron Fe_2B phase from a solid solution of boron in iron consists of two stages—incubation and kinetic. The conventional scheme of the electrolytic synthesis of the boride coating is shown in Fig. 14.1.

During the incubation period, a weak, supersaturated solid solution of boron in iron is formed in the near-surface layer of iron. The release of excess boron leads to the formation of embryos of the new boride phase. When the critical size of the nuclei in the solid solution is reached, the low-boron Fe_2B phase begins to grow. After the formation of a continuous layer of a new phase over the entire surface, the kinetic stage begins, i.e. boride growth.

The "Fe_2B—solid solution" interface moves due to the formation of thermodynamically nonequilibrium supersaturated solid solution in certain areas and the subsequent formation of a stoichiometric chemical compound.

Electron microscopy makes it possible to reliably conduct quantitative analysis and distribution of borides (Fig. 14.2).

As the thermodynamic equilibrium concentration of boron in the surface layers at the grain and block boundaries is reached, precipitates of the boride phase are formed, i.e. the system goes into an equilibrium two-phase state (Fig. 14.3). The subsequent electrolytic saturation with boron is accompanied by an increase in the grain size of the boride phase and the formation of a continuous layer of Fe_2B in the near-surface layers. Further diffusive mass transfer of boron is carried out through

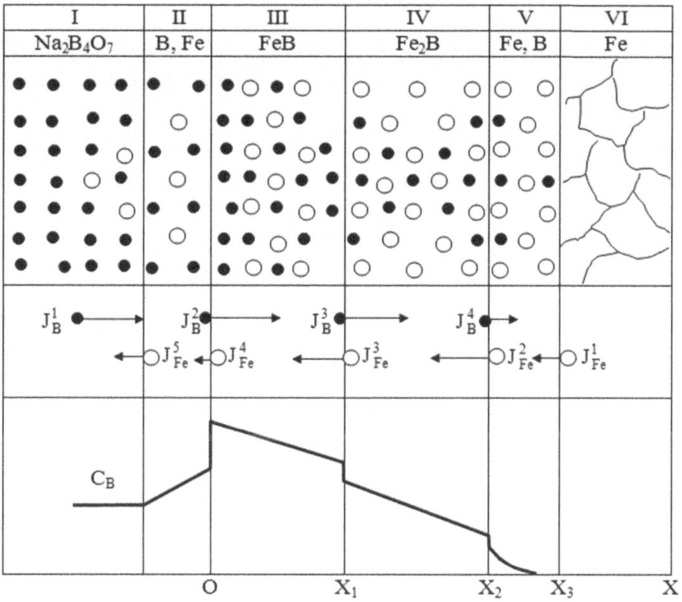

Fig. 14.1 Conventional scheme for the synthesis of diffusion coating, distribution of diffusion fluxes and boron concentration during high-temperature electrolysis boronization of iron: I layer—electrolyte, II layer—adsorption layer, III layer—FeB phase, IV layer—low-boron Fe_2B phase, V—boron solid solution in iron, VI—metal base

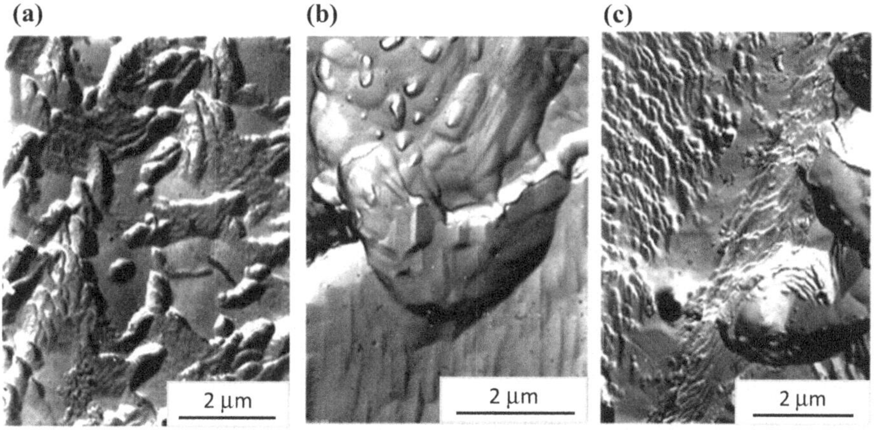

Fig. 14.2 Electron micrographs of carbon replicas of diffusion boride layers on nickel: **a** the structure of borides in the transition zone; **b** the structure of the boride column; **c** the structure of the grain boundaries in the zone of solid solution of boron in Armco iron

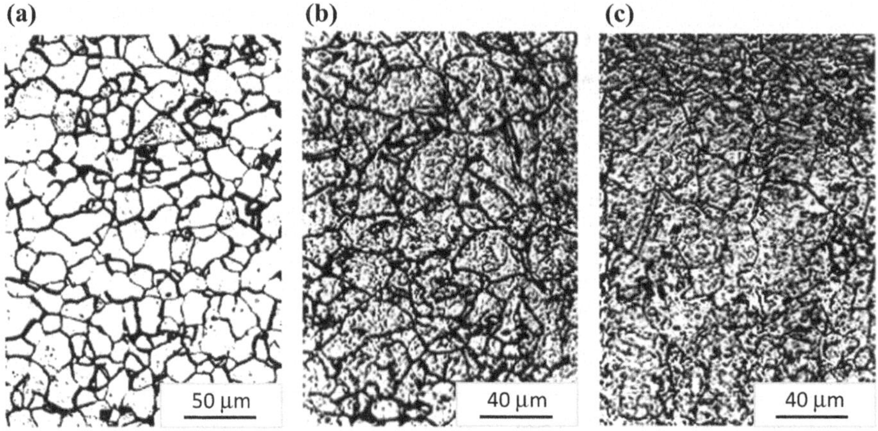

Fig. 14.3 The surface structure of Armco iron before (**a**) and after electrolysis boronation at T = 1123 K for 120 (**b**) and 240 (**c**) seconds

the layer of the Fe_2B phase. The difference between the diffusion rates of the solid solution and the lower boride of iron leads to the accumulation of boron in Fe_2B and the formation of a new FeB phase. Thus, the system becomes three phase—a solid solution of boron in iron and borides FeB and Fe_2B.

The kinetics of changes in the concentration of boride FeB is described by the equation

$$\frac{\mathrm{d}N}{\mathrm{d}t} = k(C_{\mathrm{Fe}} - N)(C_{\mathrm{B}} - N), \tag{14.1}$$

where N—is the concentration of iron boride FeB; k—is a constant characterizing the reaction rate; C_{Fe}—concentration of iron atoms; C_{B}—concentration of boron atoms.

Consider a particular case where the concentrations of the atoms of iron and boron do not change over time, i.e.

$$C_{\mathrm{Fe}} = \mathrm{const}, \ C_{\mathrm{B}} = \mathrm{const}. \tag{14.2}$$

Such a situation takes place, if we assume that in the studied time interval, the change in concentrations, N, C_{Fe}, C_{B} is insignificant.

Solving Riccatti's (14.1), we obtain an expression that establishes the dependence of the boride concentration of FeB on the time of boriding:

$$N = \frac{C_{\mathrm{Fe}}(C_{\mathrm{B}} - N_0)\mathrm{e}^{(C_{\mathrm{B}} - C_{\mathrm{Fe}})kt} - C_{\mathrm{B}}(C_{\mathrm{Fe}} - N_0)}{(C_{\mathrm{B}} - N_0)\mathrm{e}^{(C_{\mathrm{B}} - C_{\mathrm{Fe}})kt} - (C_{\mathrm{Fe}} - N_0)}, \tag{14.3}$$

where N_0 is the boride concentration at the initial time (t = 0).

At the initial time, the boride concentration is zero, from (14.3) we immediately get:

$$N = \frac{C_{\mathrm{Fe}} C_{\mathrm{B}} \left(\mathrm{e}^{(C_{\mathrm{B}} - C_{\mathrm{Fe}})kt} - 1\right)}{C_{\mathrm{B}} \mathrm{e}^{(C_{\mathrm{B}} - C_{\mathrm{Fe}})kt} - C_{\mathrm{Fe}}}. \tag{14.4}$$

To solve (14.4), it is necessary to consider two cases of the subsequent synthesis of boride over time:

1. $C_{\mathrm{B}} > C_{\mathrm{Fe}}$. In this case, as follows from (14.4):

$$\lim_{t \to \infty} = C_{\mathrm{Fe}}; \tag{14.5}$$

2. $C_{\mathrm{B}} < C_{\mathrm{Fe}}$. Then:

$$\lim_{t \to \infty} = C_{\mathrm{B}}. \tag{14.6}$$

A similar situation will be in the case if the $N_0 \neq 0$ and dependence $N(t)$ is described by the relation (14.3).

The increase in sample size in the process of boriding, i.e. the movement of the outer boundary of the diffusion layer outside the sample can be represented as the following scheme (Fig. 14.4). Here, the origin of coordinates coincides with the electrolyte interface—the product at the initial moment of time.

In the process of electrolytic boronization, the boundary moves towards the electrolyte, i.e. negative values X of according to the law $-X = y(t)$ [2].

The process of electrolysis boronization can be described by a system of counter flows

$$J_{\mathrm{B}} = -D_{\mathrm{B}} \frac{\partial \mu_{\mathrm{B}}}{\partial x} + (\dot{y} + v) C_{\mathrm{B}}, \tag{14.7}$$

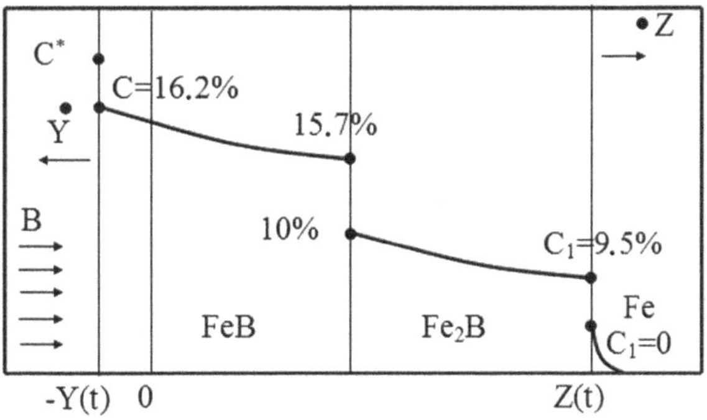

Fig. 14.4 Scheme of synthesis of boride coatings on iron with allowance for counter-currents of diffusant and saturable matrix [1]

$$J_{\mathrm{Fe}} = -D_{\mathrm{Fe}} \frac{\partial \mu_{\mathrm{Fe}}}{\partial x} + (\dot{y} + v) C_{\mathrm{Fe}}, \tag{14.8}$$

where, μ_{B} and μ_{Fe}—chemical potential of boron and iron, respectively; C is the concentration of the diffusing element, v is the velocity of the lattice flow.

From here

$$-D_{\mathrm{B}} \frac{\partial \mu_{\mathrm{B}}}{\partial x} + (\dot{y} + v) C_{\mathrm{B}} = -k_{\mathrm{B}} (C_{\mathrm{Fe}} - N)(C_{\mathrm{B}} - N) \tag{14.9}$$

$$-D_{\mathrm{Fe}} \frac{\partial \mu_{\mathrm{Fe}}}{\partial x} + (\dot{y} + v) C_{\mathrm{Fe}} = -k_{\mathrm{Fe}} (C_{\mathrm{Fe}} - N)(C_{\mathrm{B}} - N) \tag{14.10}$$

where, $k_{\mathrm{B}} = mk$, $k_{\mathrm{Fe}} = nk$, m, n—is the average molar concentration of boron and iron, respectively. Solving the system of (14.9, 14.10), the speed of movement of the external surface of the sample is defined as:

$$\dot{y}(t) = \frac{k(n - m)(C_{\mathrm{Fe}} - N)(C_{\mathrm{B}} - N) + D_{\mathrm{B}} \frac{\partial \mu_{\mathrm{B}}}{\partial x} - D_{\mathrm{Fe}} \frac{\partial \mu_{\mathrm{Fe}}}{\partial x}}{C_{\mathrm{B}} + C_{\mathrm{Fe}}} - v. \tag{14.11}$$

The kinetics of the outer surface can be described as:

$$\dot{y}(t) = \alpha k (C_{\mathrm{Fe}} - N)(C_{\mathrm{B}} - N), \tag{14.12}$$

where $\alpha = (n - m)(mC_{\mathrm{B}} + nC_{\mathrm{Fe}})$ at $v = v_0$ and $\alpha = \left(\frac{nD_{\mathrm{Fe}}}{\tilde{D}} - \frac{mD_{\mathrm{FeB}}}{\tilde{D}}\right) / (mC_{\mathrm{B}} - nC_{\mathrm{Fe}})$ at $v = 0$, here $\tilde{D}$ is the coefficient of mutual diffusion.

Diffusion processes in the positive direction $+X = y(t)$ (Fig. 14.1), associated with the formation of borides inside the sample, are due to the chemical potential gradient μ:

$$\frac{\partial \mu}{\partial t} = \frac{\partial}{\partial x} \left(D \frac{\partial \mu}{\partial x} \right) \tag{14.13}$$

The derived equations describe the kinetics of the growth of the boride layer, but the numerical values of the diffusion parameters are difficult to determine because there are no reliable data on the activity values of the iron and boron atoms in the borides. Reliable are data on the activities of iron and boron in a mixture of borides.

To solve the kinetic equations, the calculation formulas of [3] were used. It was assumed that the activity of boron and iron is equal to I, and all parameters were determined through the concentration of C.

The total thickness of the boride layer (taking into account the movement of the outer boundary) was defined as:

Table 14.1 Dependence of boron diffusion parameters in iron on electrolysis temperature

T, K	μ	a	b	D	K
1123	0.22	5.39×10^{-4}	2.72×10^{-9}	2.46×10^{-9}	4.56×10^{-6}
1173	0.23	16.5×10^{-4}	27.5×10^{-9}	25.2×10^{-9}	15.25×10^{-6}
1223	0.23	32.9×10^{-4}	63.8×10^{-9}	56.3×10^{-9}	17.7×10^{-6}

$$h(t) = y(t) + Z(t) = \left(a^2 + bt\right)^{1/2} - a, \tag{14.14}$$

where $a = \frac{D}{k}, b = 2D(1-\mu)\frac{C_1 - C_2}{C^* - C_1}, \mu = \frac{y(t)}{Z(t)}$.

The values a and b were calculated using the methods of successive approximations and least squares. Both methods showed good convergence of the obtained results, and the results of calculations are presented in Table 14.1.

The temporal and temperature dependences of the displacement of the boride phase boundary in the Z and Y directions for the case of boron eutectoid steel are presented in Fig. 14.5.

The calculated theoretical values differ from experimental data by 3–5%. The boron concentrations were determined on an X-ray microanalyzer, which made it possible to calculate the ratio

$$\frac{C^* - C_1}{C_1 - C_2} \cong 0.71 \tag{14.15}$$

The calculated coefficient of mutual diffusion is well described by the expression:

$$D = 2.6 \exp\left(-\frac{42,700}{RT}\right) \tag{14.16}$$

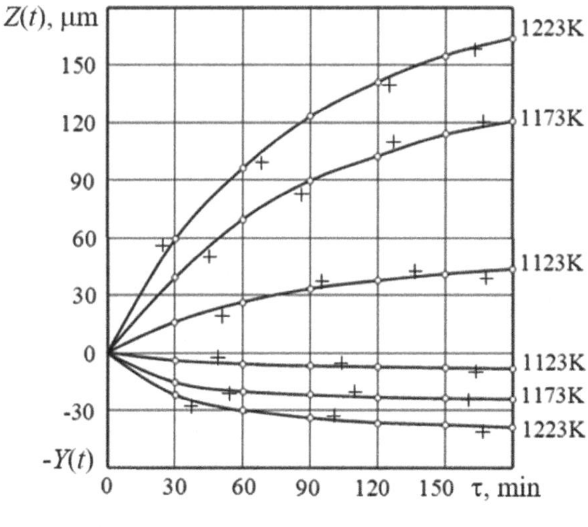

Fig. 14.5 The effect of time and temperature of electrolysis on the movement of the outer Z (t) and inner Y (t) boundaries of boride layers on eutectoid steel

The parameters of the diffusion boronization of eutectoid steel are significantly different from the calculations based on the constancy of the integral of the error function for the growth phase with a constant concentration:

$$\Phi\left(\frac{y}{2\sqrt{D\tau}}\right) = \text{const}, \quad (14.17)$$

where

$$\frac{y}{2\sqrt{D\tau}} = \text{const}, \quad (14.18)$$

i.e.

$$y = z\sqrt{D\tau}. \quad (14.19)$$

Taking the temperature dependence of the diffusion coefficient in the form

$$D = A\,\mathrm{e}^{-E/RT} \quad (14.20)$$

and substituting it in (14.19), we get

$$y^2 = z^2\mathrm{e}^{-E/RT}\tau \quad (14.21)$$

where y is the depth of the layer; E—activation energy of diffusion.

Denoting $Az^2\mathrm{e}^{-E/RT}$ by $2p$, we have:

$$y^2 = 2p\tau \quad (14.22)$$

Since (14.22) has a parabolic dependence, the method for calculating surface growth described above is valid only if the parabolic dependence of the total depth of the diffusion layer on the time of the boriding.

Figure 14.6 shows the dependences of the «full depth» of boride layers on eutectoid steel on the time and temperature of the process when boriding in the melt a drill with a cathodic current density of 2000 A/m^2.

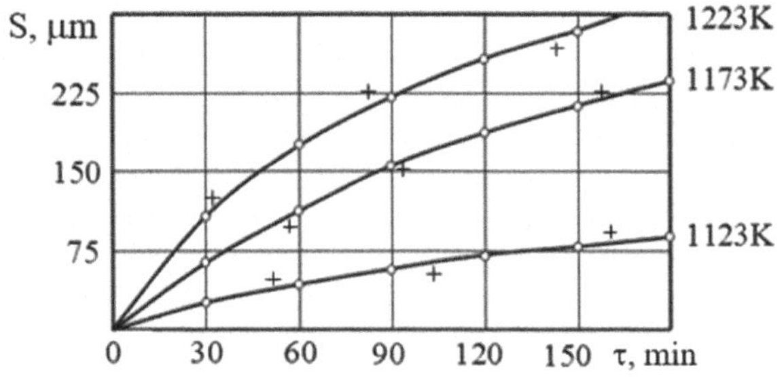

Fig. 14.6 The influence of time and temperature of electrolysis on the value of the boride layer on eutectoid steel (DK = 2000 A/m^2)

From the graphical dependence $y^2 = f(\tau)$, it is possible to calculate the parameter $2p$ in terms of the slope of the straight lines. From (14.21) and (14.22) it follows that

$$\ln 2p = \ln Az^2 - \frac{E}{2.3RT}. \tag{14.23}$$

Graphic dependence $\ln 2p = f(1/T)$ is a direct, allowing to determine the numerical value $\ln Az^2$. For our case, it is equal to 1.95. Substituting this value into (14.23), we find the activation energy of boron in eutectoid steel:

$$E = 2.3\ (\ln Az^2 - \ln 2p)RT \cong 29880\ \text{cal/mol} \tag{14.24}$$

and diffusion coefficient

$$D = 11.2 \cdot 10^{-3} \exp\left(\frac{29880}{RT}\right). \tag{14.25}$$

These parameters allow us to derive an empirical equation that determines the dependence of the boride layer depth y on temperature and electrolysis time τ:

$$y = \left(z\sqrt{A}\right) \exp\left(-\frac{E}{RT}\right) \sqrt{\tau} \tag{14.26}$$

The value $z\sqrt{A}$ can be determined from experimental data. For example, for electrolysis saturation with boron of eutectoid steel for 30 min at $T = 950$ °C, $y = 70$ μm:

$$y = 0.07 = \left(z\sqrt{A}\right) \exp\left(-\frac{29880}{RT}\right) \sqrt{0.5}, \tag{14.27}$$

from here $z\sqrt{A} = 45.19$.

Substituting the values into formula (14.27), we have:

$$y = 45.19 \exp\left(-\frac{29880}{RT}\right) \sqrt{\tau}. \tag{14.28}$$

14.3 Conclusion

The kinetic equations for the growth of the boride coating both in the depth of the iron sample and the growth of the outer layer, which increases the size of the product, are obtained.

For manufacturers engaged in chemical-thermal processing of precision parts and assemblies (dies, punches, dies, broaches, etc.), two methods of computer calculations of the required parameters have been developed and implemented: the least squares method and the method of successive approximations. Both methods showed good convergence of the results. The values found in this way are presented in our work [4]. The discrepancies between the experimental and theoretical points obtained do not exceed 3–5%.

References

1. A.A. Afanasyev, *Improving the Quality of the Surfaces of Machine Parts* (BSTU Publishing House, V.G. Shukhov, Belgorod, 2007), p. 235
2. A.A. Borsyakov, A.N. Osintsev, A.S. Borsyakov, Mathematical modeling of the process of moving the external border during electrolysis borination, in *In the Collection: Innovative Technologies and Equipment of the Machine-Building Complex* (Interuniversity Sat. scientific tr. Feder. Education Agency, Voronezh State Technical University, Obl. board NTO machine builders. Voronezh, 2005), pp. 20–22
3. Y.M. Lakhtin, B.N. Arzamasov, *Chemical Heat Treatment of Metals* (M: Metallurgy, 1985), p. 256
4. A.S. Borsyakov, A.A. *Afanasyev, Scientific and Practical Aspects of Obtaining Multicomponent Boron-Containing Coatings with High Service Properties* (BSTU Publishing House, V.G. Shukhov, Belgorod, 2017), p. 432

Index

E. V. Levchenko et al. (eds.), *Theory and Simulation in Physics for Materials Applications*, Springer Series in Materials Science 296,
https://doi.org/10.1007/978-3-030-37790-8

The manufacturer's authorised representative in the EU is Springer Nature Customer Service Centre GmbH, Europaplatz 3, 69115 Heidelberg, Germany. If you have any concerns regarding our products, please contact ProductSafety@springernature.com

Printed and bound by CPI Group (UK) Ltd, Croydon, CR0 4YY
30/09/2026
02225591-0010